Manufacturing Process Selection Handbook

Manufacturing Process Selection Handbook

K. G. Swift

Department of Engineering, University of Hull

J. D. Booker

Department of Mechanical Engineering, University of Bristol

ELSEVIER

Amsterdam • Boston • Heidelberg • London • New York • Oxford • Paris
San Diego • San Francisco • Singapore • Sydney • Tokyo
Butterworth-Heinemann is an imprint of Elsevier

B
H

Butterworth-Heinemann is an imprint of Elsevier
The Boulevard, Langford Lane, Kidlington, Oxford, OX5 1GB
225 Wyman Street, Waltham, MA 02451, USA

First published 2013

British Library Cataloguing in Publication Data
A catalogue record for this book is available from the British Library

Library of Congress Cataloguing in Publication Data
A catalog record for this book is available from the Library of Congress

ISBN: 978-0-08-099360-7

Printed and bound in the United Kingdom

13 14 15 16 10 9 8 7 6 5 4 3 2 1

Contents

Preface

As we write this handbook, most nations of the world are facing economic pressures not seen since the 1930s and consequently, it has never been more important to design products that can be manufactured economically and at high levels of quality.

In order to facilitate the achievement of the required quality and cost objectives for the manufacture of a component/product design solution, it is necessary to carry out the interrelated activities of selecting candidate processes and tuning a design to get the best out of a chosen manufacturing route. These are difficult decision-making tasks that few experts do well, particularly in the situation of new product introduction.

Failure to get this right often results in late engineering change, with its associated problems of high cost and lead time protraction, or having to live with components that are of poor quality and/or expensive to make.

There is a need for specialist knowledge across a range of manufacturing technologies to enable the correct design decisions to be made from the breadth of possibilities. The difficulties faced by businesses in this area are frequently due to a lack of the necessary process selection strategies, knowledge and data.

The Handbook addresses this gap through the provision of a comprehensive range of process groups; extensive knowledge and data on specific processes; and a set of process selection strategies. The selection strategies guide the reader quickly to find the most appropriate set of candidate processes for a design. The selection strategies are each supported by a number of industrial case studies.

The Handbook title, rather than a standard new edition of our earlier textbook in this field, was felt to be more appropriate, given the nature of the contents and our plans for substantial growth in the reference material (number of processes) to be included. The Handbook holds on to our notion of the PRocess Information MAp (PRIMA) as the means of providing the required technological and economic data on specific processes. The PRIMAs provide detailed data on the characteristics and capabilities of each process in a standard format under headings including: material suitability, design considerations, quality issues, economics, and process fundamentals and variations. A distinctive feature is the inclusion of tolerance capability charts for processing key material types and process-capable geometric tolerancing data for some key processes and geometric characteristics. The Handbook contains some 85 PRIMAs covering primary and secondary shaping processes, rapid prototyping, surface engineering, assembly processes and joining techniques.

The inclusion of assembly is very deliberate – assembly issues are too often neglected in product engineering. Through consideration of assembly many strategically important issues can be addressed. For example, Design for Assembly (DFA) impacts much more than assembly itself for, in addition to reducing component fitting and handling costs, DFA encourages part-count optimisation, variety reduction and standardisation.

Another distinctive feature of the Handbook is the inclusion of easily applicable methods for estimating product costs, based on both design characteristics and process routes. The cost associated with processing a component design is based on the notion of a design-independent basic processing cost and a set of relative cost coefficients for taking account of the design characteristics including material, geometry, tolerance, etc. The overall component cost is logically based on the sum of the material processing and material purchase cost elements. A simple method for estimating manual assembly costs has also been included. The effects of design characteristics on handling and fitting process costs are enumerated and are used in conjunction with a standard labour rate (cost per second) associated with a standard (most simple) assembly operation to estimate assembly costs.

While, where possible, the costing methods should be used with company-specific data, approximate data has been included on a sample of common manufacturing and material groups and assembly processes. This can be used to illustrate the design costing process and show, at least in relative terms, the effects of design choices and alternative processing routes on manufacturing cost.

The Handbook is primarily intended to be useful to practising engineers as an aid to the problem of selecting processes and costing design alternatives in the context of concurrent engineering. The work will also be useful as an introduction to manufacturing processes and their selection for students of engineering and management.

Acknowledgements

We would like to thank the Engineering and Physical Sciences Research Council (EPSRC) of the UK for support of our research in engineering for manufacture, including research under grant number GR/J97922 concerned with probabilistic design and process capability analysis and under GR/M53103 and GR/M55145 on the Designers' Sandpit project. Also, we are greatly indebted to our industrial collaborators on the various EPSRC projects, not least Phil Baker, Graham Hird, Duncan Law and Brian Miles while at CSC Manufacturing Ltd, and to Richard Batchelor, formerly of TRW.

We are immensely grateful to the engineering students and researchers at the Universities of Bristol and Hull, who have been involved over many years in supporting the research in engineering for manufacture. In addition, special thanks are given to Bob Swain for help in the preparation of many of the diagrams used in the Handbook and to Nathan Brown for his research in joining process selection.

K. G. Swift and J. D. Booker
September 2012

Notation

List of Terms

A	Total average cost of setting up and operating a specific process, including plant, labour, supervision and overheads, per second in the chosen country.
B	Average annual cost of tooling for processing an ideal component, including maintenance.
A_h	Basic handling index for an ideal design using a given handling process.
A_f	Basic fitting index for an ideal design using a given assembly process.
C_c	Relative cost associated with producing components of different geometrical complexity.
C_f	Relative cost associated with obtaining a specified surface finish.
C_{ft}	Value of C_t or C_f (whichever is greatest).
C_l	Labour rate.
C_{ma}	Total cost of manual assembly.
C_{mp}	Relative cost associated with material-process suitability.
C_{mt}	Cost of the material per unit volume in the required form.
C_s	Relative cost associated with size considerations and achieving component section reductions/thickness.
C_t	Relative cost associated with obtaining a specified tolerance.
F	Component fitting index.
H	Component handling index.
M_c	Material cost.
M_i	Manufacturing cost (pence).
n	Number of operations required to achieve the finished component.
N	Total production quantity per annum.
P_a	Penalty for additional assembly processes on parts in place.
P_c	Basic processing cost for an ideal design of component by a specific process.
P_f	Insertion penalty for the component design.
P_g	General handling property penalty.

P_o	Orientation penalty for the component design.
Ra	Roughness average (surface finish).
R_c	Relative cost coefficient assigned to a component design.
T	Process time in seconds for processing an ideal design of component by a specific process.
V	Volume of material required in order to produce the component.
V_f	Finished volume of the component.
W_c	Waste coefficient.
α	Cost of setting up and operating a specific process, including plant, labour, supervision and overheads, per second.
β	Process-specific total tooling cost for an ideal design.

Units

m	metre
μm	micron/micrometre
mm	millimetre
t	tonne (metric)
kg	kilogramme
ℓ	litre
g	gramme
h	hour
min	minute
s	second
rpm	revolutions per minute

Abbreviations – General

CAD	Computer-aided Design
C_{pk}	Process Capability Index
DFA	Design for Assembly
DFM	Design for Manufacture
HV	Vickers Hardness
JIT	Just in Time
NDT	Non-destructive Testing
OEM	Original Equipment Manufacturer
PDS	Product Design Specification
ppm	parts per million
PRIMA	Process Information Map

Abbreviations – Manufacturing Processes

3DP	3D Printing
AJM	Abrasive Jet Machining
ATB	Automated Torch Brazing
ATS	Automated Torch Soldering
CM	Chemical Machining
CNC	Computer Numerical Control
CVD	Chemical Vapour Deposition
CW	Cold Welding
DB	Dip Brazing
DS	Dip Soldering
DFW	Diffusion Bonding (Welding)
DFB	Diffusion Brazing
DMLS	Direct Metal Laser Sintering
EBM	Electron Beam Machining
EBW	Electron Beam Welding
ECG	Electrochemical Grinding
ECM	Electrochemical Machining
EDG	Electrical Discharge Grinding
EDM	Electrical Discharge Machining
EGW	Electrogas Welding
ESW	Electroslag Welding
EXW	Explosive Welding
FB	Furnace Brazing
FS	Furnace Soldering
FCAW	Flux Cored Arc Welding
FDM	Fused Deposition Modelling
FRW	Friction Welding
FW	Flash Welding
GW	Gas Welding
IB	Induction Brazing
INS	Iron Soldering
IRB	Infrared Brazing
IRS	Infrared Soldering
IS	Induction Soldering
JPS	Jetted Photopolymer System
LBM	Laser Beam Machining
LBW	Laser Beam Welding
LOM	Laminated Object Manufacturing

MIG	Metal Inert-gas Welding
MIM	Metal Injection Moulding
MMA	Manual Metal Arc Welding
NDT	Non-destructive Testing
NTM	Non-traditional Machining
PAW	Plasma Arc Welding
PVD	Physical Vapour Deposition
RB	Resistance Brazing
RP	Rapid Prototyping
RPW	Resistance Projection Welding
RS	Resistance Soldering
RSEW	Resistance Seam Welding
RSW	Resistance Spot Welding
SAW	Submerged Arc Welding
SGC	Solid Ground Curing
SLA	Stereolithography
SLS	Selecting Laser Sintering
SOUP	Solid Object Ultraviolet Laser Plotter
SW	Stud Arc Welding
TB	Manual Torch Brazing
TIG	Tungsten Inert-gas Welding
TS	Manual Torch Soldering
TW	Thermit Welding
USM	Ultrasonic Machining
USW	Ultrasonic Welding
USEW	Ultrasonic Seam Welding
WS	Wave Soldering

Manufacturing Process Key (for Chapter 12)

AM	Automatic Machining
CCEM	Cold Continuous Extrusion (Metals)
CDF	Closed Die Forging
CEP	Continuous Extrusion (Plastics)
CF	Cold Forming
CH	Cold Heading
CM2.5	Chemical Milling (2.5 mm depth)
CM5	Chemical Milling (5 mm depth)
CMC	Ceramic Mould Casting
CNC	Computer Numerical Controlled Machining

CPM	Compression Moulding
GDC	Gravity Die Casting
HCEM	Hot Continuous Extrusion (Metals)
IC	Investment Casting
IM	Injection Moulding
MM	Manual Machining
PDC	Pressure Die Casting
PM	Powder Metallurgy
SM	Shell Moulding
SC	Sand Casting
SMW	Sheet-metal Work
VF	Vacuum Forming

Materials Key (for Plastics Processing)

ABS	Acrylonitrile Butadiene Styrene
BMC	Bulk Moulding Compound
CA	Cellulose Acetate
CP	Cellulose Propionate
PF	Phenolic
PA	Polyamide
PBTP	Polybutylene Terephthalate
PC	Polycarbonate
PCTFE	Polychlorotrifluoroethylene
PE	Polyethylene
PESU	Polyethersulphone
PETP	Polyethylene Terephthalate
PMMA	Polymethylmethacrylate
POM	Polyoxymethylene
PPS	Polyphenylenesulphone
PP	Polypropylene
PS	Polystyrene
PSU	Polysulphone
PU	Polyurethane
PVC-U	Poly Vinyl Chloride – Unplasticised
SAN	Styrene Acrylonitrile
SMC	Sheet Moulding Compound
UP	Polyester

CHAPTER 1

Introduction to the Handbook

1.1 The Economic Case for Manufacturing Process Selection

A productive manufacturing engineering sector comes from making the best use of our manufacturing processes – understanding and utilising their technological capabilities and the economic opportunities they can offer. If we can get the quality and cost of our products right for our customers, then our businesses will reap the financial rewards.
It is often thought that selecting manufacturing processes for making our products is the responsibility of the production people. In reality, this can be highly influenced by design – the processes used can be largely predetermined by the design alternative selected: choice of materials, sizes, shapes, finishes and tolerances, etc. Therefore, process options should feature at the concept selection stage. There are always design alternatives, with some being more economic and robust than others. All other things being equal, the design professional should select the most competitive design in terms of quality and cost while ensuring that the functional requirements of the design are met.

So the designer has the huge responsibility of guaranteeing that the product will conform to customer requirements, comply with specification and ensure quality in every aspect of the product, including its manufacture and assembly, all within compressed timescales. Cost and quality are essentially designed into products (or not!) in the early stages of the product introduction process and thus consideration of manufacturing problems at the design stage is the major means available for reducing manufacturing costs, improving quality and increasing productivity. Such considerations are particularly important in the current economic climate, where the vast majority of (if not all) manufacturing businesses are facing pressure on their margins. Therefore, this Handbook has been designed to make process selection easy, efficient and effective, providing a ready opportunity to engineer a product that gives customers the functional performance they want at a competitive price and with a minimum of design risk.

Regarding design risk, the company that waits until the product is at the end of the line to measure its conformity and cost will not be competitive. The need to understand and quantify the consequences of design decisions on product manufacture and quality of conformance has never been greater. It has been found that more than 30% of product development effort can be wasted on rework [1] and it is not uncommon for manufacturing operations to have a 'cost of quality' equal to 25% of total sales revenues [2].

Manufacturing Process Selection Handbook. http://dx.doi.org/10.1016/B978-0-08-099360-7.00001-X

Why do we continually face these difficulties? The costs 'fixed' at the planning and design stages in product development are between 60% and 85%, while the costs actually incurred at that stage range from only 5% to 7% [3]. Therefore, the more problems prevented early on, through careful engineering of a design for manufacture, the fewer problems that have to be corrected later when they are difficult and expensive to change.

Increasing manufacturing efficiency, improving quality and reducing costs does not only accrue from investment in automation and advanced machine tools. The benefits of picking the right process can be enormous. This point is illustrated in Figure 1.1, which shows the relative costs and technical merits for a number of components, and resulting from alternative processing routes.

We know from Design for Manufacture (DFM) and Design for Assembly (DFA) research that huge savings can also be made at the product level. The results of numerous applications of these approaches, carried out across a wide variety of industries, show average part-count

Component	Material	Number Per Annum	Manufacturing Process	Relative Economic and Technical Merits	Relative Cost
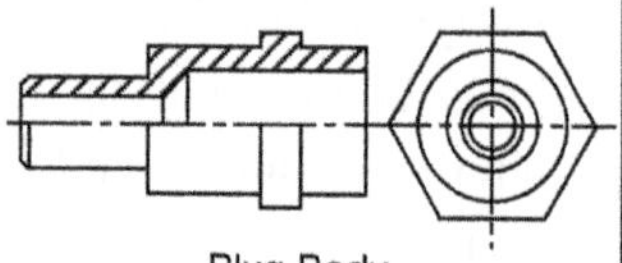 Plug Body	Low Carbon Steel	1,000,000	Machining	High waste Low to medium production rates Poor strength	4.2
			Cold Forming	Little waste Very high production rates High strength	1
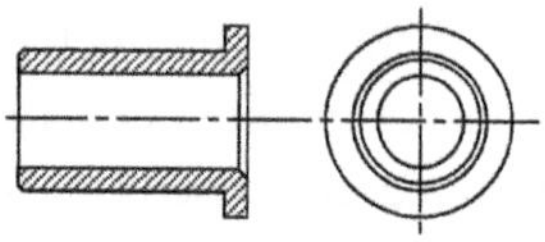 Plain Bearing	Bronze	50,000	Machining	High waste Low to medium production rates Non-porous properties	2.2
			Powder Metal Sintering	No waste High production rates Porous product	1
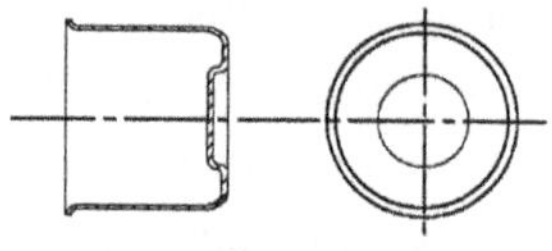 Cover	Alum. Alloy	5,000	Spinning	High labour costs Low production rates Limited detail and accuracy	1.8
			Deep Drawing	Low labour costs High production rates High detail and accuracy	1
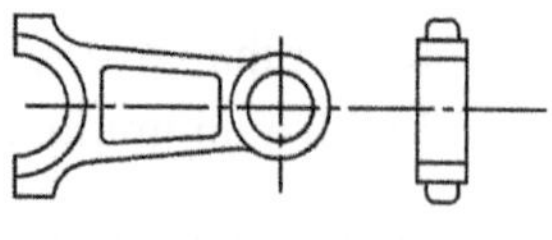 Connecting Rod	Medium Carbon Steel	100,000	Closed Die Forging	Long lead times High tooling costs High equipment costs	1.3
			Sand Casting	Short lead times Low tooling costs Low equipment costs	1
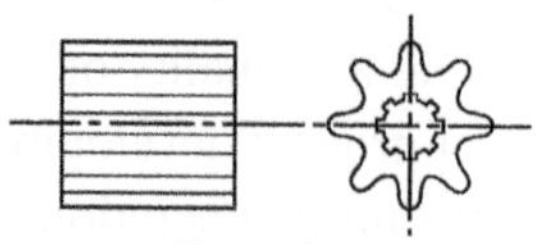 Pump Gear	Low Carbon Steel	5,000	Machining	High waste Low to medium production rates Poor strength	2.6
			Cold Extrusion	Little waste Very high production rates High strength	1

Figure 1.1: Contrast in Component Cost for Different Processing Routes.

reductions of almost 50% and average assembly cost savings of 45%. Associated savings in product cost of the order of 30% are not uncommon [4–7].

Where do these product savings come from? DFA is particularly interesting in the context of this Handbook, since its main benefits result from systematically reviewing functional requirements and replacing component clusters by single integrated pieces and by selecting alternative joining processes [4]. Therefore, invariably the proposed design solutions rely heavily on adopting different manufacturing processes – material combinations as shown in the part-count reduction examples in Figure 1.2. (A number of guidelines for manufacturing and assembly-oriented design are provided for the reader in Appendices A and B.)

Example 1

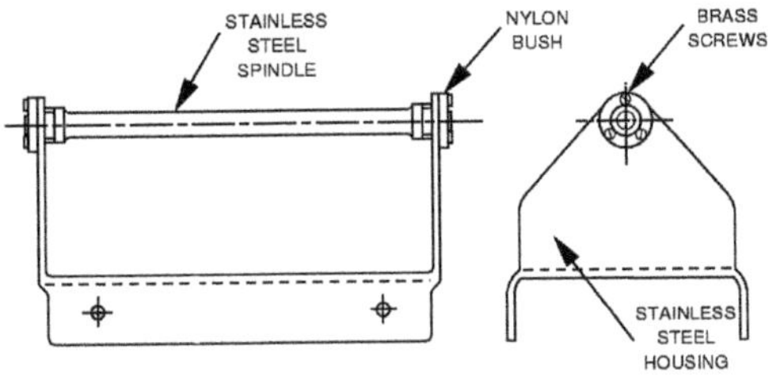

This spindle/housing assembly (sheet metal housing) has ten separate parts and an assembly efficiency rating of 7%.

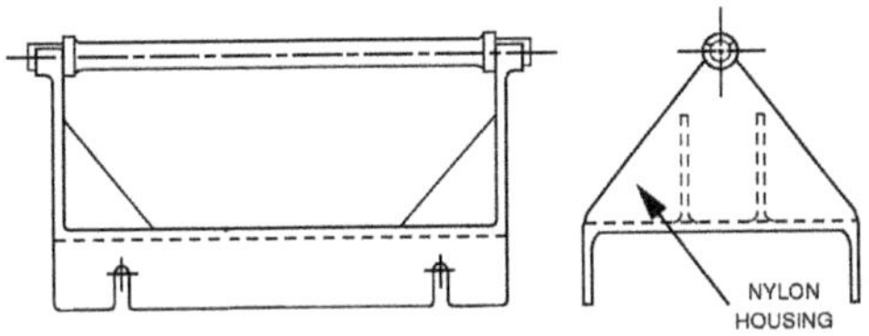

The two-part design, utilising an injection-moulded nylon housing, has an assembly efficiency rating of 93%.

Example 2

a)

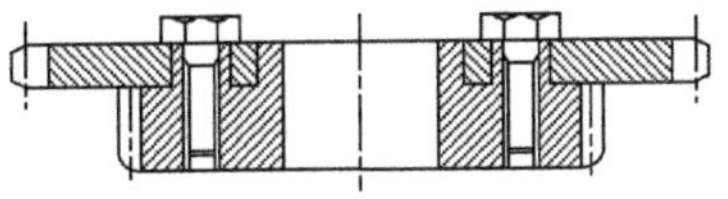

b)

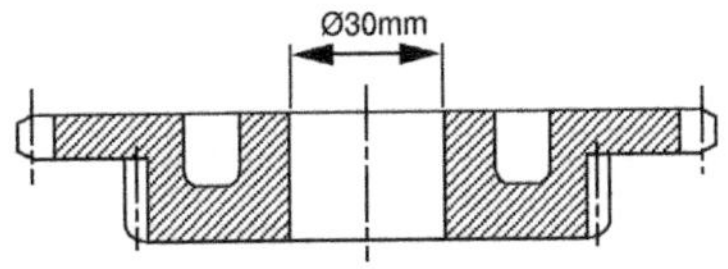

Two different types of sprocket and gear wheel: (a) includes assembly: this is made of steel and produced by machining. The individual teeth are cut. It is necessary to divide it into two elements due to production technique reasons. (b) does not include assembly: this is produced from sintered metal, the teeth being sintered in accordance with the required tolerance and surface quality; the advantages being no waste material, short processing time and no assembly.

Figure 1.2: Sample DFA Case Studies to Illustrate the Power of Process Selection (Example 1 after Ref. [4], Example 2 after Ref. [7]).

So what's the underlying problem? While some designers have practical experience of production processes and understand the limitations and capabilities they must work within, there are many more that do not [4]. The designer needs to be aware of the importance of manufacturing and assembly, and understand the processes and capabilities they are designing for in order to mitigate problems and potentially save money. Furthermore, the effects of assigning tolerances and specifying geometry and materials in design have far-reaching implications on manufacturing operations and service life, and the associated risks are rarely (if ever) properly understood. Understanding the effects of variability and the severity/cost of failure is key to risk assessment and its management.

The use of design techniques such as DFM and DFA early in the product development process has emerged as an effective way of reducing costs and improving competitiveness as they help measure the performance of designs and support the experience of the designer. In order to achieve the required quality and cost objectives for the manufacture of a design, it is necessary to carry out the interrelated activities of selecting candidate processes and tuning a design to get the most out of a chosen manufacturing route. This is not always so easy, particularly in the situation of new product development. In most cases there are several processes that can be used and selection depends on a large number of factors.

Different manufacturing technologies such as primary shape-generating processes, joining techniques and assembly systems require that selection takes place based on the factors relevant to that particular technology. Although there may be many important selection drivers with respect to each process technology, a simple and effective strategy for selection must be sought for the general situation and for usability.

Selection strategies based on key economic and technical factors interpreted from the Product Design Specification (PDS), or other requirements, are necessary. The selection strategies, together with the information provided by design guidelines, in-house data and handbooks, must complement business strategy and the costing of designs in order to provide a procedure that fully justifies final selection.

1.2 Manufacturing Process Information for Designers

The need to provide the design activity with information regarding manufacturing process capabilities and costs has been recognised for many years and some of the work that has been done to address this problem will be touched up on. However, there is relatively little published work in this area. The texts on design rarely include relevant data and while a few of the volumes on manufacturing processes do provide some aid in terms of process selection and costing [8–13], the information is seldom sufficiently detailed and systematically presented to do more than indicate the apparent enormity of the problem. Typically, the facts tend to be process specific and described in different formats in each case, making the engineer's task more difficult. There is a considerable amount of data available but precious little knowledge of how it can be applied to the problem of manufacturing process selection. The available information tends to be inconsistent: some processes are described in great detail, whilst others are perhaps neglected. This may give a disproportionate impression of the processes and their availability.

Information in manufacturing texts can also be found displayed in a tabulated and comparative form on the basis of specific process criteria. While useful, the design-related data tends to be somewhat limited in scope and detail. Such forms may be adequate if the designer has expertise in the respective processes, but otherwise, gaps in the detail leave room for misconceptions and may be a poor foundation for decision-making. Manufacturing catalogues and information on the Internet can be helpful; however, they tend to be sales orientated and, again, data is presented in different formats and at various levels of detail. Suppliers rarely provide much on design considerations or information on process capability. In addition, there are often differences in language between the process experts and the users.

In recent years a number of research groups have concentrated specifically on the design/manufacture interface. Processes and systems for cost estimation have been under development in areas such as machining, powder metallurgy, die casting and plastic moulding, and on broader techniques with the goal of providing DFM and cost-related information for the designer. A selection of cost estimation techniques for the early stages of design and a method for relating product cost to material cost, total batch size and level of underlying technology can be found in Refs [14–16].

Companies recognising the importance of design for manufacture have also searched for many years for a solution to this problem, with most opting for some kind of product 'team' approach, involving a multitude of persons supposedly providing the necessary breadth of experience in order to obtain 'production-friendly products'. While sometimes obtaining reasonable results, this approach often faces a number of obstacles, such as: assembling the persons with the relevant experience; lack of formal structure (typically such meetings tend to be unstructured and often involve ad hoc attacks on various 'pet' themes); the location of the persons required in the team can also present problems (not only can designers and production engineers be found in different functional departments, but they can frequently be on different sites, in foreign countries and are, in the case of subcontractors, in different companies). In addition, the chances are that the expertise in the team will only cover the primary activities of the business and hence the opportunity to exploit any benefits from alternative processes may be lost.

1.3 This Handbook, its Objectives and Strategy

When considering alternative design solutions for cost and quality, it is necessary to explore candidate materials, geometries and tolerances etc., against possible manufacturing routes. This requires some means of selecting appropriate processes and estimating the costs of manufacture early on in product development, across a whole range of options. In addition, the costs of non-conformance [17] need to be understood – that is, appraisal (inspection and testing) and failure, both internal (rework, scrap, design changes) and external (warranty claims, liability claims and product recall). Therefore, we also need a way of exploring conformance levels before a process is selected. For more information on this important aspect of design, the reader is directed to Ref. [18].

The primary objective of this Handbook is to provide support for manufacturing process selection in terms of technological feasibility, quality of conformance and manufacturing cost. The strategy adopted for the satisfaction of this objective is based on:

- Selection strategies, for each generic process group, that can be used to focus attention on those processes that are likely candidates based on top-level product and process requirements.
- Information on the characteristics and capabilities of a range of important manufacturing, rapid prototyping, surface engineering, joining and assembly processes. The intention is to promote the generation of design ideas and facilitate the matching and tuning of a design to a process.
- Methods and data to enable the exploration of design solutions for component manufacturing and assembly costs in the early stages of the product design and development process.

To provide for the first point, simple selection strategies have been formulated that enable key economic and technical factors that may be readily interpreted from the PDS and other important requirements, to be used to refine the process search space. In this way, processes can be filtered out and smaller grouping of candidates created. The candidates may then be further refined by considerations of more detailed technological and cost factors based on the data provided under the second and third points discussed below.

To provide for the second point, a set of manufacturing PRocess Information Maps (PRIMAs) have been developed. All in a standard format for each process, the PRIMAs present knowledge and data on areas including: material suitability, design considerations, quality issues, economics, and process fundamentals and process variants. The information includes not only design considerations relevant for the respective processes, but quite purposefully, an overview of the functional characteristics of the process so that a greater overall understanding may be achieved. Within the standard format, a similar level of detail is provided on each of the processes included. The format is very deliberate. Firstly, an outline of the process itself – how it works and under what conditions it functions best. Secondly, a summary of what it can do – the limitations and opportunities it presents – and finally an overview of quality considerations, including process capability charts for relating tolerances to characteristic dimensions.

To provide for the third point, techniques are put forward that can be used to estimate the costs of component manufacture and assembly for designs at an early stage of development.

It enables the effects of product structure, design geometry and materials to be explored against various manufacturing and assembly routes. A sample data set is included that enables the techniques to be used to predict component manufacturing and assembly costs for a wide range of processes and materials. The process of cost estimation is illustrated through a number of case studies and the scope for, and importance of application with company-specific data is discussed.

The body of the Handbook begins with the strategies employed for PRIMA selection, where attention is focused on identification of candidate processes based on strategic criteria such as material, process technology and production quantity. Having identified the possible targets, the data in the PRIMAs is used to do the main work of selection. The PRIMAs include the five main manufacturing process groups: casting; plastic and composite processing methods; forming; machining, and non-traditional processes. In addition, rapid prototyping, surface engineering process, the main assembly systems and the majority of commercially available joining processes are covered. In all, 85 PRIMAs are presented, giving reference to well over 100 manufacturing, assembly and joining processes. Note that the PRIMAs have been designed such that they can also be used independently – as a source of reference for process operational characteristics, process capabilities and key design considerations.

Following on from the PRIMAs, the Handbook concentrates on the cost estimation methodologies for components and assemblies, their background, theoretical development and application. In practice, the selection strategies and PRIMAs facilitate the selection of the likely candidate processes for a design from the whole range of possibilities, while cost estimation techniques provide a means of gauging the manufacturing and assembly costs of the candidate processes.

1.3.1 Outline Process for Design for Manufacture and Assembly

For process selection (or DFA for that matter) to be most effective, it all needs to be part of the product introduction process. The placing in the product design cycle of process selection, in the context of engineering for manufacture and assembly is illustrated in Figure 1.3. The selection of an appropriate set of candidate processes for a product is difficult to undertake without a sound PDS. A well-constructed PDS lists all the requirements of the customers, end-users and the business that need to be satisfied. The reader is referred to Ref. [19] for information on how to develop a PDS. It should be written and used by the Product Team and provide a reference point for any emerging design or prototype. Any conflict between customer needs and product functionality should be referred back to the PDS.

The first step in the process is to analyse the design or prototype with the aim of simplifying the product structure and optimising part-count. As shown earlier, without proper analysis,

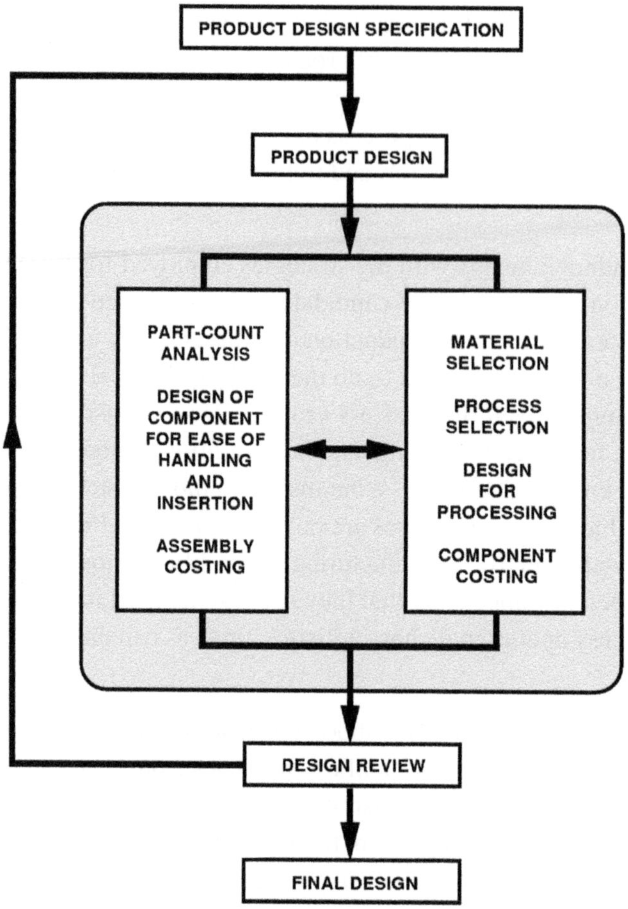

Figure 1.3: Outline Process for Design for Manufacture and Assembly.

design solutions invariably tend to have too many parts. Therefore, it is important to identify components that are candidates for elimination or integration with mating parts (every component part must be there for a reason and the reason must be in the PDS). This must be done with due regard for the feasibility of material process combinations and joining technology. A useful and popular approach for material selection in mechanical design is provided in Ref. [20] and related software [21]. Materials selection in terms of environmental issues can be found in Ref. [22]. Also, Ref. [23] provides coverage and information of the design/materials/manufacture interface. Figure 1.4 provides a general classification of engineering materials.

The next steps give consideration to the problems of component handling and fitting processes, the selection of appropriate manufacturing processes and ensuring that components are tuned to the manufacturing technology selected. Estimation of component manufacture

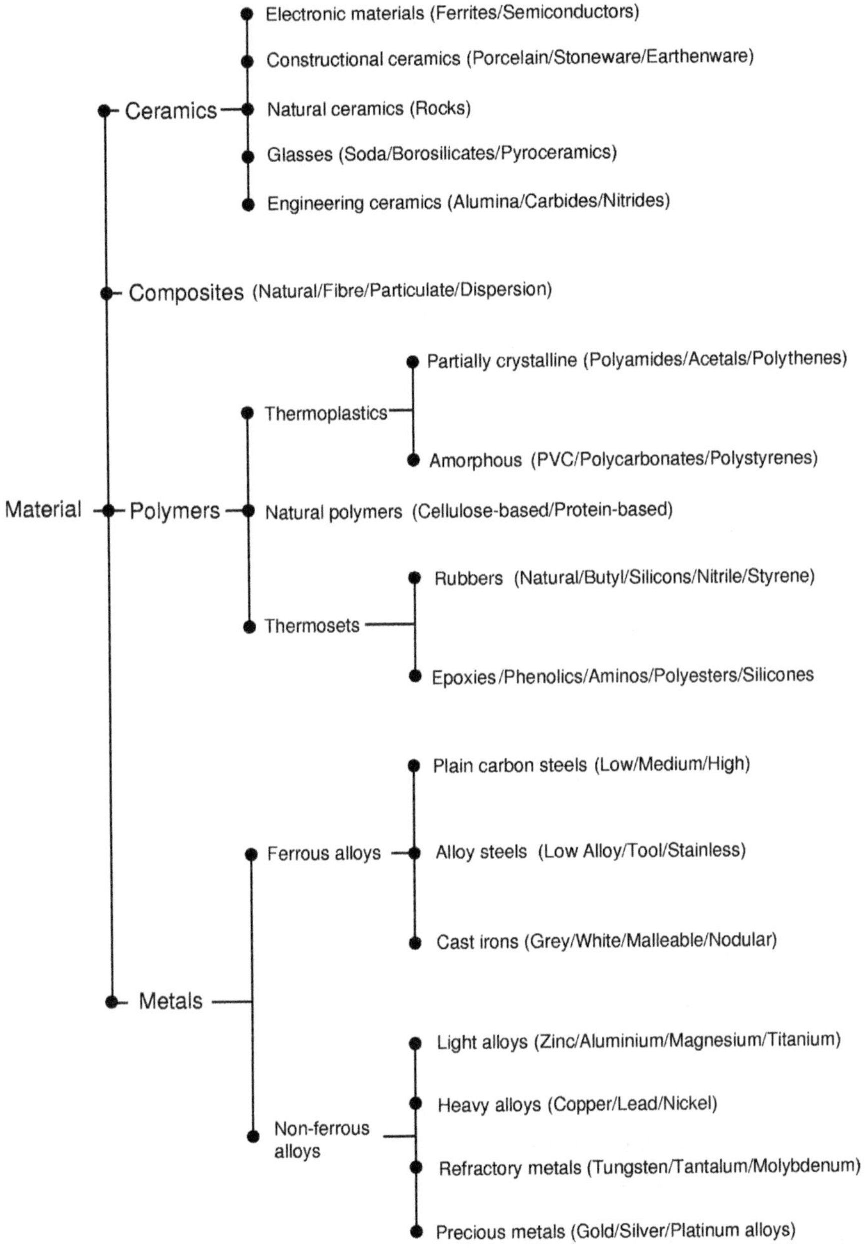

Figure 1.4: General Classification of Materials.

and assembly costs during the design process are important for both assessing a design against target costs and in trade-off analysis. Overall, the left-hand side of Figure 1.3 is closely related to DFA while the right-hand side involves material selection, process selection and component design for processing – essentially considerations of DFM.

1.3.2 Classification of Materials and Processes

As mentioned previously, selecting the right manufacturing process is not always simple and obvious. In the vast majority of cases there are several processes and materials that can be used for a component design. The problem is compounded by the range of manufacturing processes and wide variety of material types commonly in use. Figures 1.5–1.10 provide a general classification, hierarchy and guide to the range of processes – component manufacturing, rapid prototyping, assembly, joining, and bulk and surface engineering respectively – that are widely available. All of these processes are discussed in detail in the Handbook except bulk surface engineering treatments, and the reader is referred to Refs [8–13] for more information on these. This is primarily because only surface treatments and coatings modify component surfaces to improve wear resistance, fatigue life and corrosion resistance, which are all key selection drivers.

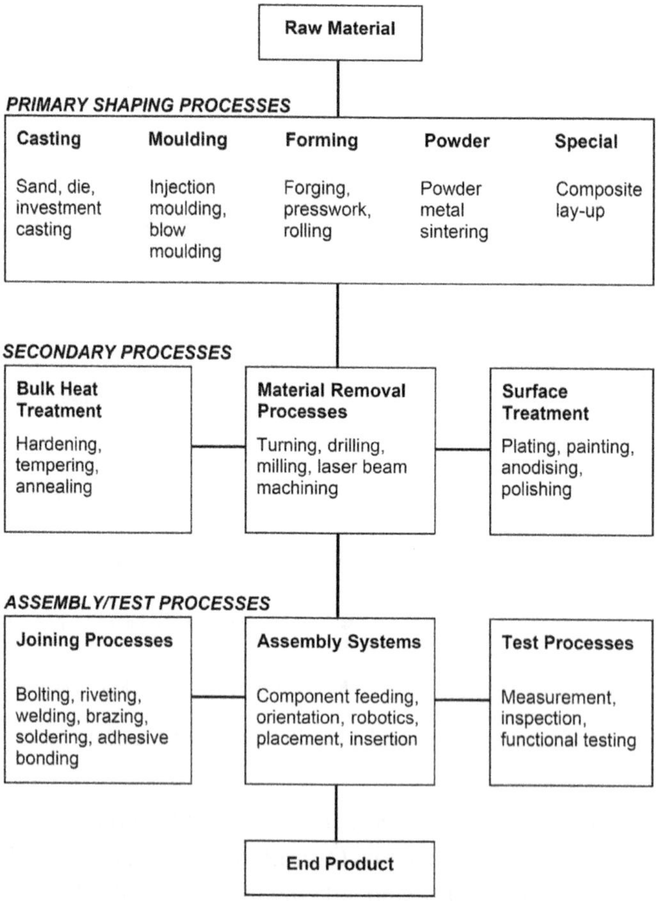

Figure 1.5: Hierarchy of Manufacturing Processes.

1.3.3 Process Selection Strategy Overview

Different manufacturing technologies such as primary shape-generating processes, rapid prototyping, joining techniques, assembly systems and surface engineering processes require that selection takes place based on the factors relevant to that particular class of process technology. For example, the selection of a joining technique may be heavily reliant on the ability of the process to join dissimilar materials of different thickness. This is a particular

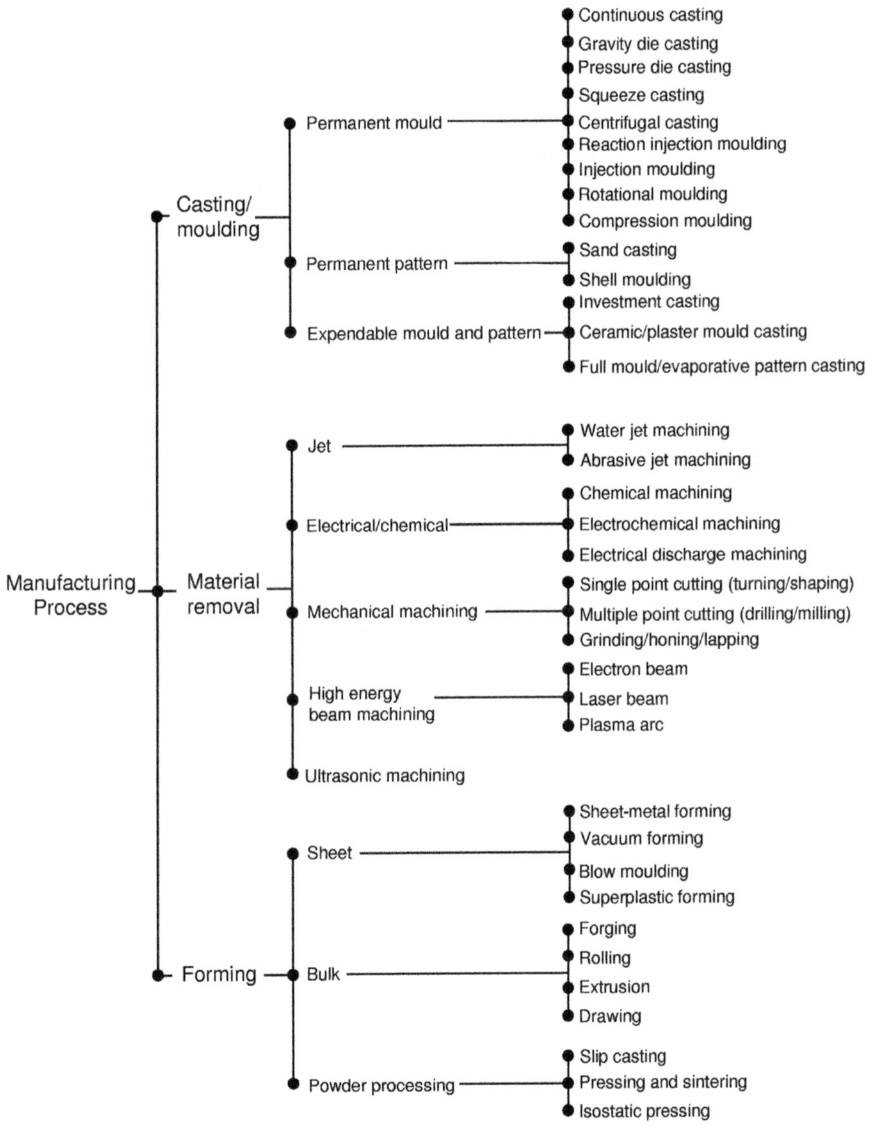

Figure 1.6: General Classification of Manufacturing Processes.

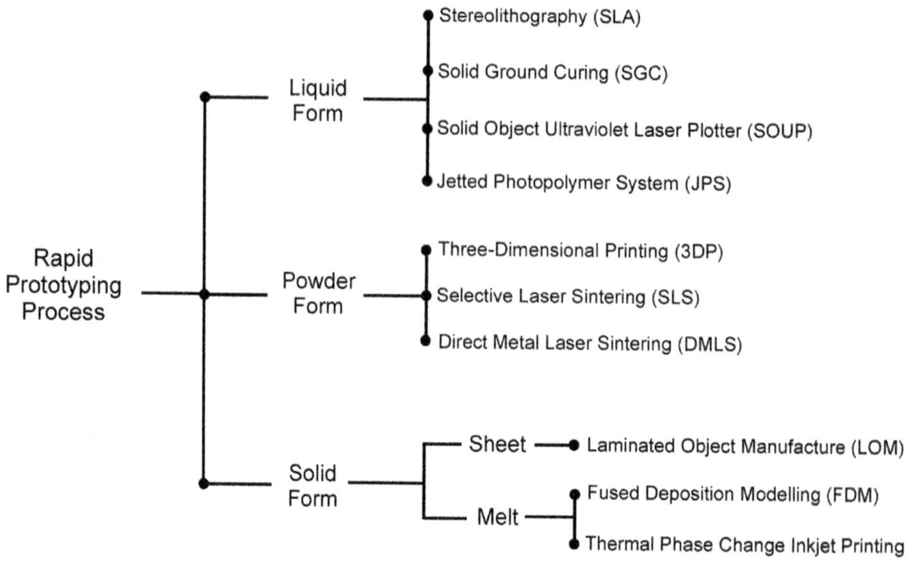

Figure 1.7: General Classification of Selected Rapid Prototyping Processes.

requirement not necessarily defined by the PDS, but one that has been arrived at through previous design decisions, perhaps based on spatial or functional requirements, whereas assembly system selection may simply be dictated by a low labour rate in the country of manufacture and therefore manual assembly becomes viable for even relatively large production volumes.

Although there may be many important selection drivers with respect to each process technology, a simple and effective strategy for selection must be sought that is widely applicable and usable. Selection strategies can be formulated by concentrating on the key economic and technical factors that may be readily interpreted from the PDS and other important requirements. Put in a wider context, the selection strategies, together with the information provided in the PRIMAs, must complement business strategy and the costing of designs in order to provide a procedure that fully justifies final selection. A flowchart is shown in Figure 1.11 relating the general factors and activity stages relevant to the process selection strategies discussed in Chapter 2 of the Handbook.

However, for any process, design considerations that should be taken into account are highlighted below:

- What shapes can be manufactured?
- What constraints on dimensional parameters are imposed by the process?
- What accuracy can be produced by the process?
- What are the implications of the process for surface condition?

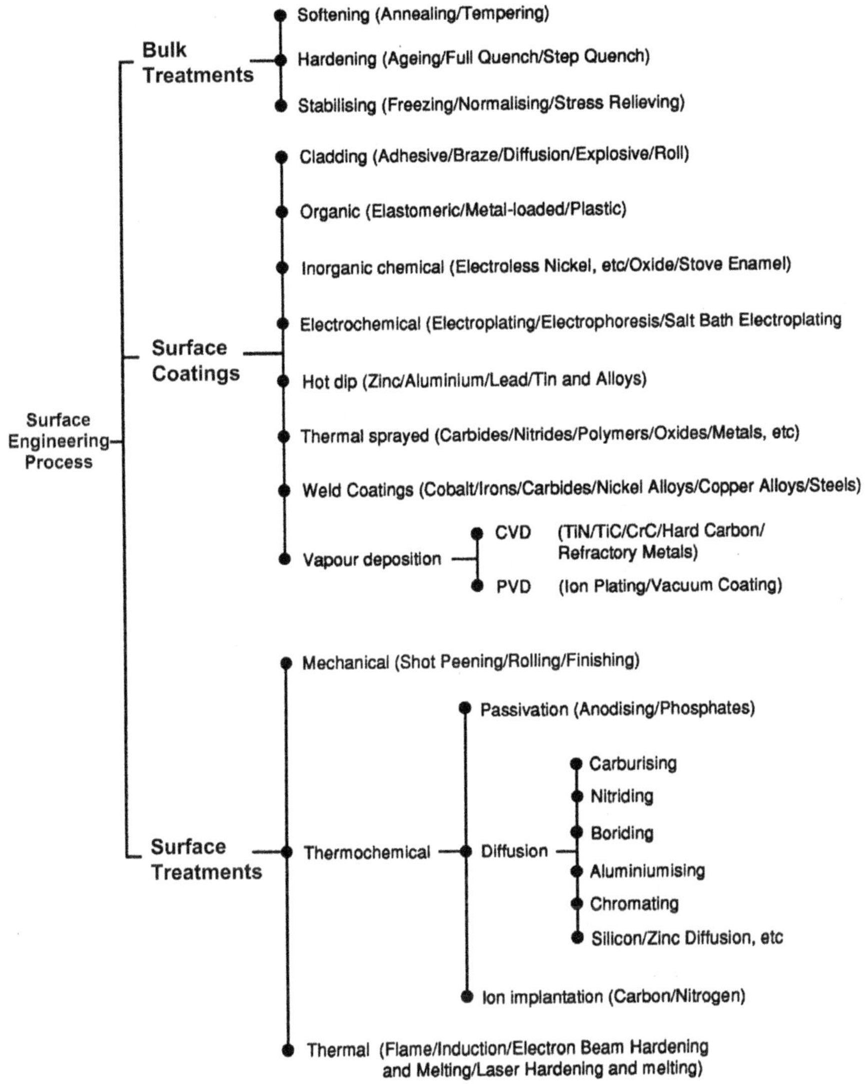

Figure 1.8: General Classification of Surface Engineering Processes.

- What are the costs of the process and of the required tooling?
- What are the component integrity implications?

Remember, in all design for manufacture projects:

- Think about manufacturing upfront.
- Consult the manufacturer, whether in-house or supplier, but be informed about the process and its capabilities, in order to get the most out of the dialogue.
- Search for general design rules and specific process guidelines.

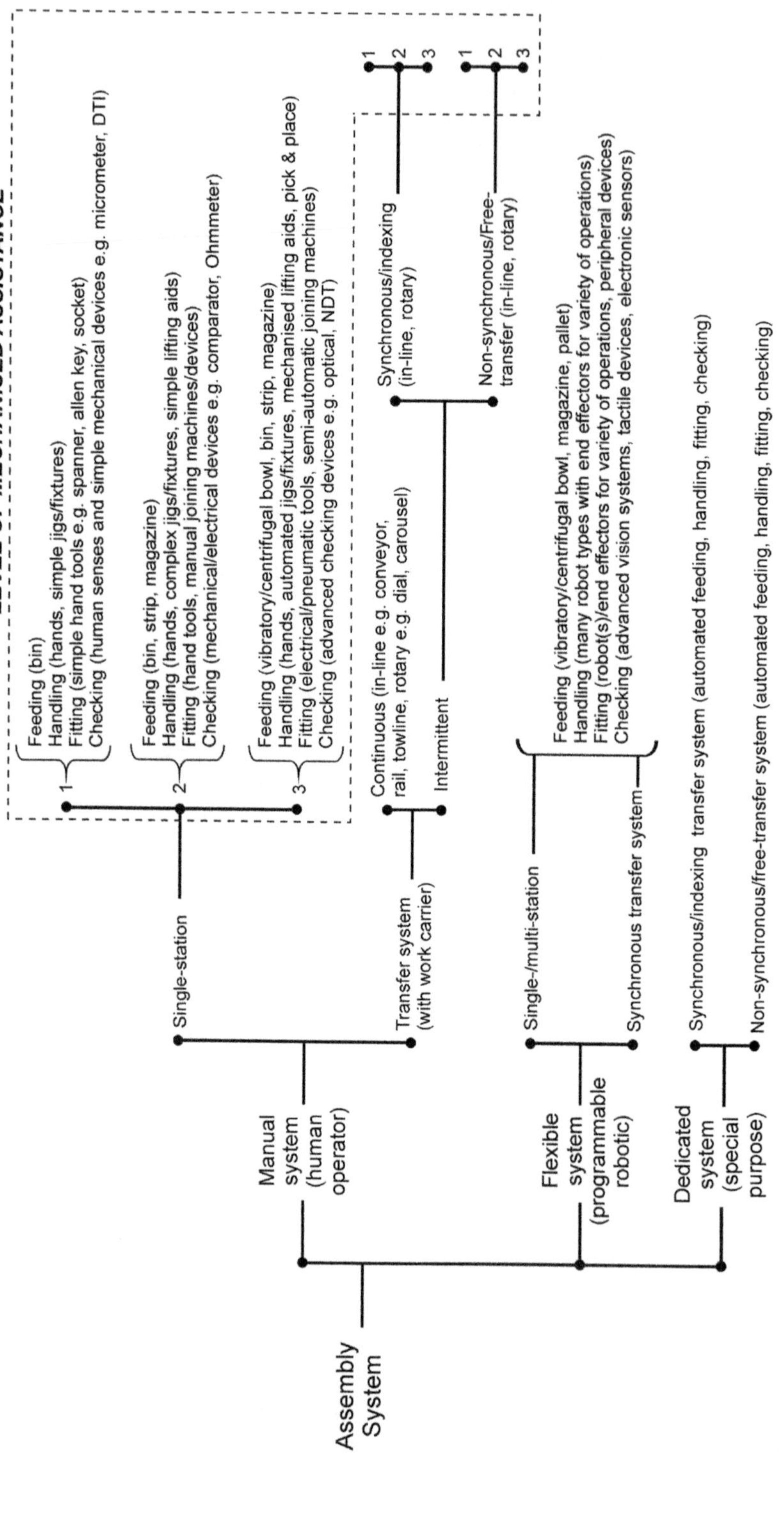

Figure 1.9: General Classification of Assembly Systems.

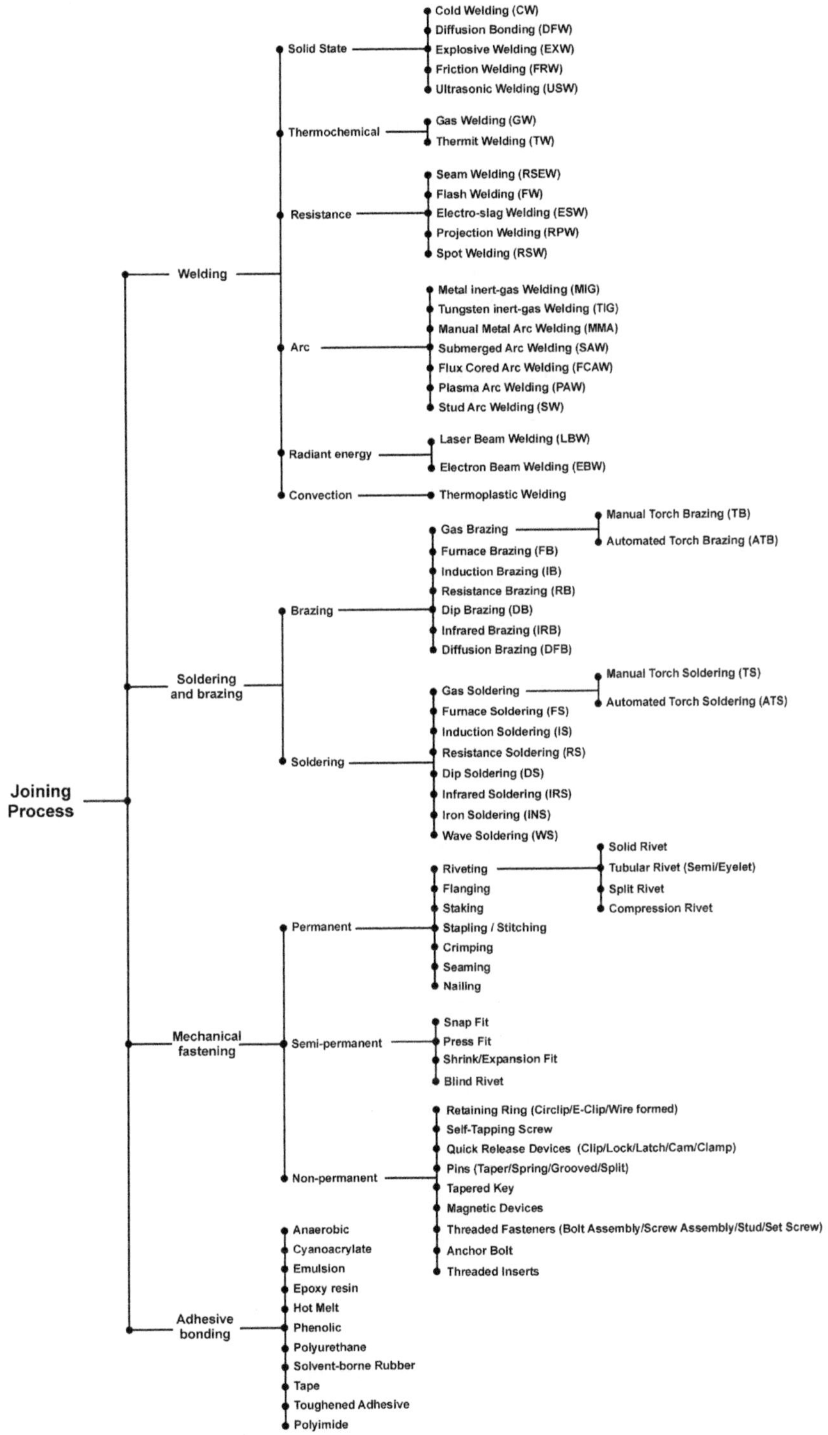

Figure 1.10: General Classification of Joining Processes.

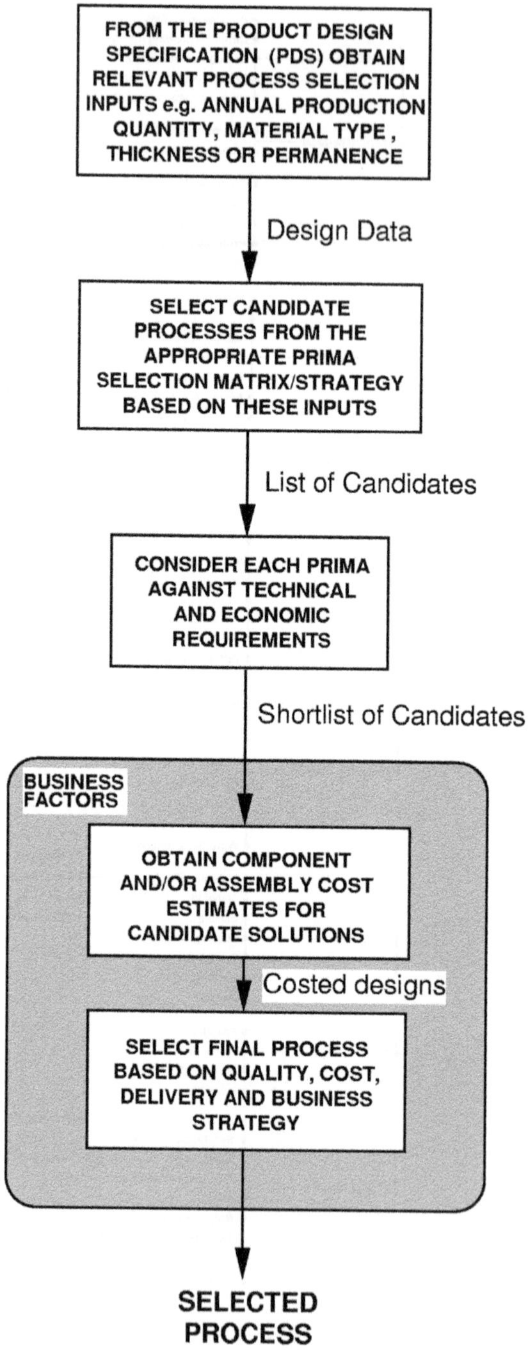

Figure 1.11: General Process Selection Flowchart.

Case studies are used to support the application of each of the process selection strategies. The studies are examples of real problems that provide for validation and show how to go about the application of the strategic concepts. They provide the opportunity for the reader to try out the strategies for themselves and bridge the gap between theory and practice. Case studies are also employed later in the Handbook to contextualise component and assembly costing methodologies (see Chapters 12 and 13).

1.3.4 PRIMA Structure

As touched on previously, selecting the right process and optimising the design to suit the process selected involves a series of decisions that exert considerable influence on the quality and cost of components and assemblies. Such decisions can significantly affect the success of a product in the marketplace. Again, in selecting processes and tuning designs for processing many factors need to be taken into consideration. The PRIMAs presented in this Handbook attempt to provide the knowledge and data required to refine the decisions coming out of the various process selection strategies. It is the PRIMAs that provide the means of making more detailed assessments regarding the technological and economic feasibility of a process.

Design considerations are provided to enable the designer to understand more about the technical feasibility of the design decisions made. The process quality considerations give valuable information on process conformance, including data on process tolerance capability associated with characteristic dimensions. A good proportion of the PRIMAs are associated with quality considerations as non-conformance often represents a large quality cost in a business. Such losses result from rework, order exchange, warranty claims, legal actions and product recall. The goal is to provide data that enables the selection of processes that have the capability to satisfy the engineering needs of the application, including those associated with conformance to quality requirements.

Each PRIMA is divided into seven categories as listed and defined below, covering the characteristics and capabilities of the process:

- **Process Description**: an explanation of the fundamentals of the process together with a diagrammatic representation of its operation and a finished part.
- **Materials**: describes the materials currently suitable for the given process.
- **Process Variations**: a description of any variations of the basic process and any special points related to those variations.
- **Economic Considerations**: a list of several important points, including: production rate, minimum production quantity, tooling costs, labour costs, lead times and any other points that may be of specific relevance to the process.
- **Typical Applications**: a list of components or assemblies that have been successfully manufactured or fabricated using the process.

- **Design Aspects**: any points, opportunities or limitations that are relevant to the design of the part as well as standard information on minimum section, size range and general configuration.
- **Quality Issues**: standard information includes a process capability chart (where relevant), typical surface roughness and detail, geometrical tolerance data (where relevant), as well as any information on common process faults.

A key feature of the PRIMAs is the inclusion of process capability charts for many of the primary shaping and material removal processes. Tolerances tend to be dependent on the overall dimension of the component characteristic and the relationship is specific and largely non-linear. The charts have been developed to provide a simple means of understanding the influence of dimension on tolerance capability. The regions of the charts are divided by two contours. The region bounded by these two contours represents a spectrum of tolerance–dimension combinations where the process capability index $C_{pk} \geq 1.33$ is achievable. If the process characteristic is a normal distribution, C_{pk} can be related to a parts per million (ppm) defect rate. $C_{pk} = 1.33$ equates to a defect rate of 30 ppm at the nearest limit. At $C_{pk} = 1$, the defect rate equates to approximately 1350 ppm. See Ref. [24] for more information about process capability indices. Below the lowest contour, tolerance–dimension combinations are likely to require special control or secondary processing if $C_{pk} = 1.33$ is to be realised.

In the preparation of the process capability charts it has been assumed that the component shape is well suited to the process and that all operational requirements are satisfied. Where the material under consideration is not mentioned on the charts, care should be taken. Any adverse affects due to this or shape-driven component variation should be taken into consideration. For more information the reader is referred to Ref. [18]. The data used in the charts has been compiled from contacts in industry and from published work. Although attempts have been made to standardise the data as far as possible, difficulties were faced in this connection since it was not always easy to obtain a consensus view. Consequently, as many as 20 different data sources have been used in the compilation of the individual process capability charts to provide an understanding of the general tolerance capability range offered by each manufacturing process.

The process capability charts refer to dimensional tolerances and do not explicitly relate to geometric tolerance capabilities such as roundness, flatness and parallelism, etc. However, it can be argued that the process capability charts in the PRIMAs may sometimes provide clues to the geometric tolerance capability of the process. As is the case with dimensional tolerances, the geometric tolerance capability tends to vary widely with the manufacturing process. For many processes, such as moulding, casting and forming, it might well be assumed that the geometric tolerance capabilities of the process are no better than that associated with the corresponding dimensional tolerances.

Conversely, for machining processes, many of the geometric capabilities can substantially exceed those associated with their linear counterparts. Frequently, given the nature of machining processes, it is the case that the machines used tend to be set up to produce linear dimensions, with geometric properties being implicit (dependent on the machine type, e.g. solids of revolution) and the resulting geometric capabilities are closer to 'machine' rather than 'process' capabilities.

Several of the machining process PRIMAs provide explicit data on likely geometric variation, but this is not extensive due to the scarcity of published work in the field of process capable geometric tolerancing. Given the widespread use of geometric tolerances and their importance in engineering design, this is clearly an opportunity for future research.

In the following, we will briefly define the terms used in the Handbook in connection with the data given in the PRIMAs on manufacturing variation associated with geometric characteristics. The geometric data specifies the likely variation in form or position from true geometry. The geometric variation data gives the width or diameter of the tolerance zone and are not ± values.

1.3.5 Definitions for Geometrical Characteristics

Taper: Variation in the conical form of a hole based on the differences in diameter over a given conical length.
Ovality: Ovality or non-circularity is the degree of deviation from perfect circularity of the cross-section of a hole, a condition in which a hole that should be round has two opposing lobes, resulting in an oval shape.
Roundness: Describes the condition of a hole characterised by having the same diameter relative to a central common axis.
Flatness: A three-dimensional geometric characteristic that controls how much a feature can deviate from a flat plane.
Cylindricity: A three-dimensional geometric tolerance that controls how much a feature can deviate from a perfect cylinder.
Straightness (cylinders or cones): A two-dimensional geometric characteristic that defines how much a characteristic deviates from a straight line.
Parallelism of flat surfaces: A three-dimensional geometric characteristic that defines how much a flat surface deviates from an orientation parallel to the specified datum.
Parallelism of cylinders: A three-dimensional geometric characteristic that defines how much a cylinder deviates from an orientation parallel to the specified datum.

For further information and resources on geometric dimensioning and tolerancing, the reader is directed to Refs [25–27].

References

[1] A. Parker, Engineering Is Not Enough. Manufacturing Engineer, December, 1997, pp. 267–271.
[2] D.F. Kehoe, The Fundamentals of Quality Management, Chapman & Hall, London, 1996.
[3] W.J. Fabrycky, Modeling and indirect experimentation in system design and evaluation, Journal of Systems Engineering 1 (1) (1994) 133–144.
[4] G. Boothroyd, P. Dewhurst, W. Knight, Product Design for Manufacture and Assembly, second ed., Marcel Dekker, New York, 2002.
[5] J. Shimada, S. Mikakawa, T. Ohashi, Design for manufacture, tools and methods: the assemblability evaluation method (AEM), in: Proc. FISITA'92 Congress, London, 1992, No. C389/460.
[6] B. Miles, K.G. Swift, Design for Manufacture and Assembly. Manufacturing Engineer, October, 1998, pp. 221–224.
[7] M. Andreasen, S. Kahler, T. Lund, Design for Assembly, second ed., IFS Publications/Springer-Verlag, New York, 1988.
[8] R.H. Todd, D.K. Allen, L. Alting, Manufacturing Processes Reference Guide, Industrial Press, New York, 1994.
[9] S. Kalpakyian, S.R. Schmid, Manufacturing Engineering and Technology, fourth ed., Prentice Hall, New York, 2009.
[10] D. Koshal, Manufacturing Engineer's Reference Book, Butterworth-Heinemann, Oxford, 1993.
[11] E.P. Degarmo, R.A. Kohser, Materials and Processes in Manufacturing, tenth ed., Wiley, New York, 2007.
[12] J.M. Walker (Ed.), Handbook of Manufacturing Engineering, Marcel Dekker, New York, 1996.
[13] J. Schey, Introduction to Manufacturing Processes, third ed., McGraw-Hill, New York, 1999.
[14] K. Schreve, H.R. Schuster, A.H. Basson, Manufacturing cost estimation during design of fabricated parts, in: Proc. EDC'98, Brunel University, UK, 1998 July, pp. 437–444.
[15] A.M.K. Esawi, M.F. Ashby, Cost estimates to guide pre-selection of processes, Materials and Design 24 (2005) 605–616.
[16] P. Liebl, M. Hundal, G. Hoehne, Cost calculation with a feature-based CAD system using modules for calculation, comparison and forecast, Journal of Engineering Design 10 (1) (1999) 93–102.
[17] J.J. Plunkett, B.G. Dale, Quality Costing, Chapman & Hall, London, 1991.
[18] J.D. Booker, M. Raines, K.G. Swift, Designing Capable and Reliable Products, Butterworth-Heinemann, Oxford, 2001.
[19] K.N. Otto, K. Wood, Product Design: Techniques in Reverse Engineering and New Product Development, Prentice-Hall, New York, 2001.
[20] M.F. Ashby, Material Selection in Mechanical Design, third ed., Butterworth-Heinemann, Oxford, 2004.
[21] www.grantadesign.com/products/ces, accesssed 17 July 2012.
[22] M.F. Ashby, Materials and the Environment: Eco-informed Material Choice, second ed., Butterworth-Heinemann, Oxford, 2012.
[23] G.E. Dieter (Ed.), tenth ed., ASM Handbook – Materials Selection and Design, Vol.20, ASM International, Ohio, 1997.
[24] S. Kotz, C.R. Lovelace, Process Capability Indices in Theory and Practice, Arnold, London, 1998.
[25] www.tec-ease.com, accessed 17 July 2012.
[26] P.J. Drake, Dimensioning and Tolerancing Handbook, McGraw-Hill, New York, 1999.
[27] B. Griffiths, Engineering Drawing for Manufacture, Penton Press, London, 2002.

CHAPTER 2

Process Selection Strategies and Case Studies

2.1 Manufacturing Process Selection

2.1.1 Selection Strategy

The term manufacturing processes used here represents the main shape-generating methods such as casting, moulding and forming processes, as well as traditional and non-traditional machining processes. The processes specific to this section are classified in Figure 1.6. The purpose is to provide a guide for the selection of the manufacturing processes that may be suitable candidates for a component. In most cases there are several processes that can be used for a component and final selection depends on a large number of factors, mainly associated with a range of technical capabilities and process economics, not least component size, geometry, tolerances, surface finish, capital equipment and labour costs [1]. Some of the main process selection drivers are listed below. The intention is not to infer that these are necessarily of equal importance or occur in this fixed sequence.

Important process selection drivers are:

- Product quantity.
- Equipment costs.
- Tooling costs.
- Processing times.
- Labour intensity and work patterns.
- Process supervision.
- Maintenance.
- Energy consumption and other overhead costs.
- Material costs and availability.
- Material to process compatibility.
- Component form and dimensions.
- Tolerance requirements.
- Surface finish needs.
- Bulk treatment and surface engineering.
- Process to component variability.
- Process waste.
- Component recycling.

Manufacturing Process Selection Handbook. http://dx.doi.org/10.1016/B978-0-08-099360-7.00002-1

The strategy for selecting manufacturing processes is described below. Points 1, 2 and 3 are specific to this category of processes, but points 4, 5 and 6 apply to all selection strategies.

1. Obtain an estimate of the annual production quantity.
2. Choose a material type to satisfy the PDS.
3. Refer to Figure 2.1 to select candidate PRIMAs.
4. Consider each PRIMA against the engineering and economic requirements:
   - Understand the process and its variations.
   - Consider the material compatibility.
   - Assess conformance of component concept with design rules.
   - Compare tolerance and surface finish requirements with process capability data.
5. Consider the economic positioning of the process and obtain component cost estimates for alternatives.
6. Review the selected manufacturing process against business requirements.

The intention is that the candidate processes are selected before the component design is finalised, so that any specific constraints and/or opportunities may be borne in mind. To this end, the manufacturing process PRIMA selection matrix (see Figure 2.1) has been devised based on just two basic requirements:

- *Material type* – Accounts for the compatibility of the parent material with the manufacturing process and is therefore a key technical selection factor. A large proportion of the materials used in engineering manufacture have been included in the selection methodology, from ferrous alloys to precious metals.
- *Production quantity per annum* – The number of components to be produced to account for the economic feasibility of the manufacturing process. The quantities specified for selection purposes are in the following ranges:
  - Very low volume = 1–100.
  - Low volume = 100–1,000.
  - Medium volume = 1,000–10,000.
  - Medium to high volume = 10,000–100,000.
  - High volume = 100,000+.
  - All quantities.

The justification for basing the matrix on material type and production quantity alone is that it combines technological and economic issues of prime importance in a simple manner. Many manufacturing processes are only viable for low-volume production due to the time and labour involved. On the other hand, some processes require expensive equipment and are therefore unsuitable for low production volumes. By considering production quantities in the early stages, a process can be selected that proves to be the most economical later in the development process. The boundaries of economic production, however, can be vague when so many factors are relevant; therefore, the matrix concentrates rather more on the use of

QUANTITY \ MATERIAL	IRONS	STEEL (carbon)	STEEL (tool, alloy)	STAINLESS STEEL	COPPER & ALLOYS	ALUMINIUM & ALLOYS	MAGNESIUM & ALLOYS	ZINC & ALLOYS	TIN & ALLOYS
VERY LOW 1 TO 100	[3.5][3.6] [3.7][6M] [7.1][7.4]	[3.5][3.7] [4.10] [6M][7.1] [7.5][7.6]	[3.1][3.5] [3.7][4.10] [6M][7.1] [7.5][7.6][7.7]	[3.5][3.7] [4.7][4.10] [6M][7.1] [7.5][7.6]	[3.5][3.7] [4.10][6M] [7.1]	[3.5][3.7] [4.7][4.10] [6M][7.5]	[3.6][3.7] [4.10][6M] [7.1][7.5]	[3.1][3.7] [4.10][6M] [7.5]	[3.1][3.7] [4.10][6M] [7.5]
LOW 100 TO 1,000	[3.2][3.5] [3.6][3.7] [6M][7.1] [7.3][7.4]	[3.2][3.5] [3.7][4.10] [6M][7.1] [7.3][7.4] [7.5]	[3.1][3.2] [3.7][6M][7.1] [7.3][7.4][7.5] [7.6][7.7]	[3.2][3.7] [4.7][4.10] [6M][7.1] [7.3][7.4][7.5]	[3.2][3.5][3.7] [3.8][4.5] [4.10][6M] [7.1][7.3][7.5]	[3.2][3.5][3.7] [3.8][4.7] [4.10][6M] [7.3][7.4][7.5]	[3.6][3.7] [3.8][4.10] [6M][7.5]	[3.1][3.7] [3.8][4.10] [6M][7.5]	[3.1][3.7] [3.8][4.10] [6M][7.5]
LOW TO MEDIUM 1,000 TO 10,000	[3.2][3.3] [3.5][3.6] [3.7][4.11] [6A][7.2]	[3.2][3.3][3.5] [3.7][4.1][4.3] [4.10][4.11] [6A][7.2][7.3] [7.4][7.5]	[3.2][3.5] [3.7][4.1][4.4] [4.11][6A][7.2] [7.3][7.4][7.5]	[3.2][3.5][3.7] [4.1][4.3][4.7] [4.10][4.11][6A] [7.2][7.3] [7.4][7.5]	[3.2][3.3][3.5] [3.8][4.1][4.3] [4.10][4.11][6A] [7.2][7.3][7.4]	[3.2][3.3][3.5] [3.8][4.1][4.3] [4.7][4.10][4.11] [6A][7.3][7.4] [7.5]	[3.3][3.6][3.8] [4.1][4.3][4.4] [4.10][6A][7.5]	[3.3][3.8] [4.3][4.10] [6A][7.5]	[3.3][3.8] [4.3][4.10]
MEDIUM TO HIGH 10,000 TO 100,000	[3.2] [3.3] [4.11] [6A]	[3.9][4.1] [4.3][4.4] [4.5][4.11] [4.12][4.13] [6A][72][7.5]	[4.1][4.4] [4.5][4.11] [4.12][4.13] [6A][7.2]	[3.9][4.1][4.3] [4.4][4.5][4.11] [4.12][4.13] [6A]	[3.2][3.4][3.9] [4.1][4.3][4.4] [4.5][4.11] [4.12][4.13][6A]	[3.2][3.3][3.4] [3.9][4.1][4.3] [4.4][4.5][4.11] [4.12][4.13] [6A][7.5]	[3.3][3.4] [4.1][4.3] [4.4][4.5] [4.13][6A]	[3.3][3.4] [4.3][4.4] [4.5][4.13] [6A]	[3.3][3.4] [4.3][4.4] [4.13]
HIGH 100,000+	[3.2] [3.3] [4.11] [6A]	[3.9][4.1] [4.2][4.3] [4.4][4.5] [4.12] [4.13][6A]	[4.12] [6A]	[3.9][4.2] [4.3][4.12] [6A]	[3.2][3.9][4.1] [4.2][4.3][4.4] [4.5][4.7][4.8] [4.11][4.12] [4.13][6A]	[3.2][3.3][3.4] [3.9][4.1][4.2] [4.3][4.4][4.5] [4.8][4.12][4.13] [6A]	[3.3][3.4] [4.1][4.3] [4.4][4.8] [4.13][6A]	[3.4][4.2] [4.3][4.4] [4.5][6A]	[3.4][4.3] [4.4][6A]
ALL QUANTITIES	[3.1]	[3.1][3.6] [4.6][4.8] [4.9]	[3.6][4.6]	[3.1][3.6] [4.6][4.8] [4.9]	[3.1][3.6] [4.6][4.8] [4.9][7.5]	[3.1][3.6] [4.6][4.8] [4.9]	[3.1][4.6] [4.8][4.9]	[4.6][4.8] [4.9]	

QUANTITY \ MATERIAL	LEAD & ALLOYS	NICKEL & ALLOYS	TITANIUM & ALLOYS	THERMOPLASTICS	THERMOSETS	FR COMPOSITES	CERAMICS	REFRACTORY METALS	PRECIOUS METALS
VERY LOW 1 TO 100	[3.1][4.10] [6M][7.5]	[3.5][3.7] [4.10][6M] [7.1][7.5][7.6]	[3.1][3.6][4.7] [4.10][6M] [7.1][7.5] [7.6][7.7]	[5.5] [5.7]	[5.5] [7.7]	[5.2] [5.8] [7.7]	[3.5] [7.1] [7.5] [7.6] [7.7]	[3.1] [7.7]	[7.5]
LOW 100 TO 1,000	[3.1][3.8] [4.10][6M] [7.5]	[3.2][3.5][3.7] [4.10][6M] [7.1][7.3] [7.4][7.5]	[3.1][3.6][4.7] [4.10][6M] [7.1][7.3][7.4] [7.5][7.6][7.7]	[5.3] [5.5] [5.7]	[5.2] [5.3]	[5.2] [5.3] [5.8] [7.7]	[7.1] [7.3] [7.5] [7.6] [7.7]	[7.7]	[7.5]
LOW TO MEDIUM 1,000 TO 10,000	[3.3][3.8] [4.3][4.10]	[3.2][3.3][3.5] [3.7][4.1][4.3] [4.10][4.11][6A] [7.2][7.3][7.4] [7.5]	[4.1][4.7] [4.10][4.11] [6A][7.2][7.3] [7.4][7.5]	[5.3] [5.5] [5.6] [5.7]	[5.2] [5.3] [5.4]	[5.1] [5.2] [5.3]	[7.2] [7.3] [7.4] [7.5]		[7.5]
MEDIUM TO HIGH 10,000 TO 100,000	[3.3][3.4] [4.3][4.4] [4.5][4.13] [6A]	[3.1][3.3][3.4] [3.5][4.11] [4.12][4.13] [6A][7.2][7.5]	[4.1][4.4] [4.11][4.12] [4.13][6A] [7.2][7.5]	[5.1] [5.3] [5.5] [5.6] [5.10]	[5.1] [5.3] [5.10]	[5.1] [5.3] [5.9]	[4.11] [4.12]	[4.12] [4.13]	[4.5] [4.12]
HIGH 100,000+	[3.4][4.2] [4.3][4.4] [6A]	[4.2][4.3] [4.12][6A]	[4.12] [6A]	[5.1] [5.6] [5.10]	[5.1] [5.3] [5.4] [5.10]	[5.9]	[4.7] [4.11] [4.12]	[4.12]	[4.5] [4.12]
ALL QUANTITIES	[4.6]	[3.1][3.6] [4.6][4.8] [4.9]	[4.8][4.9]				[7.5]	[3.6]	[3.6]

KEY TO MANUFACTURING PROCESS PRIMA SELECTION MATRIX:

CASTING PROCESSES
[3.1] SAND CASTING
[3.2] SHELL MOULDING
[3.3] GRAVITY DIE CASTING
[3.4] PRESSURE DIE CASTING
[3.5] CENTRIFUGAL CASTING
[3.6] INVESTMENT CASTING
[3.7] CERAMIC MOULD CASTING
[3.8] PLASTER MOULD CASTING
[3.9] SQUEEZE CASTING

FORMING PROCESSES
[4.1] FORGING
[4.2] ROLLING
[4.3] DRAWING
[4.4] COLD FORMING
[4.5] COLD HEADING
[4.6] SWAGING
[4.7] SUPERPLASTIC FORMING
[4.8] SHEET-METAL SHEARING
[4.9] SHEET-METAL FORMING
[4.10] SPINNING
[4.11] POWDER METALLURGY
[4.12] METAL INJECTION MOULDING
[4.13] CONTINUOUS EXT (METALS)

PLASTIC & COMPOSITE PROCESSING
[5.1] INJECTION MOULDING
[5.2] REACTION INJECTION MOULDING
[5.3] COMPRESSION MOULDING
[5.4] RESIN TRANSFER MOULDING
[5.5] VACUUM FORMING
[5.6] BLOW MOULDING
[5.7] ROTATIONAL MOULDING
[5.8] CONTACT MOULDING
[5.9] PULTRUSION
[5.10] CONTINUOUS EXTRUSION (PLASTICS)

MACHINING PROCESSES
[6A] AUTOMATIC MACHINING
[6M] MANUAL MACHINING

NON-TRADITIONAL MACHINING PROCESSES
[7.1] ELECTRICAL DISCHARGE MACHINING (EDM)
[7.2] ELECTROCHEMICAL MACHINING (ECM)
[7.3] ELECTRON BEAM MACHINING (EBM)
[7.4] LASER BEAM MACHINING (LBM)
[7.5] CHEMICAL MACHINING (CM)
[7.6] ULTRASONIC MACHINING (USM)
[7.7] ABRASIVE JET MACHINING (AJM)

Figure 2.1: Manufacturing Process PRIMA Selection Matrix.

materials. By limiting itself in this way the matrix cannot be regarded as comprehensive and should not be taken as such. It represents the main common industrial practice, but there will always be exceptions at this level of detail. It is not intended to represent a process selection methodology itself. It is essentially a first-level filter. The matrix is aimed at focusing attention on those PRIMAs that are most appropriate based on the important considerations of material and production quantity. It is the PRIMAs that do the task of guiding final manufacturing process selection.

The reader interested in other approaches to manufacturing process selection is referred to Refs [2,3], and for a review of research associated with computer-based process selection approaches, see Ref. [1].

With further reference to the manufacturing process PRIMA selection matrix in Figure 2.1, it can be seen that the requirement to process carbon steel in low to medium volumes (1,000–10,000), for example, returns 13 candidate processes. This is a large number of processes from which to select a frontrunner. However, some processes can be eliminated very quickly – for example, those that are on the border of economic viability for the production volume requested.

The next step involves reference to the PRIMAs. The process of elimination is also aided by the consideration of several of the key process selection drivers in parallel.

For example:

- For the required major or critical dimension, does the tolerance capability of the process achieve specification and avoid secondary processing?
- What is the labour intensity and skill level required to operate the process, and will labour costs be high as dictated by geographical location?
- Is the initial material costly and can any waste produced be easily recycled?
- Is the lead time high, together with initial equipment investment, indicating a long time before a return on expenditure?

In this manner, a process of elimination can be observed that gives full justification to the decisions made. An overriding requirement is the component cost, and the methodology provided in Chapter 12 of this book may be used in conjunction with the selection process when deciding the most suitable process from a number of candidates. However, not all processes are included in the component costing analysis and in this case it must be left to the designer to gather all the detailed requirements for the product and relate these to the data in the relevant PRIMAs.

Due to page size constraints and the number of processes involved, each manufacturing process has been assigned an identification code rather than using process names, as shown at the bottom of Figure 2.1. There may be just one or a dozen processes at each node in the selection matrix representing the possible candidates for final selection. The PRIMA selection matrix cannot be regarded as comprehensive. Figure 2.1 reflects the process position (material and quantity) under normal industrial practice, but there will always be exceptions at this level of selection and detail. Also, the order in which the PRIMAs are listed in the nodes of the matrix has no significance in terms of preference. Note that conventional machining and Non-Traditional Machining (NTM) processes are often considered as secondary rather than primary manufacturing processes, although they can be applicable to both situations. The user should be aware of this when using the PRIMA selection matrix. Also note that conventional machining processes are grouped under just two headings in the matrix: manual (M) and automatic (A) machining. Reference should be made to the individual machining processes and their variants for more detail.

2.1.2 Case Studies

2.1.2.1 Chemical Tank

Consider the problem of specifying a manufacturing process for a small chemical tank. It is made from thermoplastic, which is inert to most chemicals and has major dimensions 1 m length, 0.5 m depth and 0.5 m width. A uniform thickness of 2 mm is considered initially with the requirement of a thicker section if needed. The likely annual requirement is 5000 units, but this may increase over time. The manufacturing process PRIMA selection matrix in Figure 2.1 shows that there are four possible processes considered economically viable for a thermoplastic material with a production volume of 1,000–10,000, and these are:

5.3 Compression Moulding.
5.5 Vacuum Forming.
5.6 Blow Moulding.
5.7 Rotational Moulding.

Next, we proceed to compare relatively the data in each PRIMA for the candidate processes against product requirements. Figure 2.2 provides a summary of the key data for each process upon which a decision for final selection should be based. A symbol '**x**' next to certain process data indicates that they should be eliminated as candidates. Vacuum forming is found to be the prime candidate as it is suited to the manufacture of tub-shaped parts of uniform thickness within the size range required. Vacuum forming is also relatively inexpensive compared with the other processes and has low to moderate tooling, equipment and labour costs, with a reasonably high production rate achievable. Production volumes over 10,000 make it a very competitive process.

2.1.2.2 Generator Laminations

The development of an experimental wind turbine generator based on a radial permanent magnet machine topology requires the use of laminations as part of the active material on the rotor, to accommodate coils of insulated copper wire. Each lamination is 0.5 mm thick, fabricated from supplied sheets of iron–silicon alloy (coated with thin polymer coating to

5.3 Compression Moulding

Economic Considerations

- Production rates from 20 to 140/h.
- The greater the thickness of the part, the longer the curing time.
- Lead times may be several weeks according to die complexity.
- Material utilisation is high. No sprues or runners.
- Tooling costs are moderate to high. ✗
- Equipment costs are moderate.
- Direct labour costs are low to moderate.
- Finishing costs are generally low. Flash removal required.

Typical Applications

- Dishes.

Design Aspects

- Shape complexity is limited to relatively simple forms. Moulding in one plane only.
- Threads, ribs inserts, lettering, holes and bosses possible.
- Thin walled parts with minimum warping and dimensional deviation may be moulded.
- Maximum section, typically = 13mm.
- Minimum section = 0.8mm.
- Maximum dimension, typically = 450mm. ✗
- Maximum area = $1.5m^2$.

Quality Issues

- Variation in raw material charge weight results in variation of part thickness and scrap.
- Dimensions in the direction of the mould opening and the product density will tend to vary more than those perpendicular to the mould opening.

5.5 Vacuum Forming

Economic Considerations

- Production rates from 60 to 360/h commonly.
- Lead times of a few days typically.
- Material utilisation is moderate to low. Unformed parts of the sheet are lost and cannot be directly recycled.
- Set-up times and change-over times are low.
- Sheet material much more expensive than raw pellet material.
- Tooling costs are low to moderate, depending on complexity.
- Equipment costs are low to moderate, but can be high if automated.
- Labour costs are low to moderate.
- Finishing costs are low. Some trimming of unformed material after moulding.

Typical Applications

- Open plastic containers and panels.
- Bath tubs, sink units and shower panels.

Design Aspects

- Shape complexity limited to mouldings in one plane.
- Open forms of constant thickness.
- Undercuts possible with a split mould.
- Cannot produce parts with large surface areas.
- Bosses, ribs and lettering possible, but at large added cost.
- Maximum section = 3mm.
- Minimum section = 0.05mm to 0.5mm, depending on material used.
- Sizes range from $25mm^2$ to 7.5m x 2.5m in area.

Quality Issues

- Thermoplastic material must possess a high uniform elongation otherwise tearing at critical points in the mould may occur.
- Sheet material will have a plastic memory and so at high temperatures the formed part will revert back to original sheet profile. Operating temperature therefore important.

5.6 Blow Moulding

Economic Considerations

- Production rates between 100 and 2,500/h, depending on size.
- Lead times are a few days.
- There is generally little material waste, but can increase with some complex geometries using extrusion blow moulding.
- Set-up times and change-over times are relatively short.
- Tooling costs are moderate to high.
- Equipment costs are moderate to high, especially for full automation.
- Direct labour costs are low. One operator can manage several machines.
- Finishing costs are low. Some trimming required.

Typical Applications

- Hollow plastic parts with relatively thin walls. ✗

Design Aspects

- Complexity limited to hollow, well rounded, thin walled parts with low degree of asymmetry. ✗
- Undercuts, bosses, ribs, lettering, inserts and threads possible.
- Maximum section = 6mm.
- Minimum section = 0.25mm.
- Sizes range from 12mm in length to volumes upto $3m^3$.

Quality Issues

- Poor control of wall thickness, typically ±50% of nominal. ✗
- Good surface detail and finish possible.

5.7 Rotational Moulding

Economic Considerations

- Production rates of 3 to 50/h, but dependent on size.
- Lead time several days.
- Material utilisation very high. Little waste material.
- Tooling costs are low.
- Equipment costs are low to moderate.
- Labour costs are moderate.
- Finishing costs are low. Little finishing required.

Typical Applications

- Water tanks.
- Buckets.
- Drums.

Design Aspects

- Complexity limited to large, hollow parts of uniform wall thickness. ✗
- Large flat surfaces should be avoided due to distortion and difficulty to form. Use stiffening ribs.
- Maximum section = 13mm.
- Minimum section is typically 2mm, but can be as low as 0.5mm for certain applications. ✗
- Sizes up to $4m^3$.

Quality Issues

- Control of inside surface finish is not possible. ✗
- Wall thickness is determined by the close control of the amount of raw material used.
- Wall thickness tolerances are generally between ±5 to ±20% of the nominal.

Figure 2.2: Chemical Tank Case Study – Comparison of Key PRIMA Data for the Candidate Processes.

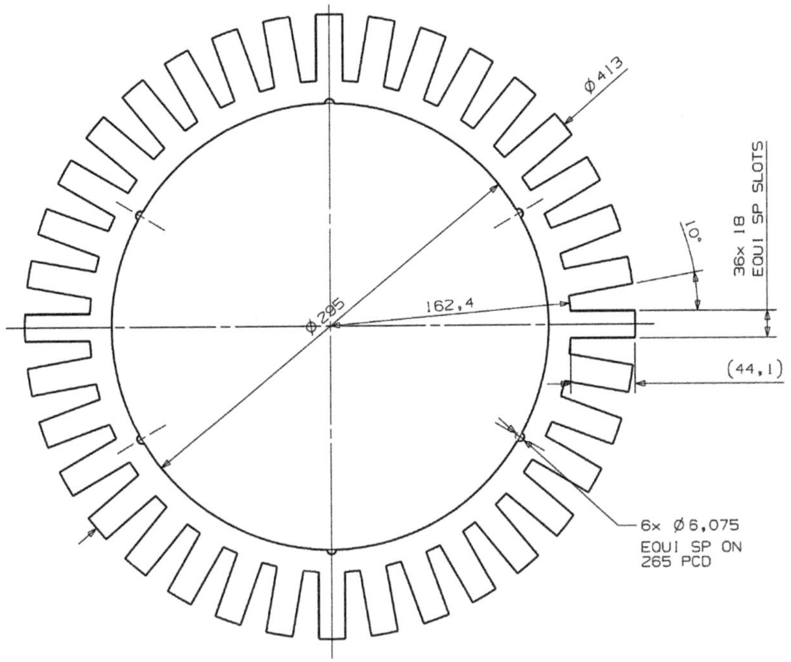

Figure 2.3: Generator Lamination Case Study – Detailed Component Drawing.

effectively reduce eddy currents). In total, 180 laminations of the design shown in Figure 2.3 are required. A very short lead time is required for the prototype batch, with high accuracy and good surface finish and minimal burring to allow multiple laminations to stack together without accumulated error on the width.

From the manufacturing process selection matrix shown in Figure 2.1, there are eight processes initially considered viable for an iron material and a low production volume (100–1,000).

The candidate processes are:

3.2 Shell Moulding.
3.5 Centrifugal Casting.
3.6 Investment Casting.
3.7 Ceramic Mould Casting.
6M Manual Machining.
7.1 Electrical Discharge Machining (EDM).
7.3 Electron Beam Machining (EBM).
7.4 Laser Beam Machining (LBM).

The four casting processes (PRIMAs 3.2, 3.5, 3.6 and 3.7) can quickly be eliminated as the iron–silicon alloy material is already supplied in 0.5-mm-thick sheet, therefore no primary shape forming to this thickness is required.

Manual machining, specifically vertical milling, is an obvious candidate initially (PRIMA 6.2), but on further examination is not actually valid. Moving to automatic machining (CNC milling) at these low volumes does not improve the situation either. It is mainly due to the complexities of machining the 2D lamination geometry itself, and the special tooling required, that manual machining must be rejected. Machining individual components would require sacrificial tooling to avoid component deformation due to the low stiffness of the sheet material and the high machining forces. It would also be slow and laborious to machine individual components this way. It is possible that a stack of laminations could be machined simultaneously, again using special (and substantial) tooling, and reducing the lead time; however, error propagation throughout multiple laminations is a possibility, which means scrapping many components rather than one. Milling would also produce a large radius at the slot profile root formed by the milling tool diameter, when in fact to accommodate the preformed coils, a square corner is preferred. In addition, edge burrs from milling would need to be manually removed from each component.

From these arguments, manual machining (PRIMA 6M) can be eliminated, leaving EDM, EBM and LBM, three similarly accurate NTM processes that will now be compared with reference to some key PRIMA data related to requirements.

EDM (more specifically wire EDM) has no cutting forces, but is relatively slow. EDM also produces a recast layer with a surface roughness less than 25 μm Ra, which would indicate a finishing process was necessary. The polymer coating (non-conductive) on each lamination could be problematic using this process. This is especially the case when multiple sheets are processed simultaneously in stacks in an attempt to improve lead time. Sharp corners are possible though, and are related to the wire diameter used.

Laser beam cutting, a variant of LBM, is widely used for rapidly (up to 70 mm/s) producing complex 2D profiles from sheet, where sharp corners are possible, producing surfaces with a finish less than 6.3 μm Ra. It does produce localised thermal stresses giving very small heat-affected zones, a small recast layer with the possibility of low distortion of thin parts. Recast layers can be easily removed if undesirable though. Individual sheets are processed using simple tooling only, and the polymer coating is not a problem as it would simply be vaporised during cutting.

EBM cutting also produces localised thermal stresses, giving very small heat-affected zones, a small recast layer and the possibility of low distortion of thin parts. Although it is capable of cutting speeds up to 10 mm/s and comparable surface finishes to LBM, the tooling costs are high, lead times are long and sharp corners are difficult, with reference to its PRIMA data.

Considering the above arguments and with reference to several key technical and economic capability requirements, LBM was finally selected, and 180 laminations to high accuracy and a short lead time were delivered for the prototype generator, with very little post-manufacture cleaning required before rotor assembly.

2.2 Rapid Prototyping Process Selection

2.2.1 Selection Strategy

Rapid prototyping covers a wide range of relatively new manufacturing technologies, developed to take full advantage of modern 3D CAD modelling capabilities to produce physical components with a minimum lead time. The aim is to shorten product realisation cycles and communicate designs earlier to customers, whether internal or external to a business, by visualising a physical component or, increasingly, to produce tooling and patterns for other manufacturing processes, e.g. investment casting, an application termed rapid tooling.

In this Handbook, only five of the most established technologies and most commonly used processes are covered in depth using PRIMAs, although over 30 commercial processes exist. The following are included in the selection strategy for rapid prototyping: Stereolithography (SLA), Three-Dimensional Printing (3DP), Selective Laser Sintering (SLS), Laminated Object Manufacturing (LOM) and Fused Deposition Modelling (FDM). They are all based on additive layer manufacturing principles. However, a number of related process variants are also briefly described in the PRIMAs and are shown in Figure 1.7, the general classification of rapid prototyping processes based on the initial material form. Processes based on subtractive material removal, e.g. CNC machining, are sometimes grouped within rapid prototyping; however, they are treated separately in this Handbook under Section 2.1.

In general, all additive layer rapid prototyping processes share a number of basic steps for component manufacture:

1. Create a 3D CAD model of the component.
2. Convert the CAD model to the '.stl' file format (a standard file format for rapid prototyping models and the original abbreviation for stereolithography).
3. Slice the .stl model file into thin layers (usually about 0.1 mm thick).
4. Build the prototype component up one layer on top of another using one of the various manufacturing technologies.
5. Post-process the rapid prototyped component, e.g. remove supports (where applicable), machine, clean, etc.

Rapid prototyping processes are not suited to any level of volume production. They are more applicable for the generation of one-off components, i.e. prototypes (as the name suggests), and models. However, very small batches of fewer than 50 components can be economically viable for some processes depending on size and complexity. Therefore, these processes are not bound by the same large-volume manufacture-related economic issues as the more traditional processes when considering their selection. Although cost should always be important in engineering decision-making, the niche application domain of these processes in product development suggests that final component cost may not be a key selection criterion. In addition, rapid prototyping does not necessarily mean instantaneous prototypes and the time it takes to make a component depends on the technology used, and the size and complexity of the model. The time can vary from a few hours to a day or more; therefore, speed is a tangible economic selection criterion related to how quickly the 3D design model can be physically realised.

The material for the prototype is a key technical requirement, and should have the same level of importance placed on it as in the selection strategy used for traditional manufacturing processes, but for different reasons. In the more traditional manufacturing processes, the material choice is often directed by the specification, manufacturing compatibility and/or structural integrity issues. For rapid prototyping, the prototyping material may not be the same type as the final larger production volume component material. Even if the same material is chosen, the material processed using rapid prototyping technologies will rarely achieve or exhibit the same physical properties as the material processed, perhaps for a mass-produced component, using a more economically viable manufacturing process. A key design objective when specifying rapid prototyping processes should be that the prototype component possesses sufficient strength, stiffness, dimensional stability and environmental protection characteristics commensurate with its purpose. Significant strength would make a rapid prototyped component certainly handlable, able to be assembled with other components to demonstrate component qualities such as fit and form together with other components in a product, and even structurally load bearing to sustain loads in a wind tunnel or test rig, for example.

The material choice is also dependent on its raw cost, of course, and the final volume of material used. The final volume could be greater than the component volume as two of the main rapid prototyping processes (SLA and FDM) require extra support material for undercuts and overhanging features.

Tolerances and surface roughness, two elements of form and fit [4], could also be useful selection drivers, and vary between all technologies [5]. Models made by some rapid prototyping routes require extensive finishing of the surfaces, which adds cost and time to deliver the models. A typical feature with most rapid prototyping processes, for example, is the 'stair-stepping' marks evident on walls in the vertical build direction, which may need to be post-processed. With respect to shape and size, a distinct advantage of all rapid prototyping processes over the more traditional manufacturing processes is their ability to create complex features and geometries faster and usually more inexpensively due to the layered build-up of the component. However, another variation is the maximum component size possible, or more specifically component volumes, together with a variation between achievable width, height and depth for any individual process.

Summarising, the main selection drivers for the rapid prototyping processes discussed are:

- Prototyping material type/cost.
- Speed of component production.

- Achievable tolerances.
- Achievable surface finish.
- Overall component size/volume.

In reality, the selection of rapid prototyping processes is a difficult and complex task and here the focus is on those issues that will largely satisfy product requirements [6]. It is also important to maintain a simple and consistent approach for all the selection strategies in the initial stages to identify candidate processes. Combining just two factors, one technical (material type) and one economic (production volume), was discussed in Section 2.1 for manufacturing processes. A combination of material type and production speed are selection drivers that again combine technical and economic issues for rapid prototyping processes, and can vary significantly from process to process to create an effective selection strategy for the five processes included. A PRIMA selection matrix showing the relevance of rapid prototyping processes with respect to material type and production speed is provided in Figure 2.4. The matrix defines the following requirements or inputs:

- *Material type* – The selection of the material is mainly based on the purpose of the physical model, e.g. functional prototype, form and fit part, concept model, rapid tooling, etc. The materials compatible for rapid prototyping processes include the categories:
  - Natural materials, e.g. plaster, wax, paper.
  - Elastomers, thermosetting plastics and thermoplastics, e.g. epoxy, ABS, PC, PA.
  - Metals, e.g. steels, titanium and other alloys.
  - Ceramics and ceramic–metal composites.
- *Production speed* – Based on build speed of the component in terms of volume of material in ℓ/h (litres per hour), or in terms of vertical build speed in mm/h (millimetres per hour). The production speed does not include pre- and post-processing time for producing the final component, which can be significant. The quantities specified for selection purposes are broadly in the ranges:
  - Slow = Volume of material 0.05 ℓ/h, vertical build speed 1.5 mm/h.
  - Medium = Volume of material 0.1 ℓ/h, vertical build speed 5 mm/h.
  - Fast = Volume of material 0.25 ℓ/h, vertical build speed 20 mm/h.

The information in Figure 2.4 represents a wide range of industrial and commercial practices, and includes several new material combinations with existing rapid prototyping technologies. In the case of 3D printing, both the binder and powder material have been highlighted as relevant in the selection process, each having an approximate 50% contribution by volume to the component. There will always be exceptions at this level of detail though, and more so than in other process categories, the developments in rapid prototyping are numerous and often. The matrix is not intended to represent a process selection methodology itself either. It is essentially a first-level filter to identify candidates. Consistent with the other process selection strategies presented, it is the PRIMAs that guide final selection.

MATERIAL / SPEED	PAPER	SAND	PLASTER	CELLULOSE	STARCH	WAX	ELAST-OMER	EPOXY	ABS	PA	PC	PE	PPS	PS	PU	PVC	UP	IRON - COPPER ALLOYS	STEEL (carbon)	STEEL (tool, alloy)	STAINLESS STEEL	COPPER ALLOYS	ALUMINIUM	NICKEL ALLOYS	TITANIUM ALLOYS	TUNGSTEN	CERAMICS	METAL-CERAMIC COMPOSITES
SLOW **Volume 0.05 ℓ/h** **Vertical Build 1.5mm/h**						[8.5]	[8.5]		[8.5]	[8.5]	[8.5]	[8.5]	[8.5]				[8.5]										[8.5]	
MEDIUM **Volume 0.1 ℓ/h** **Vertical Build 5mm/h**	[8.4]	[8.3]				[8.3]	[8.1] [8.3]	[8.1]	[8.4]	[8.3]	[8.3] [8.4]			[8.3] [8.4]	[8.1]	[8.3] [8.4]		[8.3]	[8.3] [8.4]	[8.3]	[8.3]	[8.3]	[8.3] [8.4]	[8.3]	[8.3]	[8.3]	[8.4]	
FAST **Volume 0.25 ℓ/h** **Vertical Build 20mm/h**		[8.2]	[8.2]	[8.2]	[8.2]	[8.2]	[8.2]	[8.2]							[8.2]						[8.2]	[8.2]					[8.2]	[8.2]

KEY TO RAPID PROTOTYPING PROCESS PRIMA SELECTION MATRIX:

RAPID PROTOTYPING PROCESSES

- [8.1] STEREOLITHOGRAPHY (SLA)
- [8.2] 3D PRINTING (3DP)
- [8.3] SELECTIVE LASER SINTERING (SLS)
- [8.4] LAMINATED OBJECT MANUFACTURING (LOM)
- [8.5] FUSED DEPOSITION MODELLING (FDM)

MATERIALS

ABS	Acrylonitrile Butadiene Styrene
PA	Polyamide
PC	Polycarbonate
PE	Polyethylene
PPS	Polyphenylenesulphone
PS	Polystyrene
PU	Polyurethane
PVC	Polyvinylchloride
UP	Polyester

Figure 2.4: Rapid Prototyping PRIMA Selection Matrix.

2.2.2 Case Study – Prototype Turgo Turbine

Figure 2.5 shows a 3D CAD model of a Turgo turbine used in the development of a new pico-hydro power generation system. In order to assess its experimental performance, a rapid prototyped component is required for integration in a test rig to conduct extensive trials. The material choice should be a tough, lightweight material (to minimise mass moment of inertia and fly-wheeling effects during testing), which does not absorb water and can sustain its mechanical properties between a wide range of service temperatures, such as a thermoplastic. The turbine has been designed for resistance to fatigue anticipated due to the repeated loading of a water jet impacting on each cantilevered cup on rotation of the turbine. The stresses at the major changes in section have been minimised in this respect. From the CAD model, the volume of the turbine is 161,700 mm^3, with a minimum thickness of 4 mm and an outside diameter of 150 mm. The surface finish is important only inside the cups in order that frictional losses are minimised during interaction with a water jet.

Scrutinising the material requirements, ABS is an inexpensive rubber-toughened thermoplastic exhibiting good impact resistance, that can operate over a range of temperatures, and has good water and chemical resistance. Although it is damaged by sunlight, this is not a major problem during experimental testing of the turbine.

From Figure 2.4, with no speed requirement specified as yet, ABS is compatible with:

8.4 Laminated Object Manufacturing (LOM).
8.5 Fused Deposition Modelling (FDM).

FDM is slow in comparison with LOM. However, given the small volume of the Turgo turbine model, the build time will be fairly short for both processes. Assuming the least volume of

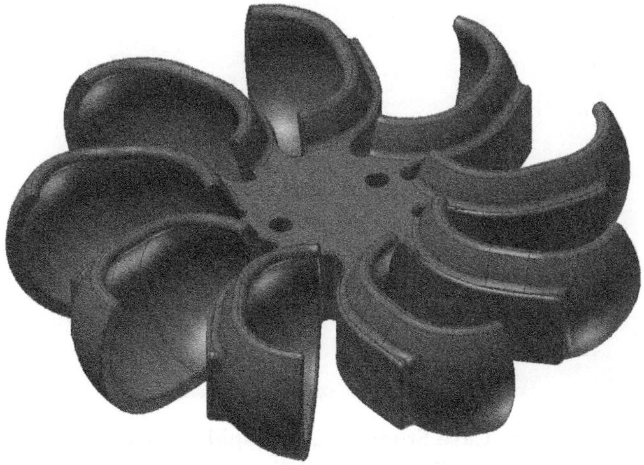

Figure 2.5: Turgo Turbine Case Study – CAD Model of Prototype.

material deposited with the addition of any pre- and post-processing time, the component could be completed within hours. With reference to the PRIMAs, both processes can easily achieve the maximum dimension of the Turgo turbine and have similar tolerance capability, which is satisfactory for this component installation. With respect to surface finish, FDM outperforms LOM, and with the requirement of a fine surface finish only being required on the inside faces of the turbine cups, these surfaces can be hand finished at very low cost and time effort.

The structural integrity of a rapid prototyped component that will be subjected to loading, as in this case, is an important issue for all rapid prototyping processes as the mechanical properties are anisotropic depending on build direction. Preference in the build direction and therefore the material strength orientation should be given to the geometry where highly stressed regions are located. This is even more of an acute problem with LOM as individual layers of ABS sheet will be built up, separated with a low interfacial shear stress adhesive. This also means that LOM produces components that are essentially porous in that plane, and post-processing would involve the addition of a water-resistant coating to avoid swelling for the pico-hydro turbine application. FDM produces little porosity in the material in comparison.

FDM was selected mainly due to its ability to create a functional water-resistant, dimensionally and structurally stable prototype in ABS with satisfactory surface finish and tolerance capability. Figure 2.6 shows the finished Turgo turbine (shown in off-white ABS material) under test conditions, surviving hundreds of hours of operation to this day.

As can be seen in Figure 2.5, the Turgo design of turbine is very complex with repeated re-entrant features and changing sections. Rapid prototyping is an ideal process choice for a

Figure 2.6: Turgo Turbine Case Study – FDM Rapid Prototyped Component under Experimental Testing.
The water jet is entering the Turgo cups from the nozzle on the left-hand side.

one-off prototype in a thermoplastic such as ABS. Looking further ahead, the economic and technical viability of the pico-hydro system needs to be evaluated, of course. A key component is the Turgo turbine component itself, as it is the power conversion interface between the water and generator. With reference to Figure 2.1, specifying an aluminium alloy material (corrosion resistant, lightweight, strong), together with a variable production volume (associated with uncertainty in the number of units manufactured per annum, initially), suggests that investment casting and sand casting are candidates for the primary shape-creating manufacturing process for the turbine. Comparing the sand and investment casting PRIMAs, it is evident that the surface roughness capability is superior for investment casting, and much thinner sections are obtainable compared to sand casting. Therefore, investment casting is a candidate process to make the turbine in low to medium volumes.

2.3 Surface Engineering Process Selection

2.3.1 Selection Strategy

The processes included in the selection strategy discussed next only include surface coatings and treatments, and not those processes associated with bulk treatments, e.g. annealing, quenching, tempering, etc. Bulk treatment processes are typically used to relieve stresses after forming or machining of a component, to improve ductility or to increase the hardness of the entire part. The exclusion of bulk treatments is intentional, as there is greater process knowledge and routine utilisation of these processes than those strictly associated with engineering just the surface. The 13 processes included in the Handbook cover a widely used subset of available surface coating and treatment technologies. A full classification of surface engineering processes is shown in Figure 1.8. Several surface coating processes, e.g. weld coating (hardfacing), diffusion bonding and rolling, are all discussed in Chapters 4 and 11, under forming and joining processes respectively.

Typically, engineering of a surface using coatings and treatments is seen as a 'band-aid' solution to a wear, corrosion or fatigue problem in-service, rather than considered at the design stage, and therefore the maximum benefit is not achieved. If a customer does know what they require, this is usually based on prior experience with a coating system on similar components. Surface engineering may be used in the design, but is seldom specified by designers. Unless specified at the design stage, surface coatings and treatments will always remain factors considered at redesign, incurring additional manufacturing costs that need to be traded against quality loss reduction [7].

Where process data does exist, typically within companies specialising in a small range of processes, it is often withheld, being regarded as commercially sensitive. Information on coatings and treatments is therefore dispersed among a variety of sources and formats, making it difficult for the designer to compare them objectively. Add to this the sheer number of possible combinations, and the designer's problem becomes substantial. Other information that exists in the public domain is either in the wrong form or purely qualitative and therefore subjective in nature. The design guidelines that are available tend to be based on cumulative experience and, consequently, design for surface engineering is still seen as an art rather than a science [8].

The provision of an effective process selection methodology is of prime importance to the designer in this respect. There is much to be gained by the specification of an appropriate surface coating or treatment for a particular working environment. However, the range of processes available and the interaction with the substrate makes this task one where considerable expertise is required, and a great deal of research has been conducted in this area [9–11].

The selection of the most appropriate surface engineering process has many guiding factors, including [12]:

- Service conditions, e.g. contact, pressure, chemicals.
- Design constraints, e.g. geometry, size, hardness, friction.
- Processing constraints, e.g. temperature, material compatibilities.
- Economic issues, e.g. cost, production volume, production rate.
- Conformance issues, e.g. coating thickness, surface finish, distortion.

The main reason for the growing importance of surface engineering (and its inclusion in this Handbook) is an increasing realisation that a component usually fails when its surface cannot withstand the external forces or environment to which it is subjected, e.g. wear, corrosion, fatigue, etc. The use of surface coatings and treatments can improve the surface properties in a cost-effective manner, without processing the whole component to improve relevant properties. Therefore, although the issues listed above broadly align with the PRIMA description categories, the adopted selection strategy is initially based on the requirement to modify a surface in order that a component can survive its anticipated operating conditions and/or improve its aesthetic qualities.

The focus for selection is limited to the improvement or increased resistance of a component to one or more of the following three common life-reducing failure modes and the requirement to provide a decorative finish as defined next:

- *Fatigue* – Failure mode of a component that is subjected to a repeated loading cycle in service, ultimately resulting in total fast fracture well below a stress that would have caused fracture under a single application of the load. Typically, cracks appear and propagate from the surface of the component, where the highest stresses are present.
- *Corrosion* – Deterioration of a component usually initiating at the surface of the material, due to the interaction with its service environment, typically through a chemical process by which the metal is oxidised.
- *Wear* – The removal and deformation of material resulting from the mechanical sliding interactions of component surfaces under pressure through abrasion and friction mechanisms.
- *Decorative* – Provision of colour and/or surface finish to a component's surface to improve aesthetic appeal.

The surface engineering processes that satisfy each of these requirements are shown in Figure 2.7 using a Venn diagram. This can be used to identify candidate processes before referring to the individual PRIMAs to aid final selection based on economic and technical requirements. Preference can be given to processes that satisfy multiple requirements, e.g. nitriding is used to improve fatigue, corrosion and wear. Otherwise, where there is a single need/functionality of the surface, all candidate processes should be rigorously evaluated and no preference given. Note that additional or secondary improvements to functionality that are not required may be considered expensive.

2.3.2 Case Study – Fatigue Life Improvement of a Shaft

A low-cost surface treatment is required to increase the fatigue resistance of a shaft made from low carbon steel, as shown in Figure 2.8. The shaft has a diameter of 10 mm and a length of 100 mm, along which are located a number of stress-raising features. The maximum temperature is 1,000°C and the hardness of the shaft cannot be altered by the treatment. An allowance of ±0.1 mm is specified on all dimensions.

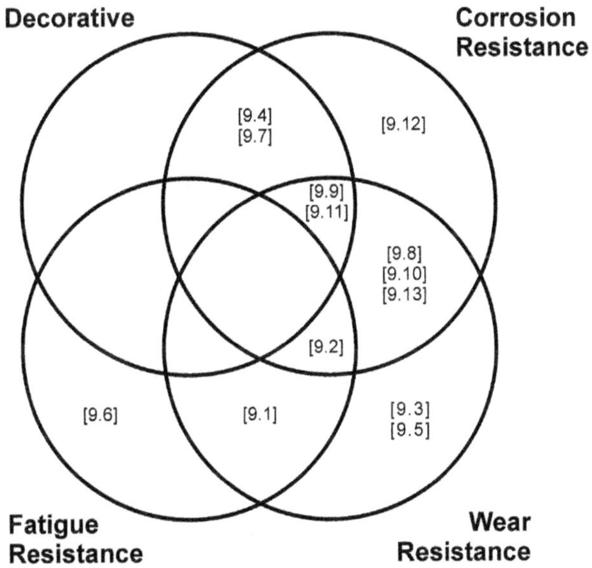

KEY TO SURFACE ENGINEERING PROCESS PRIMA SELECTION VENN DIAGRAM:

[9.1]	CARBURISING	[9.8]	CHEMICAL VAPOUR DEPOSITION (CVD)
[9.2]	NITRIDING	[9.9]	PHYSICAL VAPOUR DEPOSITION (PVD)
[9.3]	ION IMPLANTATION	[9.10]	ELECTROLESS NICKEL
[9.4]	ANODISING	[9.11]	ELECTROPLATING
[9.5]	THERMAL HARDENING	[9.12]	HOT DIP COATING
[9.6]	SHOT PEENING	[9.13]	THERMAL SPRAYING
[9.7]	CHROMATING		

Figure 2.7: Surface Engineering PRIMA Selection Venn Diagram.

Figure 2.8: Shaft Case Study.

With reference to Figure 2.7, the PRIMA selection Venn diagram indicates that there are three candidate processes for the improvement of fatigue resistance. These are:

9.1 Carburising.
9.2 Nitriding.
9.6 Shot peening.

Comparing PRIMA data, carburising is compatible with low carbon steels, but the process temperature is around the 1,000°C temperature limit. Distortion problems can occur around unsymmetrical features through the generation of residual stresses. In addition, carburising raises the surface hardness. Although nitriding is a lower temperature surface treatment with little or no distortion, it too raises surface hardness considerably. Shot peening is chosen as it does not significantly increase surface hardness and operates at ambient temperatures. It primarily improves fatigue resistance of stress-raising features and can be localised around these features without treating the whole shaft. Shot peening also has a maximum tolerance of ±0.05 mm, which is within the limits specified.

2.4 Assembly System Selection

2.4.1 Selection Strategy

Assemblies involve two or more combined components of varying degrees of build complexity and spatial configuration. The assembly technologies used range from simple manual operations through to flexible robotic operations, to fully mechanised dedicated systems. The final system or combination of systems selected has the task of reproducing the product at the volume dictated by the customer, in a cost-effective way for the producer, to be technically appropriate for the components manipulated and composed, and ultimately to satisfy the functional requirements dictated by the specification.

The assembly phase represents a significant proportion of the total production cost of a product and can outweigh manufacturing costs in some industries [13]. Through the identification of the most effective assembly technologies early in the development process, downstream activity, inefficiency and costs can be reduced. Significantly, though, assembly is a major source of engineering change, rework and production variability appearing late in product development [14]. The cost of recovering from these problems during assembly is high, estimated to be in the range of 5–10% of the final cost. In part, this is due to the fact that assembly is governed by much less controllable and less tangible issues than manufacturing, such as assembly actions and fixture design [15]. In practice, assembly selection is a very difficult task. It does not mean, however, that a sound decision cannot be made about an appropriate assembly technology to use for a given set of requirements. A number of researchers have proposed strategies for assembly system selection. The reader interested in this topic can find more information in Refs [16–18].

Three general types of assembly system are identifiable in industrial practice. Figure 1.9 showed a classification of these assembly systems, together with the variant technologies involved.

- *Manual Assembly.* Traditionally, low-volume assembly has been performed using manual assembly; however, world markets are demanding flexibility, product variety, shorter lead times and fault-free products. This is now accompanied by increasing labour costs and strict legislation from the European community. In an attempt to reduce the cost of manual assembly, many Western companies have moved their assembly plants to lower cost regions in the Far East. This is not always the ideal solution as it increases transportation costs, places a physical barrier between design and production, and quality often suffers. The need for product quality requires every product to be tested before dispatch to the customer. Manual assembly is very susceptible to quality variations because operators can easily make mistakes during assembly – for example, small parts can be forgotten, assembled in the wrong position or assembled incorrectly, e.g. screws may not be tightened to the correct torque. These demands have been met in industry through the development of manual assembly with automatic aids or semi-automatic assembly systems. Semi-automatic assembly mechanises these critical operations in the assembly sequence, such as screwing or push-fit operations. This enables the manual assembly tasks that traditionally suffer from quality variations to be controlled using automation, whilst operators perform the part feeding and positioning tasks. In the event that a wrong part is assembled or a part is forgotten, sensors within the assembly station will detect the faults and alert the operator. In this way, assembly faults are corrected 'on-line' when they occur and are not buried under layers of parts, which are assembled afterwards.
- *Flexible Assembly.* During the early 1980s a new assembly concept was born, called flexible assembly. Flexible assembly utilises robots, a flexible materials handling system and flexible part feeders in order to create a hybrid of manual, semi-automatic and

dedicated assembly that is capable of assembling different products in small batches, without suffering from the variability of manual and semi-automatic assembly and the high cost of dedicated assembly equipment. A flexible assembly system can be compared with a CNC machining station. Part programs and raw components are the system input and finished products emerge. It was argued that such systems would be used to assemble the middle range of production volume between manual assembly and dedicated assembly. However, no such systems were successfully developed due to the cost and limited capability of robot technology in this era and low-volume assembly remained a manual or semi-automatic process. The flexible assembly machine can be considered as two basic mechanical systems that operate in parallel: the assembly robot, which performs the actual assembly tasks; and the materials handling equipment, which ensures that the manipulator is fed with the correct parts, fixtures and tools at the correct time and place, whilst performing other functions such as finished product removal from the assembly area. The materials handling system can be further divided in two: small part flexible/low-cost feeders, and the pallet and fixture handling system. The assembly robot is equipped with a compliance device that is used to accommodate programming and part tolerances, and quick-change tools that allow the robot to change its gripper fingers and pick up special tools. During a product change the product-specific pallets, fixtures, low-cost small parts feeders and gripper fingers are changed and the new product's flexible-feeder and assembly robot programs are loaded.

- *Dedicated Assembly.* Dedicated assembly automates the assembly task by breaking it down into simple operations that can be conducted by a series of work-heads, the assembly being built up as it passes down the line. Parts are supplied in bulk, placed in individual parts feeders and presented to an automatic work-head, which inserts them into the part assembly at high speed. This form of assembly can achieve cycle times as low as 1 second per assembly. By and large, dedicated assembly machines are only suitable for a single product. Any significant product design change will result in considerable assembly machine redesign costs and lengthy reconfiguration time. It is also clear that such equipment can only be justified for large production volumes, as the equipment cost is spread over the life of a single product. For this reason the application of dedicated assembly has traditionally been restricted to high-volume production. There are situations where flexible system notions can be engineered into high-speed machines to offer economic assembly with high flexibility at mass production volumes – essentially assembly systems capable of mass customisation. The assembly system is built up from a number of automated stations, which are linked together using a free transfer system. At each automated station, vision system-based flexible feeders are used to feed those parts that are used to create different product variants. Part insertion and handling is conducted using a combination of robots and programmable assembly stations. Standard fixtures and gripping locations have been defined so that gripper and fixture variants are not required. In this way the cost of design changes and the introduction of new product variants has been minimised. The modular construction of the assembly line enables new assembly stations to be introduced to increase capacity or replace the few remaining manual assembly stations.

Prior to the selection of an assembly system, a number of activities should be undertaken and factors considered, some of which also help drive the final quality of the assembly:

- *Business level* – Identification and availability of assembly technologies/expertise in-house; integration into business practices/strategy; geographical location and future competitive issues, such as investment in equipment.
- *Product level* – Anticipated lead times, product life, investment return time-scale, product families/variants and product volumes required.
- *Supplier level* – Component quality (process capability, gross defects) and timely supply of bought-in and in-house manufactured parts.

The final point is of particular importance. A substantial proportion of a finished product, typically two-thirds, consists of components or subassemblies produced by suppliers [19]. The original equipment manufacturer is fast becoming purely an assembler of these bought-in parts and therefore it is important to realise the key role suppliers have in developing products that are also 'assembly friendly'. Consideration must be given to the tolerances and process variability associated with component parts from a very early stage, especially when using automated assembly technologies, because production variability is detrimental to an assembly process.

There are a number of factors or drivers that influence the selection of assembly systems. The main issues include:

- *Labour cost, availability and skill* – Very dependent on geographical location.
- *Geographical location* – For example, Boeing intends to build the new 7E7 Dream Liner in a location that gives it the best opportunity to be successful, and has identified the issues that it will consider in choosing the final assembly location, including: transportation; facilities and amenities; land and support services; workforce training infrastructure; environmental and local natural disaster conditions; and community and government support.
- *Production quantity* – A minimum of 2 years future demand forecast information for all product types is required for sufficient volume.
- *Production rate* – Higher rates are obtainable with systems that have multiple assembly stations. Also dependent on size, weight of parts and complexity of assembly operations needed, although the latter is heavily influenced by design.
- *Joining methods employed* – Studies indicate that in many designs, large proportions of the excess components identified through DFA are used only for fastening purposes. In many cases, excessive or incorrect joining processes are used, possibly due to a lack of knowledge of the availability, cost implications and functional performance of alternatives.

- *Capital cost of assembly system* – System costs range from a few thousand to many millions of pounds, depending on the degree of automation and number of parts to be assembled. Justifying the investment is a key factor when considering a new assembly technology and the ability to manage the technology.
- *System operating costs* – Costs associated with the day-to-day running of the system chosen. Can be substantial if manually centred, but less so with automated systems.
- *Number of components in the assembly* – DFA aims to reduce the number of components in an assembly design by identifying redundancy and consolidation of parts where possible; however, there still remains a strong relationship between part-count and cost.
- *Number of variants to be accommodated* – Flexibility of the system chosen is crucial to accommodate variants and product families in order to minimise cost.
- *Component quality* – Process capability, potential defects and tolerances required.
- *Component supply logistics* – Outside, as well as in-house suppliers, are an integral element in the production management system, for example Just in Time (JIT).
- *Complexity/time of assembly operations* – Increased complexity may be due to the design of components (again, DFA can help here by improving component geometry for feeding, handling, fitting and checking), interfaces or the joining method chosen. Some operations may be beyond capabilities of human operators and therefore some degree of mechanised assistance or automation is required.
- *Handling characteristics of components* – Size, weight and environmental hazards of parts to be assembled are important factors for operator fatigue or health and safety.

It is not the intention to imply that the above factors are of equal importance, or should be considered in the fixed sequence listed above. There are too many factors to consider in a single appraisal leading to the selection of one system and its variants. The assembly system selection chart devised is shown in Figure 2.9. The chart is a general guide only and merely aims to support the selection of the appropriate assembly system PRIMA. It is based on the following key drivers:

- Number of product variants (flexibility requirement).
- Production rate, or
- Production quantity per annum, or
- Capital cost of assembly equipment (though this is more of an outcome than a requirement).

The form of the curves at the boundaries shown between the system types reflects the economic consequences associated with the introduction and management of variants in the assembly system. To accommodate variants at all, requires a significant investment in the capital cost of the system, which rapidly ramps up the production volume requirement at the transition to flexible and dedicated, as variants begin to appear at a *low* variant level. As the number of variants climbs to a *high* level, the management of the variants is much more

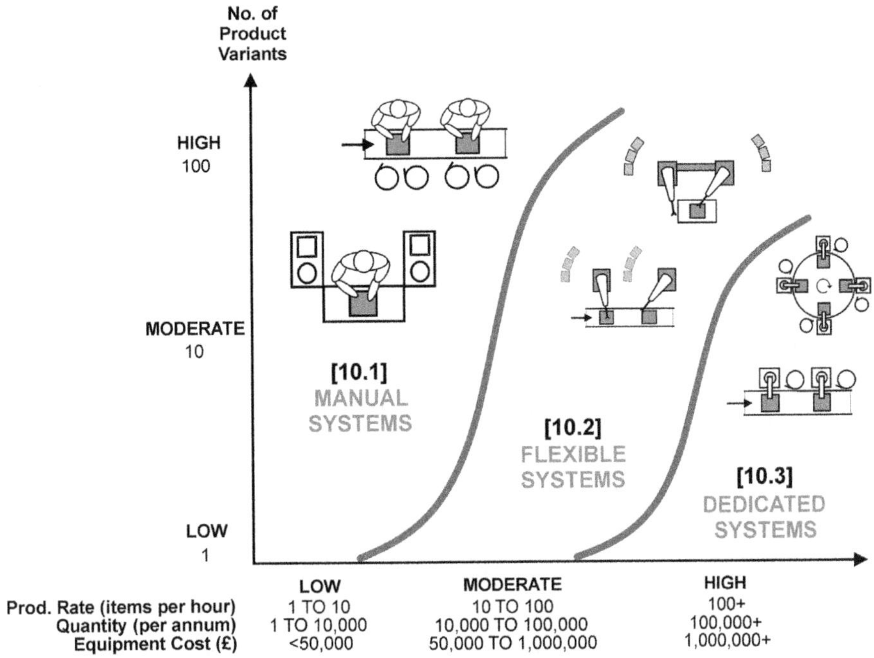

Figure 2.9: Assembly System PRIMA Selection Chart.

problematical (tooling, variability management, jams, etc.) and the running cost of the system grows rapidly. Higher production quantities are again required to justify the transition from manual to flexible, and flexible to dedicated systems.

Three basic assembly systems are summarised below with their respective PRIMA number:

10.1 *Manual assembly system* – Extremely flexible, easy to train operators, low capital cost, short development times, low to medium production rate, high-level sensing and checking capabilities. Also includes mechanised aids for the operator such as feeders and work heads.

10.2 *Flexible assembly system* – Programmable with robots. Reasonably flexible, moderate to high capital cost, long development times, medium production rate, involves complex and lengthy programming/teaching, limited sensing capabilities.

10.3 *Dedicated assembly system* – High-speed special purpose machines. Generally inflexible, high capital cost, low number of defects, long development times, high production rate, minimal sensing capabilities.

Upon assembly system selection, the technical and economic implications of the final decision must be understood, including cost estimate(s) for the product (see Chapter 13). This is particularly advantageous when Figure 2.9 shows that a set of requirements is on the boundary of two assembly system types.

In the design of the assembly system, further questions should be considered:

- What is the main objective: quality, cost or production rate?
- Care must be taken in the way we divide a product into subassemblies: How many and what are the criteria?
- How should we determine the sequence for the assembly?
- How should any necessary testing be carried out?

These issues can influence the degree of automation that can be implemented, for example. A review of the selected assembly system against business requirements should also be undertaken, as discussed earlier.

2.4.2 Case Studies

The case studies that follow describe where an automation technology has been successfully implemented as an economic and high-quality alternative to manual assembly. The intention is to illustrate the application of the selection criteria and also to indicate some of the opportunities for businesses associated with the implementation of assembly automation in industry. In the design of assembly systems, machine manufacturers have tended to adopt, where at all possible, a modular philosophy, coupled with the application of a well-trusted technology. This enables the suppliers to create systems for their customers that can be realistically priced, be effective and highly reliable. The case studies used illustrate what might be considered to be applications of automation, but with differing forms and degree of flexibility. The cases are discussed under the headings of products and customer requirements; the assembly process and machine design; and selection considerations. The case studies are all in the public domain and for more information on the studies the reader is directed to Ref. [20].

2.4.2.1 Assembly of Medical Non-return Valves

2.4.2.1.1 Product and customer requirements

The product to be assembled was a non-return check-valve used in medical equipment including catheters and tracheotomy tubes. The requirement was for a highly process-capable system with a defect rate (valve failure rate) of less than 1 part per million. Therefore, there was a requirement for checks to be built in to the assembly system to reject any part that does not conform to the process capability standard. The valve comprises six very small components and was configured in four different versions. The variants result from the requirement for the use of different material types and differences in the diameter of the caps that seal the valves. The demand for the product necessitated a production rate of 200 items/minute and cleanliness was a critical requirement for the assembly process.

2.4.2.1.2 Assembly process and machine design

To achieve the level of reliability needed at the required production rate, a linear assembly system was specially developed to assemble the six components of the valve. The cell was equipped with six vibratory bowl feeders of different sizes to feed and orient the valve's components on to pallets containing four sets of nests. The assembly system was designed with 21 stations and to enable the operator to select random samples for inspection from each of four nests. The system was configured with an operating speed of 50 cycles/minute to realise the required overall production output of 200 items per minute, and the flexible cell was capable of producing the four different versions of the product. Despite this high rate of production the valves produced were of the required quality, and displayed no surface faults (damage to the plastic components) that could have led to rejects. To meet the cleanliness requirements the parts of the assembly system that come into contact with the valve's components were made from stainless steel, and the machine was carefully designed to operate without traces of dust or particulates. In addition, precise component fitting operations were required by the product design, with some of the items having to be inserted into the body of the valve within a tolerance of 0.05 mm.

2.4.2.1.3 Selection considerations

Factors driving the selection of the assembly technology adopted for the application could be considered to include:

- High production volumes and continuous demand.
- Four different product variants.
- Very high levels of process capability (<1 ppm).
- Clean assembly process environment free from contamination.

The product volume, number of variants and process capability requirements support the application of flexible assembly system for the product as indicated in Figure 2.9.

2.4.2.2 Assembly and Test of Diesel Injector Units

2.4.2.2.1 Product and customer requirements

The requirement was for a flexible system to assemble a family of diesel unit injectors that could yield economic operation at fluctuating demand volumes. To realise the demanding tolerances necessitated by the product technology the injector unit makes use of precision shims to compensate for machining variation and the inevitable variation in the characteristics of the spring embedded within the injector body. By choosing a shim of the correct characteristic thickness and capability the business can vary the opening pressure of the valve to achieve an injector unit assembly that operates correctly first time. The customer's 'lean manufacturing' philosophy required that automation should only be introduced where there is a clear quality and economic case to do so. The automation project had to respect the customer's principle of balancing the relative benefits of automation against that of well-known manual assembly processes.

2.4.2.2.2 Assembly process and machine design

The system created by the assembly machine supplier operated on the 'Negari' principle, which readily allows production volumes to be varied depending on the number of operators allocated to the system at any one time. The machine was designed such that a single operator could operate all machine stations in sequence; however, up to four operators could work on the same machine system to create a proportionate increase in production rates. The system was designed to enable assembled injectors to be 'wet tested' to verify the functional performance of the unit. The system provides the business with a means of directly responding to fluctuations in demand for the product. The system was also designed so that when the product is eventually withdrawn from service the Negari facility will be able to provide 'service' components to reflect demand with the minimum of downtime.

2.4.2.2.3 Selection considerations

Considerations driving the selection of the assembly technology adopted include:

- Medium/high production volumes.
- Fluctuating demand patterns.
- Very high levels of process capability.
- Integrate product testing.

In order to meet the requirement for volume flexibility, the assembly system needs flexibilities in areas including: parts handling and fitting processes, machine capacity and processing routes. Adopting the Negari machine layout with multi-stations and manual handling and loading of parts, provides a natural way of dealing with this problem.

2.4.2.3 Accelerator Pedal Sensor Assembly

2.4.2.3.1 Product and customer requirements

The electronic pedal sensor provides a means of throttle control that is more accurate and more reliable than cables, and provides a product that is essentially maintenance free. The sensor design is supplied to a leading (tier 1) manufacturer whose generic throttle pedal design places it well to meet the requirements of many major Original Equipment Manufacturers (OEM). Given the safety-critical nature of the accelerator pedal sensor it is essential to electronically test each completed assembly to make sure is works correctly. The product comprised eight components and was to be assembled on a 9-second cycle time.

2.4.2.3.2 Process and assembly machine design

The process is essentially automatic, but requires two operators to load critical components. Each operation is checked to make sure that it took place correctly (any incorrect assemblies are flagged on the pallet and pass through without further work). Laser trimming calibrates the resistance of the unit and an electronic test also ensures that each completed assembly works properly. A modular approach was adopted for the design of the machine. The operator first loads a housing and rotor on to a flagged pallet. The first automatic station then loads the substrate, ensures that it is laid flat and heat stakes it into position. The system checks only that the substrate is present, not that it performs correctly. Electronic testing of the final unit is carried out later in the process. Further operations load the spring, rotor and cover, which is heat staked into position to complete the assembly.

Three of the assembly operations are particularly technically demanding: wire bonding, spring contact assembly and laser trimming. A proprietary wire-bonding station is used to weld the thin wire contacts into position to link the substrate with the electrical contacts moulded in the sensor body. The spring contact assembly positions three small twin-spoked contacts and heat stakes them in position. The contacts must be secured without deformation and a force gauge is used to measure the pressure exerted by every spoke of the contact on the substrate track to ensure proper connections are made. Laser trimming of the substrate track calibrates the final assembly to ensure it has the correct resistance at a reference position. The system checks the resistance before and after the trimming process. This is a critical operation that ensures the correct operation of the sensor.

2.4.2.3.3 Selection considerations

The assembly technology adopted for the application could be considered as driven by factors including:

- High production volumes and continuous demand.
- Very high levels of process capability.
- Complex assembly processes.
- Integrated testing processes.

The product volume and the safety-critical nature of the process, coupled with complex assembly processes, point to the need for a special purpose automatic machine with operator loading of critical components, as supported by Figure 2.9.

2.5 Joining Process Selection

2.5.1 Selection Strategy

There is extensive evidence to suggest that many industrial products are designed with far too many parts. DFA case studies indicate that in many designs large proportions of excess components are only used for fastening [21]. These non-value-added components increase part-count and production costs without contributing to the product's functionality. In many cases, incorrect joining processes are used due to a lack of knowledge of such factors as availability, cost and functional performance of alternatives. As with primary and secondary manufacturing processes, selecting the most suitable joining process greatly influences the manufacturability of a design, but the selection of the joining technology to be used can also greatly influence the ease of assembly of a design. The method chosen can also have a significant influence on the product architecture and assembly sequence, and it is well known that complicated joining processes lead to incorrect, incomplete and faulty assemblies [22].

Selecting the most appropriate joining technique requires consideration of many factors relating to joint design, material properties and service conditions. During the selection procedure the designer is required to scrutinise large quantities of data relating to many different technologies. Several selection methods exist for the selection of the process variants within individual joining technologies. However, selecting the most appropriate technology itself remains a design-oriented task that often does not get the attention it deserves. It can be concluded that a selection methodology that incorporates joining processes and technologies that can be applied at an early stage in the design process is a useful tool to support design and particularly DFA. Considering joining processes prior to the development of detailed geometry enables components to be tailored to the selected process rather than limiting the number of suitable processes. Addressing such issues during the early stages of product development actively encourages designers to employ DFA practice and reduces the need for costly redesign work.

The aim of the joining process selection methodology presented here is to provide a means of identifying feasible methods of joining, regardless of their fundamental technology. The methodology is not intended to select a specific joining method, for example torch brazing or tubular rivet, but to highlight candidate processes that are capable of joining under the given conditions. The final selection can be made after considering process-specific data and detailed data against design requirements from the PRIMAs. The reader interested in other approaches to this problem is referred to Refs [23–25].

Due to the large number of different joining processes and variants, only the most commonly used and well-established processes in industry are included. Investigations highlighted 73 major joining techniques, as shown in Figure 1.10 [26]. In order to classify them, a common factor is used based on technology and process. Technology class refers to the collective group that a process belongs to, for example welding or adhesive bonding. The process class refers to the specific joining technique, for example Metal Inert-gas Welding (MIG) or anaerobic adhesive. Each process is derived from a particular fundamental technology providing a means for classification. From this, the joining processes have been divided into five main categories: welding, brazing, soldering, mechanical fastening and adhesive bonding.

Technical classes can be separated into subcategories based on distinct differences in underlying technology. Although the basic premise of all welding processes is the same, specific techniques differ considerably due to the particular processes involved in generating heat and/or enabling the fusion process. This can be used as a means of classifying subsets. Both brazing and soldering have a number of different processes so they have been split into two subsets. Mechanical fasteners can be divided in two ways, by group technologies and degree of permanence. The latter has been chosen as it relates to the functionality of the fastener in service and therefore product requirements. Due to the large number of specific adhesives, which in many cases are exclusive to the producer, adhesive bonding has been viewed from a generic level; therefore, only the adhesive group can be selected.

In order to select the most appropriate joining process, it is necessary to consider all processes available within the methodology. As technology-specific selection criteria tend to be non-transportable between domains, evaluating the merits of joining processes that are based on fundamentally dissimilar technologies requires a different approach. Differentiating between technology classes and process classes requires the comparison of specifically selected parameters. In order to evaluate a joint, consideration must be given to its functional, technical, spatial and economic requirements. A review of important joining requirements has identified a number of possible selection criteria as shown in Table 2.1, which are discussed below.

Table 2.1: Classification of Joint Requirements.

Category	Criteria	Description
	Technological	
Functional	• Degree of permanence • Loading type (static, cyclic, impact) • Strength	Functional requirements define the working characteristics of the joint
Technical	• Joint configuration/design • Operating temperature • Material type • Material compatibility • Accuracy	Specific needs of components to be joined are categorised by a joint's technical requirements
Spatial	• Material thickness • Size, weight • Geometry	Geometric characteristics of the joint are accounted for by the spatial requirements
Miscellaneous	• Complexity • Flexibility (assembly/orientation) • Safety • Joint accessibility • Quality	Other important issues not considered by the above groups
	Economic	
	• Production quantity • Production rate • Availability of equipment • Ease of automation • Skill required • Tooling requirements • Cost	The economics of joining processes align the design with the business needs of the product

- *Functional.* Functional requirements define the working characteristics of the joint. The functional considerations for a joint are degree of permanence, load type and strength. Degree of permanence identifies whether a joint needs to be dismantled or not. In most cases the permanence of a joining process is independent of its technology class. Degree of permanence provides a suitable high-level selection criterion that is not reliant on detailed geometry. Load type and strength are often mutually dependent and can be influenced by the geometric characteristics of the joint interface. As joint design is dissimilar for different technology classes, it is difficult to use load type or strength as universal selection criteria. However, these considerations must be considered when evaluating suitable joining processes for final selection when appropriate.
- *Technical.* Specific needs of components to be joined are categorised by the joint's technical requirements. The technical considerations for a joint are material type, joint design and operating temperature. Material type is selected based on parameters defined by the product's operating environment, such as corrosion resistance. The material type is relevant to all joining technologies because they need to be compatible. Joint design is often defined by the geometry. However, if joining is considered prior to detailed geometry, the selected process can influence the design. Due to the fundamental differences in joint configurations, it is not suitable as a selection criterion for non-technology-specific selection. Operating temperature influences the performance of most joining processes, although it should be considered during material selection. While an important aspect, its effect varies for different joining technologies. Therefore, consideration of operating temperature is more appropriate during final selection.
- *Spatial.* Geometric characteristics of the joint are accounted for by the spatial requirements. The spatial requirements identified are size, weight, geometry and material thickness. The size and weight of components to be joined is considered and determined when their material is selected. As the selection methodology is intended for use prior to the development of detailed geometry, using geometry as a selection criterion would be contradictory. Material thickness has already proven to be a successful criterion in other selection methodologies and the suitability of joining processes is easily classified for different thicknesses of material.
- *Economic.* The economics of joining processes align the design with the business needs of the product. Economic considerations can be split into two sections: tooling and product. Tooling refers to the ease of automation, availability of equipment, skill required, tooling requirements and cost. Product economics relate to production rates and quantity. These business considerations are driven by the product economics as they determine the need for tooling and its complexity, levels of automation and labour requirements. Production rate and quantity are very closely linked. They can both be used to determine the assembly speed and the need for and feasibility of automation. However, as the selection methodology is to be used in the early stages of product development it is more likely that quantity will be known from customer requirements or market demand.

In order for the selection methodology to be effective at the early stages of design appraisal, the chosen parameters must apply to all joining processes. Also, it is essential that the parameters relate to knowledge that is readily available and appropriate to the level of selection. Having reviewed the requirement against the joining processes, four selection parameters have been chosen for the initial stages of the methodology:

- *Material Type* – Accounts for the compatibility of the parent material with the joining process. A large proportion of the materials used in engineering manufacture have been included in the selection methodology, from ferrous alloys to precious metals. In situations involving multiple material types the selection methodology must be applied for each.
- *Material Thickness* – Divided into three ranges: thin, ≤3 mm; medium, from 3 to 19 mm; and thick, ≥19 mm. When selecting the material type and thickness, the designer considers many other factors that can be attributed to the joint requirements, such as corrosion resistance, operating temperature and strength. Consequently, the requirements should be known and can be compared to joining process design data for making the final choice at a later stage.
- *Degree of Permanence* – This is a significant factor in determining appropriate joining processes as it relates to the in-service behaviour of the joint and considers the need for a joint to be dismantled. This selection criteria is divided into three types:
  - Permanent joint – Can only be separated by causing irreparable damage to the base material, functional element or characteristic of the components joined, for example surface integrity. A permanent joint is intended for a situation where it is unlikely that a joint will be dismantled under any servicing situation.
  - Semi-permanent joint – Can be dismantled on a limited number of occasions, but may result in loss or damage to the fastening system and/or base material. Separation may require an additional process, for example re-heating a soldered joint or plastic deformation. A semi-permanent joint can be used when disassembly is not performed as part of regular servicing, but for some other need.
  - Non-permanent joint – Can be separated without special measures or damage to the fastening system and/or base material. A non-permanent joint is suited to situations where regular dismantling is required, for example at scheduled maintenance intervals.
- *Quantity* – Production quantity per annum, and consequently the number of joints to be produced, accounts for the economic feasibility of the joining process. The quantities specified for selection purposes are the same as for the manufacturing processes selection strategy.

The joining process selection methodology is based on the same matrix approach used for manufacturing process selection. Again, due to page size constraints and the number of processes to be detailed, each process has been assigned an identification code rather than using process names. The key to the joining processes used in the matrix is shown in Figure 2.10, together with the relevant PRIMA number, where information can be found regarding that individual process or joining technology. Due to size constraints, the joining process selection matrix is divided into two parts. Figure 2.11(a) and (b) together shows the complete matrix.

The matrix representation of the selection strategy provides an intuitive way of navigating a large quantity of data. This makes the selection process simple and quick to use. Supporting the selection matrix with design advice through the use of the PRIMAs completes the methodology, allowing the user to both identify and gain an understanding of suitable joining processes. Combining material thickness with material type is a logical way of allowing the user to describe the material. It is also convenient to state quantity and then permanence in the columns of the matrix.

Note that the joining process PRIMA selection matrix cannot be considered comprehensive and should not be taken as such. It represents the main common industrial practice, but there will always be exceptions at this level of detail. Also, the order in which the PRIMAs are listed in the nodes of the matrix has no significance in terms of preference. Dissimilar metals also account for joining metals with coatings.

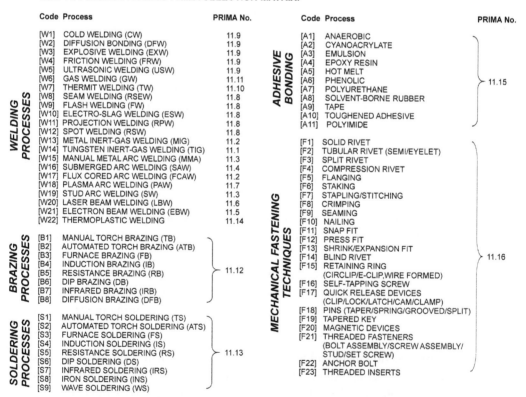

KEY TO JOINING PROCESS PRIMA SELECTION MATRIX:

	Code	Process	PRIMA No.
WELDING PROCESSES	[W1]	COLD WELDING (CW)	11.9
	[W2]	DIFFUSION BONDING (DFW)	11.9
	[W3]	EXPLOSIVE WELDING (EXW)	11.9
	[W4]	FRICTION WELDING (FRW)	11.9
	[W5]	ULTRASONIC WELDING (USW)	11.9
	[W6]	GAS WELDING (GW)	11.11
	[W7]	THERMIT WELDING (TW)	11.10
	[W8]	SEAM WELDING (RSEW)	11.8
	[W9]	FLASH WELDING (FW)	11.8
	[W10]	ELECTRO-SLAG WELDING (ESW)	11.8
	[W11]	PROJECTION WELDING (RPW)	11.8
	[W12]	SPOT WELDING (RSW)	11.8
	[W13]	METAL INERT-GAS WELDING (MIG)	11.2
	[W14]	TUNGSTEN INERT-GAS WELDING (TIG)	11.1
	[W15]	MANUAL METAL ARC WELDING (MMA)	11.3
	[W16]	SUBMERGED ARC WELDING (SAW)	11.4
	[W17]	FLUX CORED ARC WELDING (FCAW)	11.2
	[W18]	PLASMA ARC WELDING (PAW)	11.7
	[W19]	STUD ARC WELDING (SW)	11.3
	[W20]	LASER BEAM WELDING (LBW)	11.6
	[W21]	ELECTRON BEAM WELDING (EBW)	11.5
	[W22]	THERMOPLASTIC WELDING	11.14
BRAZING PROCESSES	[B1]	MANUAL TORCH BRAZING (TB)	11.12
	[B2]	AUTOMATED TORCH BRAZING (ATB)	
	[B3]	FURNACE BRAZING (FB)	
	[B4]	INDUCTION BRAZING (IB)	
	[B5]	RESISTANCE BRAZING (RB)	
	[B6]	DIP BRAZING (DB)	
	[B7]	INFRARED BRAZING (IRB)	
	[B8]	DIFFUSION BRAZING (DFB)	
SOLDERING PROCESSES	[S1]	MANUAL TORCH SOLDERING (TS)	11.13
	[S2]	AUTOMATED TORCH SOLDERING (ATS)	
	[S3]	FURNACE SOLDERING (FS)	
	[S4]	INDUCTION SOLDERING (IS)	
	[S5]	RESISTANCE SOLDERING (RS)	
	[S6]	DIP SOLDERING (DS)	
	[S7]	INFRARED SOLDERING (IRS)	
	[S8]	IRON SOLDERING (INS)	
	[S9]	WAVE SOLDERING (WS)	

	Code	Process	PRIMA No.
ADHESIVE BONDING	[A1]	ANAEROBIC	11.15
	[A2]	CYANOACRYLATE	
	[A3]	EMULSION	
	[A4]	EPOXY RESIN	
	[A5]	HOT MELT	
	[A6]	PHENOLIC	
	[A7]	POLYURETHANE	
	[A8]	SOLVENT-BORNE RUBBER	
	[A9]	TAPE	
	[A10]	TOUGHENED ADHESIVE	
	[A11]	POLYIMIDE	
MECHANICAL FASTENING TECHNIQUES	[F1]	SOLID RIVET	11.16
	[F2]	TUBULAR RIVET (SEMI/EYELET)	
	[F3]	SPLIT RIVET	
	[F4]	COMPRESSION RIVET	
	[F5]	FLANGING	
	[F6]	STAKING	
	[F7]	STAPLING/STITCHING	
	[F8]	CRIMPING	
	[F9]	SEAMING	
	[F10]	NAILING	
	[F11]	SNAP FIT	
	[F12]	PRESS FIT	
	[F13]	SHRINK/EXPANSION FIT	
	[F14]	BLIND RIVET	
	[F15]	RETAINING RING (CIRCLIP/E-CLIP,WIRE FORMED)	
	[F16]	SELF-TAPPING SCREW	
	[F17]	QUICK RELEASE DEVICES (CLIP/LOCK/LATCH/CAM/CLAMP)	
	[F18]	PINS (TAPER/SPRING/GROOVED/SPLIT)	
	[F19]	TAPERED KEY	
	[F20]	MAGNETIC DEVICES	
	[F21]	THREADED FASTENERS (BOLT ASSEMBLY/SCREW ASSEMBLY/ STUD/SET SCREW)	
	[F22]	ANCHOR BOLT	
	[F23]	THREADED INSERTS	

Figure 2.10: Key to Joining Process PRIMA Selection Matrix.

MATERIAL & THICKNESS / QUANTITY & PERMANENCE		IRONS		STEEL (carbon)			STEEL (tool, alloy)			STAINLESS STEEL			COPPER & ALLOYS			ALUMINIUM & ALLOYS			MAGNESIUM & ALLOYS			ZINC & ALLOYS			TIN & ALLOYS		
		MED. 3 to 19mm	THICK ≥19mm	THIN ≤3mm	MED. 3 to 19mm	THICK ≥19mm	THIN ≤3mm	MED. 3 to 19mm	THICK ≥19mm	THIN ≤3mm	MED. 3 to 19mm	THICK ≥19mm	THIN ≤3mm	MED. 3 to 19mm	THICK ≥19mm	THIN ≤3mm	MED. 3 to 19mm	THICK ≥19mm	THIN ≤3mm	MED. 3 to 19mm	THICK ≥19mm	THIN ≤3mm	MED. 3 to 19mm	THICK ≥19mm	THIN ≤3mm	MED. 3 to 19mm	THICK ≥19mm
VERY LOW 1 TO 100	P	[W6][W13][W15][B1]	[W6][W13][W15][B1]	[W3][W5][W8][W9][W11][W13][W14][W15][B1]	[W3][W6][W9][W11][W13][W14][W15][B1]	[W6][W7][W9][W10][W13][W15]	[W2][W3][W5][W8][W9][W11][W13][W14][W15][B1]	[W2][W3][W9][W11][W13][W14][W15][B1]	[W3][W7][W9][W10][W13][W15]	[W2][W3][W6][W8][W9][W11][W13][W14][W15][B1]	[W2][W3][W9][W11][W13][W14][W15][W17][B1]	[W9][W10][W13][W15][W17]	[W1][W2][W3][W9][W13][W14][B1]	[W1][W2][W3][W9][W13][B1]	[W3][W7][W9][W13]	[W1][W2][W3][W6][W8][W9][W11][W13][W14][B1]	[W1][W2][W3][W6][W9][W11][W13][W14][B1]	[W3][W9]	[W3][W8][W9][W13][W14][B1]	[W3][W9][W4][W13][W14]	[W13]				[W1]	[W1]	
	SP			[S1]						[S1]			[S1][S8]			[S1][S8]						[S1][S8]			[S1][S8]		
	NP	[F19][F20][F23]	[F19][F20][F23]	[F20]	[F19][F20]	[F19][F20]	[F20]	[F19][F20]	[F19][F20]	[F20]	[F19][F20]	[F19][F20]		[F19][F23]	[F19][F23]		[F23]	[F23]		[F23]	[F23]		[F23]	[F23]			
LOW 100 TO 1,000	P	[W13][W15][W16][W17][B1]	[W13][W15][W16][W17]	[W3][W8][W9][W11][W13][W14][W15][B1][B6]	[W3][W9][W11][W13][W14][W15][W16][W17][B1]	[W9][W13][W15][W16][W17]	[W2][W3][W8][W9][W11][W13][W14][W15][B1][B6]	[W2][W3][W9][W11][W13][W14][W15][W17][W16][B1]	[W3][W9][W13][W15][W16][W17]	[W2][W3][W8][W9][W11][W13][W14][W15][B1][B6]	[W2][W3][W9][W11][W13][W14][W15][W16][B1]	[W3][W9][W13][W15][W16][W17]	[W1][W2][W3][W9][W13][W14][B1]	[W2][W3][W9][W13][B1]	[W3][W9][W13]	[W1][W2][W3][W8][W9][W11][W13][W14][B1][B6]	[W1][W2][W3][W9][W11][W13][W14][B1][B6]	[W3][W9]	[W3][W8][W9][W13][W14][B1][B6]	[W4][W3][W9][W13][W14]	[W13]				[W1]	[W1]	
	SP			[S1]						[S1]			[S1][S8]			[S1][S8]						[S1][S8]			[S1][S8]		
	NP	[F19][F20][F23]	[F19][F20][F23]	[F20]	[F19][F20]	[F19][F20]	[F20]	[F19][F20]	[F19][F20]	[F20]	[F19][F20]	[F19][F20]		[F19][F23]	[F19][F23]		[F23]	[F23]		[F23]	[F23]		[F23]	[F23]			
LOW TO MEDIUM 1,000 TO 10,000	P	[W16][W17][B2][B4]	[W16][W17][B4]	[W8][W11][W13][W14][W19][B2][B4][B6]	[W11][W13][W14][W16][W17][W19][B2][B4]	[W13][W16][W17]	[W2][W8][W11][W13][W14][W19][B2][B4][B6]	[W2][W11][W13][W14][W16][W17][W19][B2][B4]	[W13][W16][W17]	[W2][W8][W11][W13][W14][W19][B2][B4][B6]	[W2][W11][W13][W14][W16][W17][W19][B2][B4]	[W13][W16][W17]	[W1][W2][W13][W14][W19][B2][B4]	[W2][W13][B2]		[W1][W2][W8][W11][W13][W14][W19][B2][B4][B6]	[W1][W2][W11][W13][W14][W19][B2][B6]		[W8][W13][W14][B2][B4][B6]	[W4][W13][W14]	[W13]	[W19]			[W1]	[W1]	
	SP			[S6]			[F11]			[F11][S6]			[S2][S6][S8]			[S2][S6][S8]						[S2][S6][S8]			[S1][S6][S8]		
	NP																										
MEDIUM TO HIGH 10,000 TO 100,000	P	[W18][B2][B3][B4][B8]	[B2][B3][B4][B8]	[W8][W11][W13][W14][W18][W19][W20][B2][B3][B4][B5][B7][B8][F7][F9]	[W4][W11][W13][W14][W16][W17][W18][W19][W20][B2][B3][B4][B8]	[W4][W13][W16][W17][B3][B8]	[W8][W11][W14][W18][W19][W20][B2][B3][B4][B5][B7][B8]	[W4][W11][W14][W16][W17][W18][W19][W20][B2][B3][B4][B8]	[W4][W16][W17][B3][B8]	[W8][W11][W13][W18][W19][W20][B2][B3][B4][B5][B7][B8][F7][F9]	[W4][W11][W14][W16][W17][W18][W19][W20][B2][B3][B4][B8]	[W4][W16][W17][B3][B8]	[W18][W19][W20][B2][B3][B4][B5][B8][F7][F9]	[W4][W18][W20][B2][B3][B8]	[B3][B8]	[W4][W8][W11][W14][W18][W19][W20][B2][B3][B4][B6][B7][B8][F7][F9]	[W4][W11][W14][W19][B2][B3][B8]	[B8]	[W8][W14][W18][W20][B2][B3][B4][B8]	[W4][W14][W18][W20][B3][B8]		[W19]					
	SP			[S3]			[F11]			[F11][S3]			[S2][S3][S4][S5][S7][S9]			[S2][S3][S4][S5][S7][S9]						[S2][S3][S4][S5][S7][S9]			[S2][S3][S4][S5][S7][S9]		
	NP						[F11]			[F11]																	
HIGH 100,000+	P	[B3][B8]	[B3][B8]	[W18][W20][W21][B3][B8][F7][F9]	[W4][W18][W20][W21][B3][B8]	[W4][W21][B3][B8]	[W18][W20][W21][B3][B8]	[W4][W18][W20][W21][B3][B8]	[W4][W21][B3][B8]	[W18][W20][W21][B3][B8][F7][F9]	[W4][W18][W20][W21][B3][B8]	[W4][W21][B3][B8]	[W18][W20][W21][B3][B8][F7][F9]	[W4][W18][W20][W21][B3][B8]	[W21][B3][B8]	[W4][W18][W20][W21][B3][B8][F7][F9]	[W4][W21][B3][B8]	[W21][B8]	[W18][W20][B3][B8]	[W18][W20][W21][B3][B8]	[W21]						
	SP			[S3]			[F11]			[F11][S3]			[S3][S9]			[S3][S9]						[S3][S9]			[S3][S9]		
	NP						[F11]			[F11]																	
ALL QUANTITIES	P	[F1][A1][A2][A4][A5][A8][A10]	[F1]	[W5][W12][A1][A2][A4][A5][A8][A10][F1][F5][F6][F8][F10]	[W12][A1][A2][A4][A5][A10][F1]	[F1]	[W5][W12][F1][A1][A2][A4][A5][A8][A10]	[F1][A1][A2][A4][A5][A10]	[F1]	[W5][W12][F1][F5][F6][F8][F10][A4][A10]	[W12][F1][A4][A10]	[F1]	[W12][F1][F5][F6][F8][F10][A1][A2][A4][A5][A10]	[W12][F1][F10][A4][A10]	[F1]	[W5][W12][F1][F2][F5][F6][F8][F10][A2][A4][A5][A7][A10]	[W5][W12][F1][F2][A2][A4][A5][A7][A10]	[F1][F2]	[W12][F1][F2][A1][A4][A5]	[W12][F1][F2][A4][A5]	[F1][F2]	[W12][F2][F10][F14]	[W12][F2]	[F2]	[W12][F2][F10][F14][A4]	[W12][F2]	[F2]
	SP	[F12][F13]	[F12][F13]	[A9][F14]	[F12][F13]	[F12][F13]	[F13][A9]	[F12][F13]	[F12][F13]	[F13][A9][F14]	[F12][F13]	[F12][F13]	[A9][F13][F14]	[F12][F13]	[F12][F13]	[F13][F14][A9]	[F12][F13]	[F12][F13]	[F13][A9]	[F12][F13]	[F12][F13]	[F14]			[F13][A9][F14]	[F13]	
	NP	[F15][F16][F18][F21]	[F15][F18][F21]	[F15][F17][F21]	[F15][F17][F18][F21]	[F15][F17][F21]	[F15][F17][F21]	[F15][F17][F21]	[F15][F17][F21]	[F15][F17][F21]	[F15][F17][F18][F21]	[F15][F17][F18][F21]	[F16][F21]	[F16][F18][F21]	[F18][F21]	[F15][F16][F17][F21]	F[15][F16][F17][F18][F21]	[F15][F17][F18][F21]	[F16][F21]	[F16][F18][F21]	[F18][F21]	[F15][F16][F21]	[F15][F16][F18][F21]	[F15][F18][F21]	[F16][F21]	[F16][F21]	[F21]

Figure 2.11: (a) Joining Process PRIMA Selection Matrix – Part A.

QUANTITY & PERMANENCE \ MATERIAL & THICKNESS		LEAD & ALLOYS			NICKEL & ALLOYS			TITANIUM & ALLOYS			THERMOPLASTICS			THERMOSETS			FR COMPOSITES			CERAMICS			REFRACTORY METALS			PREC-IOUS METALS	DISSIMILAR MATERIALS		
		THIN ≤3mm	MED. 3 to 19mm	THICK ≥19mm	THIN ≤3mm	MED. 3 to 19mm	THICK ≥19mm	THIN ≤3mm	MED. 3 to 19mm	THICK ≥19mm	THIN ≤3mm	MED. 3 to 19mm	THICK ≥19mm	THIN ≤3mm	MED. 3 to 19mm	THICK ≥19mm	THIN ≤3mm	MED. 3 to 19mm	THICK ≥19mm	THIN ≤3mm	MED. 3 to 19mm	THICK ≥19mm	THIN ≤3mm	MED. 3 to 19mm	THICK ≥19mm	THIN ≤3mm	THIN ≤3mm	MED. 3 to 19mm	THICK ≥19mm
VERY LOW 1 TO 100	P	[W1]	[W1]		[W2][W3][W6][W8][W9][W11][W13][W14][W15]	[W2][W3][W9][W11][W13][W14][W15]	[W9][W10][W13][W15]	[W2][W3][W9][W13][W14]	[W2][W3][W9][W13][W14]	[W3][W9][W13]	[W22]	[W22]	[W22]										[W2][W3][W8][W9][W13][W14][B1]	[W2][W3][W9][W13]	[W3]	[W1][W2]	[W2][W3][W8][W9][W11][B1]	[W2][W3][W9][W11][B1]	[W3][W9]
	SP	[S1][S8]			[S1][S8]															[S1]						[S1][S8]	[S1][S8]		
	NP		[F23]	[F23]	[F20]	[F20]	[F20]														[F22]	[F22]					[F20]	[F19][F20][F22]	[F19][F20][F22]
LOW 100 TO 1,000	P	[W1][W14]	[W1][W14]		[W2][W3][W8][W9][W11][W13][W14][W15][B8]	[W2][W3][W9][W11][W13][W14][W15][W16]	[W9][W13][W15]	[W2][W3][W9][W13][W14]	[W2][W3][W9][W13][W14]	[W3][W9][W13]	[W22]	[W22]	[W22]										[W2][W3][W8][W9][W13][W14][B1]	[W2][W3][W9][W13]	[W4]	[W1][W2]	[W2][W3][W9][W8][W11][B1]	[W2][W3][W9][W11][B1]	[W3][W9]
	SP	[S1][S8]			[S1][S8]															[S1]						[S1][S8]	[S1][S8]		
	NP		[F23]	[F23]	[F20]	[F20]	[F20]														[F22]	[F22]					[F20]	[F19][F20][F22]	[F19][F20][F22]
LOW TO MEDIUM 1,000 TO 10,000	P	[W1][W14]	[W1][W14]		[W2][W8][W9][W11][W13][W14][B4][B6]	[W2][W11][W13][W14][W16]	[W13]	[W2][W13][W14][W19][B4]	[W2][W13][W14][W19]	[W13]													[W2][W8][W14][B2][B4]	[W2]	[W2]	[W1]	[W2][W8][W11][W19][B2]	[W2][W11][B2]	
	SP	[S2][S6][S8]			[S2][S6][S8]															[S2][S3][S6]						[S2][S8]	[S2][S3][S6][S8][F11]	[F11]	
	NP																				[F22]	[F22]						[F22]	[F22]
MEDIUM TO HIGH 10,000 TO 100,000	P	[W14]	[W14]		[W8][W11][W14][W18][W20][B3][B4][B5][B7][B8][F9]	[W4][W11][W14][W16][W18][W20][B3][B8]	[W4][B3][B8]	[W8][W14][W18][W19][W20][B3][B4][B7][B8]	[W4][W14][W18][W19][W20][B3][B8]	[B3][B8]	[F7]			[F7]						[S2][S3][S4][S5][S7]			[W18][B2][B4][B5][B7][B8]	[W4][W18]	[W4]		[W8][W11][W19][W20][B2][B3][B8][F7][F9]	[W4][W11][W20][B2][B3][B8]	[W4][B3][B8]
	SP	[S2][S3][S4][S5][S7][S9]			[S2][S3][S4][S5] [S7][S9]			[S7]			[F11]	[F11]		[F11]	[F11]											[S2][S3][S4][S5]	[S2][S3][S4][S5][S7][F11]	[F11]	
	NP										[F11]	[F11]		[F11]	[F11]												[F11]	[F11]	
HIGH 100,000+	P				[W18][W20][W21][B3][B8][F9]	[W4][W18][W20][W21][B3] [B8]	[W4][W21][B3][B8]	[W18][W20][W21][B3][B8]	[W4][W18][W20][W21][B3][B8]	[W21][B3][B8]	[F7]			[F7]									[W18][W21][B8]	[W4][W18][W21]	[W4]	[W21]	[W20][W21][B3][B8][F7][F9]	[W4][W20][W21][B3][B8]	[W4][B3][B8]
	SP	[S3][S9]			[S3][S9]						[F11]	[F11]		[F11]	[F11]												[F11]	[F11]	
	NP										[F11]	[F11]		[F11]	[F11]												[F11]	[F11]	
ALL QUANTITIES	P	[F2][F10][A1][A4]	[F2][F10][A4]	[F2]	[W5][W12][F1][A4][A8]	[W12][F1][A4]	[F1]	[W12][A11]	[A11]		[W5][F2][F3][F1][F4] [F6][F10] [A2][A4][A8][A10]	[F2][F3][F1][F10][A2][A4][A8][A10]	[F1][F2] F3][A4][A10]	[F2][F3][F4][F10][A2][A4][A7][A8]	[F2][F3][A2][A4][A7][A8]	[F2][F3][A7]	[F2][F4][A2][A4][A7][A8][A10]	[A2][A4][A7][A8][A10]	[A4]	[F2][A2][A4][A5][A7][A8][A10]	[F10][A2][A4][A5][A7][A8][A10]	[F10][A2][A4][A5][A7][A8][A10]	[F1]	[F1]	[F1]	[W5][F1][F8][A4]	[W5][W12][F1][F2][F5][F6][F10]	[W12][F1][F2][F10]	[F1]
	SP				[F13]	[F12][F13]	[F12][F13]	[A9]			[A9]	[F12]	[F12]	[A9]	[F12]	[F12]				[F13]	[F12][F13]	[F13]	[F13]	[F12][F13]	[F12][F13]		[F13][F14]	[F12][F13]	[F12][F13]
	NP	[F16][F21]	[F16][F21]	[F21]	[F15][F16][F17][F21]	[F15][F16][F17][F18][F21]	[F15][F17][F18][F21]	[F15][F17][F21]	[F15][F17][F18][F21]	[F15][F17][F18][F21]	[F17][F21][F23]	[F17][F18][F21] [F23]	[F17][F18][F21] [F23]	[F16][F17][F21][F23]	[F15][F16][F17][F18][F21] [F23]	[F15][F17][F18][F21][F23]	[F16][F17][F21][F23]	[F16][F17][F21][F23]	[F21][F23]	[F15][F17][F21]	[F15][F17][F18][F21]	[F15][F17][F18][F21]	[F21]	[F18][F21]	[F18][F21]	[F16][F17][F18][F21]	[F15][F16][F17][F18][F21][F23]	[F15][F16][F17][F18][F21][F23]	[F15][F17][F18][F21][F23]

Figure 2.11: (b) Joining Process PRIMA Selection Matrix – Part B.

2.5.2 Case Studies

In order to illustrate the selection methodology, two sample case studies are presented. The case studies show just how many different joining processes can be used on essentially the same design and how this affects part-count, ease of assembly and functional performance in support of DFA.

2.5.2.1 Rear Windscreen Wiper Motor

The first case study shows important DFA measures and highlights where joining methods have had a detrimental effect on the design. The joining process selection methodology has been applied and the suggested joining processes compared to those used in the DFA redesigns. The architecture of the original design is shown in Figure 2.12. The DFA evaluation shows six functional parts and 23 non-functional parts, giving a DFA design efficiency of 17.8% (see Appendix B1). Twelve of the non-functional components are only present for joining and to support the joining method, two bolts and two nuts to attach the housing and four rivets (and associated spacers) to join the brush plate to the retaining plate. The motor as intended is a throwaway module – that is, if a failure occurred during operation, the motor would be replaced, not repaired. Based on this information, all joints can be stated as permanent.

The redesign based on the DFA analysis suggestions is shown in Figure 2.13. The design proposed has six functional components and no non-functional components, giving a DFA design efficiency of 100%. The redesign eliminates all 12 components used for joining. The rivets and spacers have been removed, as the components they join are not in the redesign. Integrated snap-fit fasteners have replaced the nut and bolt assemblies for fastening the housing.

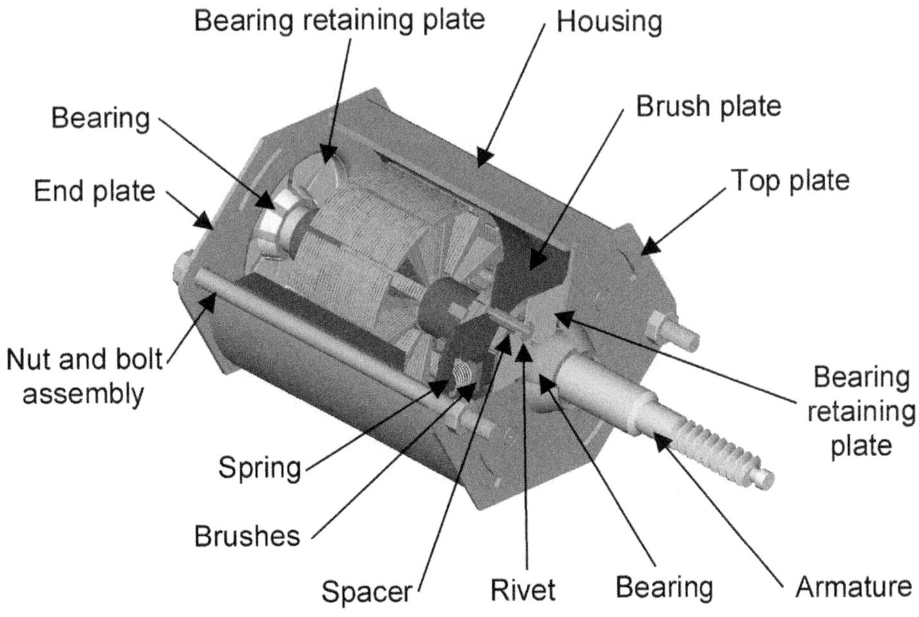

Figure 2.12: Motor Original Design.

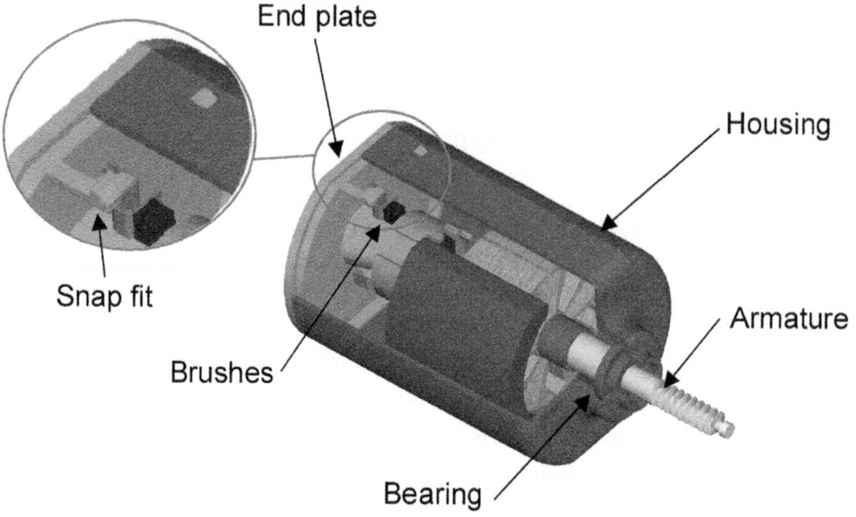

Figure 2.13: Motor Redesign.

The first step in selecting a joining process from the matrix is to determine the joint's requirements. The joint parameters for the housing are high volume (100,000+), permanent joint, thermoset material and thin (≤3 mm) material thickness. Based on these constraints the selection matrix shows the only suitable process to be a snap-fit fastener. However, the quantity column must also be evaluated for all quantities. This search identifies tubular rivets, split rivets, compression rivets, nailing, cyanoacrylate adhesives, epoxy resin adhesives, polyurethane adhesives and solvent-borne rubber adhesives as alternatives. In this case study, the geometry and material are unsuitable for riveting and nailing. A comparison of adhesives and snap-fit fasteners indicates that adhesives require more time for application, including a setting phase, and additional alignment features would need to be built into the components. Therefore, it is clear that the snap-fit fasteners are the most appropriate joining method.

Although the rivets have been removed along with the components they joined, they formed part of the assembly that held the bearing in place. Consequently, the joint between the bearing and housing needs to be considered. The joint selection parameters for the bearing to the housing are high volume, permanent joint, thermoset and steel material, and thin and medium thickness materials, respectively. For this evaluation the joining processes must match both material requirements. The search indicated two adhesive types, cyanoacrylate and epoxy resin, as candidates. A search based on the same parameters for all quantities indicates toughened adhesives as a third candidate. As all the candidate joining processes are similar, the final decision would be based on process, detailed design requirements and economic factors, such as cost and availability as provided in the PRIMAs. The proposed redesign suggested adhesive bonding for fixing the bearing into the housing.

2.5.2.2 Gas Meter Diaphragm Assembly

This case study details a sample set of designs from a case study involving 12 designs from different manufacturers. Here, three designs from different manufactures are considered.

DESIGN	DESCRIPTION	DFA MERITS*
A 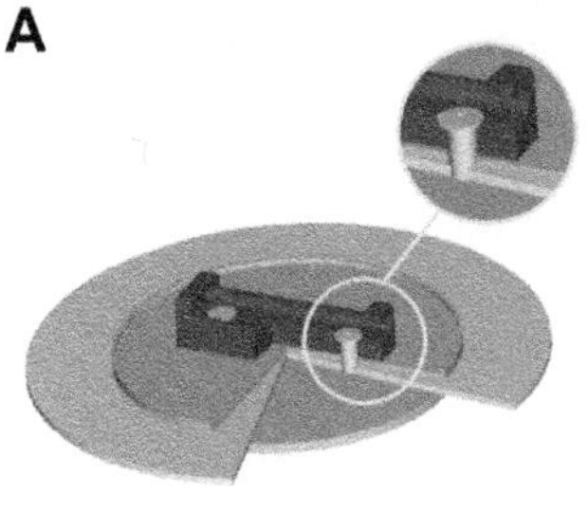	The top-plate and base-plate are steel pressings and the diaphragm is rubber. The flow measurement arm support component is a thermoplastic moulding. All four components are joined with two rivets.	**Part-count = 6** **Design Efficiency = 50%** **No. Joining Parts = 2**
B 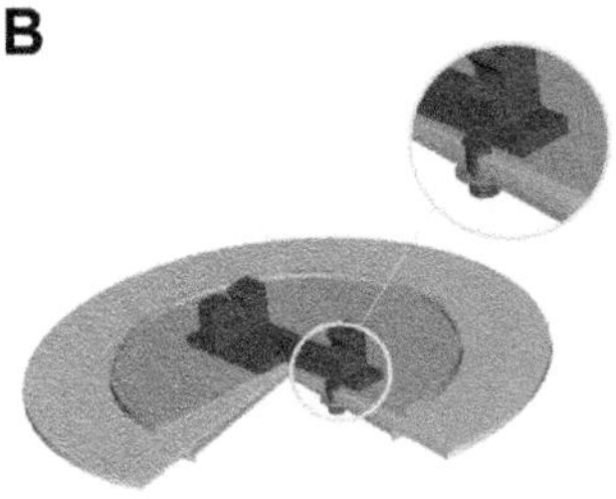	The top-plate and base-plate are steel pressings and the diaphragm is rubber. The flow measurement arm support component is a thermoplastic moulding. Two pins integrated into the support component, which are fastened with two push-on retaining rings to fasten all the parts.	Uses identical top-plates and base-plates reducing the number of unique components. Integrated pins aid with alignment during assembly. **Part-count = 6** **Design Efficiency = 50%** **No. Joining Parts = 2**
C 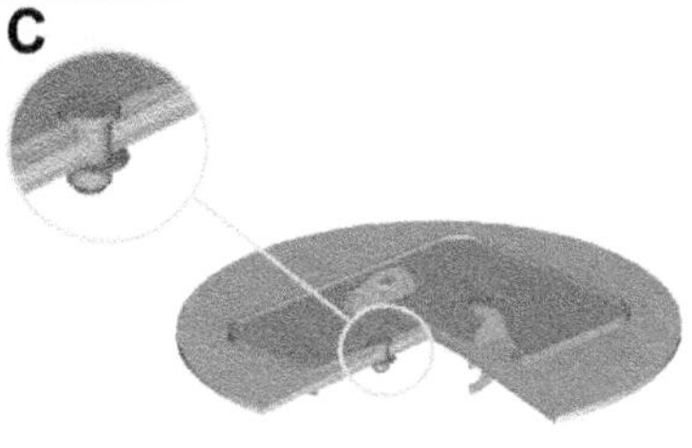	The top-plate and base-plate are steel pressing and the diaphragm is fabric. The support features are pressed into the top-plate. Two pins pushed through the assembly fastened by two push-on retaining rings.	Uses identical top-plates and base-plates reducing the number of unique components and the measurement arm support component is integrated with the plates reducing the initial part-count. **Part-count = 7** **Design Efficiency = 42.8%** **No. Joining Parts = 4**

* Part-count, design efficiency and number of joining components used to relate to this sub-assembly only and do not necessarily reflect the overall design. Design Efficiency has been determined from Appendix B1.

Figure 2.14: Diaphragm Assembly Designs.

The designs incorporate different joining processes for the same problem. Essentially, all the designs are the same with moderately different geometry, as shown in Figure 2.14. In each case there is a top plate, base plate, supports for the flow measurement arm and a rubber/fabric diaphragm. The diaphragm is sandwiched between the base plate and top plate with the flow measurement arm support on top. The joining process used fixes all components together.

Table 2.2: Diaphragm Assembly Joint Parameters and Results.

Design	Material	Quantity	Thickness	Permanence	Results
A	Steel and thermoplastic	High and all quantities	Thin to medium	Permanent	Ultrasonic welding Tubular rivet Staking Nailing Solid rivet Cyanoacrylate Epoxy resin Solvent-borne rubber Toughened adhesives
				Non-permanent	Retaining ring Quick release devices Threaded fasteners
B	Steel and thermoplastic	High and all quantities	Thin to medium	Permanent	Ultrasonic welding Tubular rivet Staking Nailing Solid rivet Cyanoacrylate Epoxy resin Solvent-borne rubber Toughened adhesives
				Non-permanent	Retaining ring Quick release devices Threaded fasteners
C	Steel	High and all quantities	Thin	Permanent	Ultrasonic welding Spot welding Plasma arc welding Laser beam welding Electron beam welding Furnace brazing Diffusion brazing Anaerobic Cyanoacrylate Epoxy resin Hot melt Solvent-borne rubber Toughened adhesives Solid rivet Flanging Stapling/stitching Seaming Staking Crimping Nailing
				Non-permanent	Retaining ring Quick release devices Threaded fasteners

The consequences of the joining process selection are highlighted by the influence on part-count and DFA design efficiency. The design processes can now be compared with the results from the joining process selection matrix. The joint parameters and results are shown in Table 2.2. It must be noted that in cases where two thicknesses are used, a match must be found for both. Also, although the quantity is high, the 'all quantities' column must be considered. While a permanent joint is required, as the joining strategy is 'through-hole', it is also necessary to consider non-permanent solutions.

The matrix results show both riveting and retaining rings (including clips), along with a number of additional processes, as candidates. This example clearly identifies the importance of selecting the most appropriate joining process. It shows that considering the impact on part-count and manufacturing processes helps to optimise the fastening of a joint. For example, both designs B and C use clips, although design C needs two extra pins to form the joint. A possible redesign would be to combine ideas from designs B and C, by integrating the top plate with the flow measurement arm support component (incorporating fastening pins as in design B) moulded from a polymer. This would eliminate the separate support component, remove the need for separate fastening pins and provide location features as part of a functional part.

The case studies show that selecting an inappropriate joining process can have a large detrimental effect on a design. It could be argued that a DFA analysis would highlight poor fastening methods and suggest the need for redesign. This point is demonstrated by the examples shown above; however, a DFA analysis requires a completed design and while highlighting the need for redesign DFA offers no support for generating redesign solutions. If a proactive DFA approach is to be realised, it is essential that joining process selection be performed. Applying the joining process selection methodology and supporting data during product development allows the geometry of components to be tailored to the selected joining process, eliminating the need for redesign.

Part-count optimisation is one of the main aims of DFA, significantly influencing economic feasibility and often the technical performance of a design. Joining has been proved to have a large influence on part-count. In many designs a significant proportion of the components are only present to support the joining process. Consequently, it can be concluded that a joining selection methodology is an important aspect of DFA. The case studies presented highlight the importance of joining process selection and its effect on the ease of assembly of a design. It can be seen that selecting an appropriate joining process in the early stages of the design process encourages a right-first-time design philosophy, reducing the need for costly redesign work.

References

[1] M. Giess, C.A. McMahon, J.D. Booker, D. Stewart, The application of faceted classification in the support of manufacturing process selection, Proc. Instn Mech. Engrs, Part B 223 (6) (2009) 597–608.

[2] M.F. Ashby, Y.J.M. Brechet, D. Cebona, L. Salvoc, Selection strategies for materials and processes, Materials and Design 25 (2004) 51–67.

[3] H.R. Shercliff, A.M. Lovatt, Selection of manufacturing process in design and the role of process modelling, Progress in Material Science 46 (2001) 429–459.
[4] M. Frank, S.B. Joshi, R.A. Wysk, Rapid prototyping as an integrated product/process development tool: an overview of issues and economics, Journal of the Chinese Institute of Industrial Engineers 20 (3) (2003) 240–246.
[5] G.D. Kim, Y.T. Oh, A benchmark study on rapid prototyping processes and machines: quantitative comparisons of mechanical properties, accuracy, roughness, speed, and material cost, Proc. Instn Mech. Engrs, Part B 222 (2008) 201–215.
[6] K. Lokesh, P.K. Jain, Selection of rapid prototyping technology, Advances in Production Engineering and Management 5 (2) (2010) 75–84.
[7] S.J. Dowey, A. Matthews, Life analysis of coated tools using statistical methods, Surface and Coatings Technology 110 (1998) 86–93.
[8] E.W. Brooman, Design for surface finishing, ASM Handbook No. 20 – Materials Selection and Design, tenth ed., ASM International, Ohio, 1997.
[9] A. Matthews, S. Franklin, K. Holmberg, Tribological coatings: contact mechanisms and selection, Journal of Physics D: Applied Physics 40 (2007) 5463–5475.
[10] R.S. Cowan, W.O. Winer, Surface engineering … an enigma of choices, Journal of Physics D: Applied Physics 25 (1992) A285–A291.
[11] C.S. Syan, Expert surface treatment and coating selection assistance in product design, Integrated Manufacturing Systems 5 (3) (1994) 29–34.
[12] A. Matthews, A. Leyland, Variability in coatings and test results, in: K.N. Strafford, C. Subramanian (Eds.), Quality Control and Assurance in Advanced Surface Engineering, Publication B678. Institute of Materials, London, 1997.
[13] A. Redford, Design for assembly, European Designer, Sept/Oct (1994) 12–14.
[14] K.G. Swift, M. Raines, J.D. Booker, Design capability and the costs of failure, Proc. Instn Mech. Engrs, Part B 211 (1997) 409–423.
[15] J.D. Booker, K.G. Swift, N.J. Brown, Designing for assembly quality: strategies, guidelines and techniques, Journal of Engineering Design 16 (3) (2005) 279–295.
[16] J. Heilala, J. Montonen, K. Helin, Selecting the right system – assembly system comparison with total cost of ownership methodology, Assembly Automation 27 (1) (2007) 44–54.
[17] A. Shtub, E.M. Dar-El, A methodology for the selection of assembly systems, International Journal of Production Research 27 (1) (1989) 175–186.
[18] A. Khan, A.J. Day, A knowledge based design methodology for manufacturing assembly lines, Computers and Industrial Engineering 41 (2002) 441–467.
[19] H. Noori, R. Radford, Production and Operations Management: Total Quality and Responsiveness, McGraw-Hill, New York, 1995.
[20] K.G. Swift, J.D. Booker, N.F. Edmondson, Strategies and case studies in assembly system selection, Proc. Instn Mech. Engrs, Part B 218 (7) (2004) 675–688.
[21] J.G. Bralla (Ed.), Design for Manufacturability Handbook, second ed., McGraw-Hill, New York, 1998.
[22] J. Wang, M. Trolio, Predicting quality in early product development, in: Proc. 3rd Annual International Conference on Industrial Engineering Theories, Applications and Practice, 28–31 December, Hong Kong, pp. 1–9, 1998.
[23] S.M. Darwish, A. Al Tamimi, S. Al-Habdan, A knowledge base for metal welding process selection, International Journal of Machine Tools and Manufacture 37 (7) (1997) 1007–1023.
[24] A.M.K. Esawi, M.F. Ashby, Computer-based selection of joining processes: methods, software and case studies, Materials and Design 25 (2004) 555–564.
[25] C. LeBacqa, Y. Brechetb, H.R. Shercliff, T. Jeggy, L. Salvob, Selection of joining methods in mechanical design, Materials and Design 23 (2002) 405–416.
[26] N.J. Brown, K.G. Swift, J.D. Booker, Joining process selection in support of proactive design for assembly, Proc. Instn Mech. Engrs, Part B 216 (2002) 1311–1324.

CHAPTER 3

Casting Processes

3.1 Sand Casting

Process Description

Moist bonding sand is packed around a pattern. The pattern is removed to create the mould and molten metal poured into the cavity. Risers supply necessary molten material during solidification. The mould is then broken to remove the part (Figure 3.1(a)).

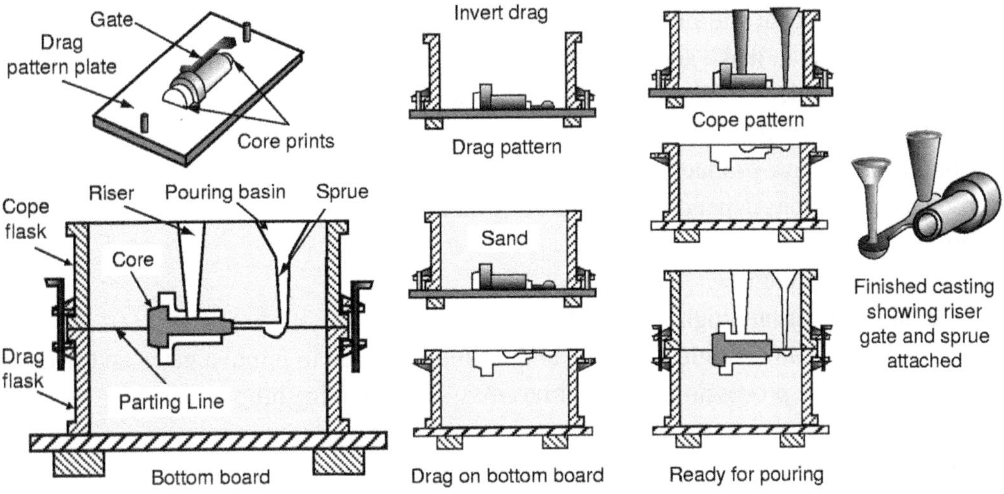

Figure 3.1(a): Sand Casting.

Materials

Most metals, particularly ferrous and aluminium alloys. Some difficulty encountered in casting: lead, tin and zinc alloys, also refractory alloys, beryllium, titanium and zirconia alloys.

Process Variations

- Green sand casting: the most common and the cheapest. Associated problems are that the mould has low strength and high moisture content.
- Dry sand: core boxes are used instead of patterns and an oven is used to cure the mould. Expensive and time consuming.

Manufacturing Process Selection Handbook. http://dx.doi.org/10.1016/B978-0-08-099360-7.00003-3

- Skin-dried sand: the mould is dried to a certain depth. Used in the casting of steels.
- Patterns: one-piece solid patterns are cheapest to make; split patterns for moderate quantities; match plate patterns for high-volume production.
- Wooden patterns: for low-volume production only.
- Metal patterns: for medium- to high-volume production. Hard plastics are also being used increasingly.
- Cosworth casting: low-pressure filling of mould used for better integrity, accuracy and porosity of casting. Longer production times and higher tooling costs, however.

Economic Considerations

- Production rates of 1–50/h, but dependent on size.
- Lead time typically days, but depends on complexity and size of casting.
- Material utilisation is low to moderate; 20–50% of material lost in runners and risers.
- Both mould material and runners and risers may be recycled.
- Patterns are easy to make and set, and are reusable.
- Pattern material dependent on the number of castings required.
- Easy to change design during production.
- Economical for low production runs of less than 100. Can be used for one-offs and high production volumes depending on degree of automation.
- Tooling costs are low.
- Equipment costs are low.
- Direct labour costs are high. Can be labour intensive.
- Finishing costs can be high. Cleaning and fettling required to remove gates and risers before secondary processing. Parting lines may also need finishing by hand.

Typical Applications

- Engine blocks.
- Manifolds.
- Machine tool bases.
- Pump housings.
- Cylinder heads.

Design Aspects

- High degree of shape complexity possible. Limited only by the pattern.
- Loose piece patterns can be used for holes and protrusions.
- All intersecting surfaces must be filleted: prevents shrinkage cracks and eliminates stress concentrations.

- Design of gating system for delivery of molten metal into mould cavity important.
- Placing of parting line important, i.e. avoid placement across critical dimensions.
- Bosses, undercuts and inserts are possible, but at added cost.
- Steel inserts can be used as heat flow barriers.
- Cored holes greater than 6 mm diameter.
- Machining allowances are usually in the range 1.5–6 mm.
- Draft angle ranges from 1° to 5°.
- Minimum section typically 3 mm for light alloys, 6 mm for ferrous alloys.
- Sizes range from 25 g to 400 t in weight.

Quality Issues

- Moulding sand must be carefully conditioned and controlled.
- Most casting defects can be traced to and rectified by sand content.
- Casting shrinkage and distortion during cooling governed by shape, especially when one dimension is much larger than the other two.
- Extensive flat surfaces are prone to sand expansion defects.
- Inspection of castings is important.
- High porosity and inclusion levels are common in castings.
- Defects in castings may be filled with weld material.
- Castings generally have rough grainy surfaces.
- Material strength is inherently poor.
- Castings have good bearing and damping properties.
- If production volumes warrant the cost of a die, close tolerances may be achieved.
- Surface detail is fair.
- Surface roughness is a function of the materials used in making the mould and is in the range 3.2–50 μm Ra.
- Not suitable for close specification of tolerances without secondary processing.
- Process capability charts showing the achievable dimensional tolerances using various materials are provided (Figure 3.1(b)). Allowances of ±0.5 to ±2 mm should be added for dimensions across the parting line.

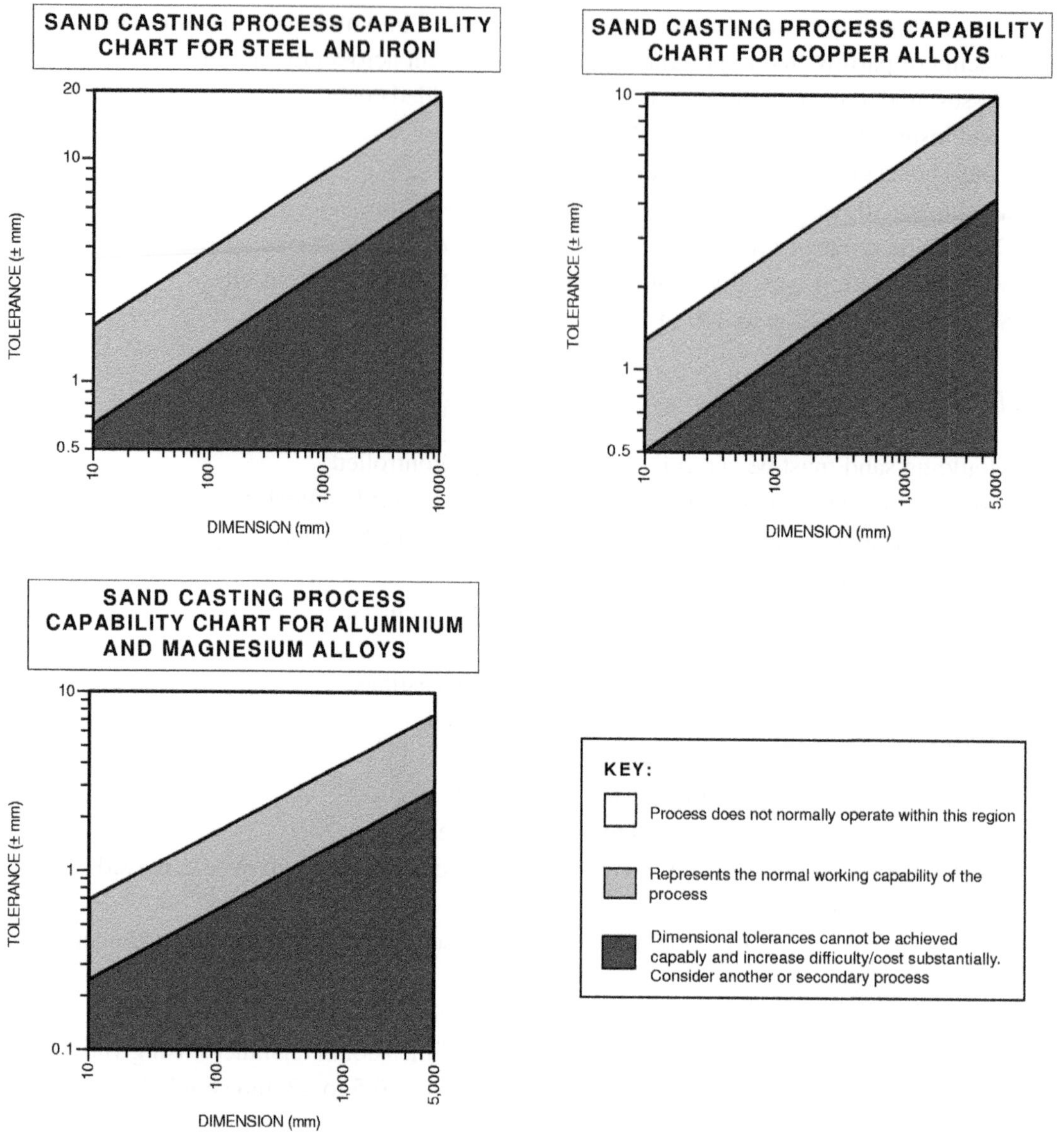

Figure 3.1(b): Sand Casting Process Capability Charts.

3.2 Shell Moulding

Process Description

A heated metal pattern is placed over a box of thermosetting resin-coated sand. The box is inverted for a fixed time to cure the sand. The box is re-inverted and the excess sand falls out. The shell is then removed from the pattern and joined with the other half (previously made). They are supported in a flask by an inert material ready for casting (Figure 3.2(a)).

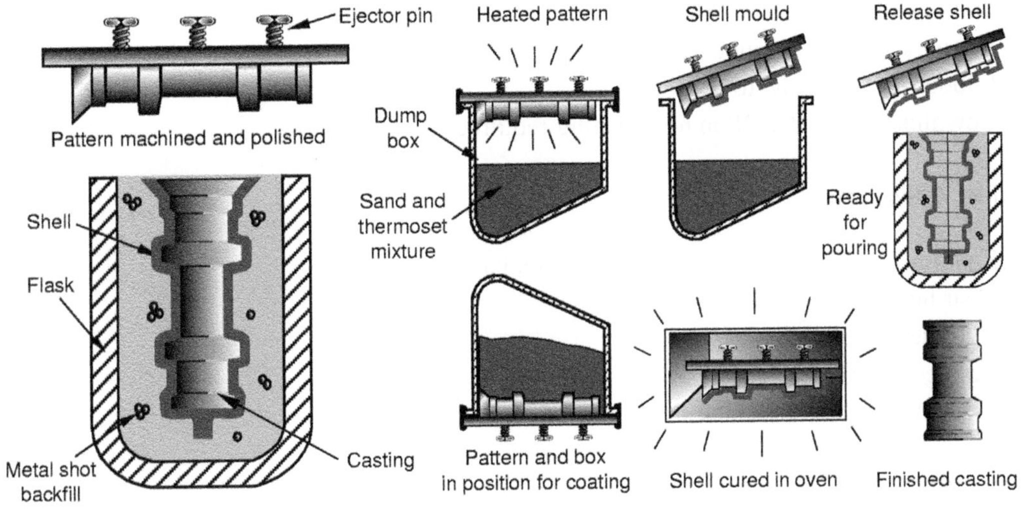

Figure 3.2(a): Shell Moulding.

Materials

Most metals, except: lead, zinc, magnesium and titanium alloys, also beryllium, refractory and zirconia alloys.

Process Variations

- Moulds produced from other casting processes may be joined with shell moulds.
- Patterns are generally made of iron or steel, giving good dimensional accuracy.
- Aluminium patterns may be used for low-volume production.
- Other pattern materials used are plaster, and graphite for reactive materials.

Economic Considerations

- Production rates of 5–200/h, but dependent on size.
- Lead time varies from several days to weeks depending on complexity and size.

- Material utilisation is high; little scrap generated.
- Potential for automation high.
- With use of gating systems several castings in a single mould are possible.
- Resin binders cost more, but only 5% as much sand is used as compared to sand casting.
- Difficult to change design during production.
- More suited to moderate- to high-volume production, but production volumes of 100–500 may be economical.
- Considered best of low-cost casting methods for large quantities.
- Tooling costs are low to moderate.
- Equipment costs are moderate to high.
- Labour costs are low to moderate.
- Low finishing costs. Often no finishing required.

Typical Applications

- Small mechanical parts requiring high precision.
- Gear housings.
- Cylinder heads.
- Connecting rods.
- Transmission components.

Design Aspects

- Good for moulding complex shapes, especially when using composite moulds.
- Great variations in cross-section possible.
- Sharper corners, thinner sections, smaller projections than possible with sand casting.
- Bosses and inserts possible.
- Undercuts difficult.
- Placing of parting line important, i.e. avoid placement across critical dimensions.
- Cored holes greater than 3 mm diameter.
- Draft angle ranges from 0.25° to 1°, depending on section depth.
- Maximum section = 50 mm.
- Minimum section = 1.5 mm.
- Sizes range from 10 g to 100 kg in weight. Better for small parts less than 20 kg.

Quality Issues

- Blowing sand on to pattern makes deposition more uniform, especially good for intricate forms.
- Few castings are scrapped due to blowholes or pockets. Gases are able to escape through thin shells or venting.

- Composite cores may include chills and cores to control solidification rate in critical areas.
- Moderate porosity and inclusions.
- Mechanical properties are better than sand casting.
- Uniform grain structure.
- Surface detail good.
- Surface roughness is in the range 0.8–12.5 μm Ra.
- Process capability charts showing the achievable dimensional tolerances using various materials are provided (Figure 3.2(b)). Allowances of ±0.25 to ±0.5 mm should be added for dimensions across the parting line.

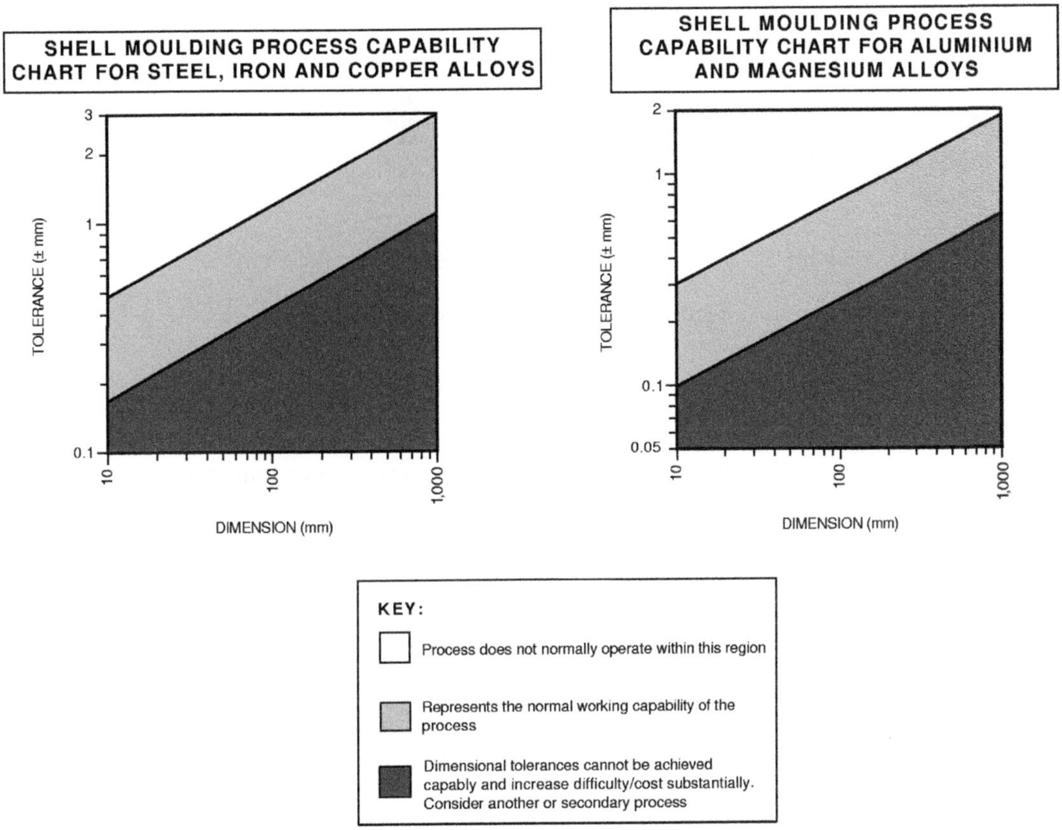

Figure 3.2(b): Shell Moulding Process Capability Charts.

3.3 Gravity Die Casting

Process Description

Molten metal is poured under gravity into a pre-heated die, where it solidifies. The die is then opened and the casting ejected. Also known as permanent mould casting (Figure 3.3(a)).

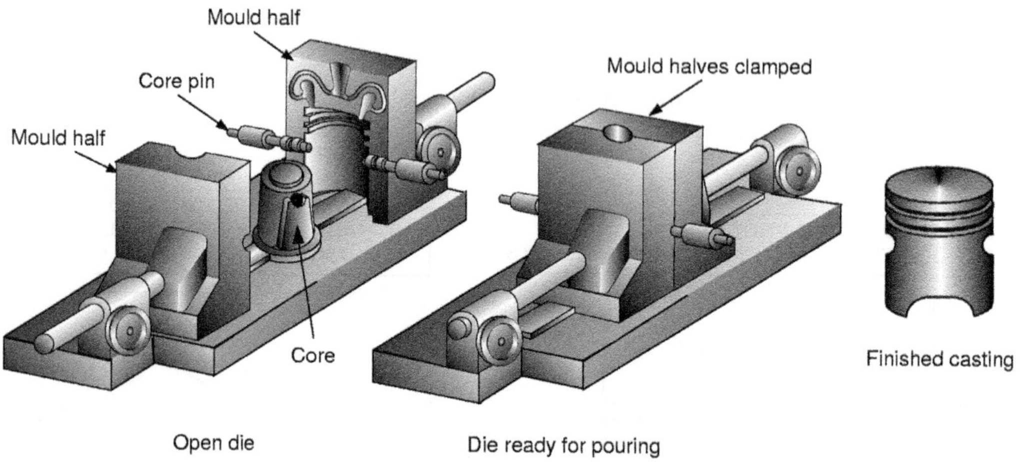

Figure 3.3(a): Gravity Die-Casting.

Materials

Usually non-ferrous metals, for example: copper, aluminium, magnesium, but sometimes iron, lead, nickel, tin and zinc alloys. Carbon steel can be cast with graphite dies.

Process Variations

- Dies typically cast iron, graphite or refractory material.
- Metal or sand cores can be used although surface finish can be poor.
- Low pressure die-casting: uses low-pressure (1 bar) air to force the molten metal into the die cavity. Less popular than gravity die-casting, and tends to be used purely for the production of car wheels. Gives lower production rates.
- Slush casting: for creating hollow parts without cores in low-melting-point metals such as lead, zinc and copper alloys.

Economic Considerations

- Production rates of 5–50/h, but dependent on size.
- Lead times can be many weeks.
- Material utilisation is moderate to high (10–40% lost in scrap, but can be recycled).

- If accuracy and surface finish is not an issue, can use sand cores instead of metallic or graphite for greater economy.
- Production volumes of 500–1,000 may be viable, but suited to higher volume production.
- Tooling costs are moderate.
- Equipment costs are moderate.
- Labour costs are low to moderate.
- Finishing costs are low to moderate. Gates need to be removed.

Typical Applications

- Cylinder heads.
- Engine connecting rods.
- Pistons.
- Gear and die blanks.
- Kitchen utensils.
- Gear blanks.
- Gear housings.
- Pipe fittings.
- Wheels.

Design Aspects

- Shape complexity limited by that obtained in die halves.
- Undercuts are possible with large added cost.
- Inserts are possible with small added cost.
- Machining allowances are usually in the range 0.8–1.5 mm.
- Vertical parting lines commonly used.
- Placing of parting line important, i.e. avoid placement across critical dimensions.
- Cored holes greater than 5 mm diameter.
- Draft angle ranges from 2° to 3°.
- Maximum section = 50 mm.
- Minimum section = 2 mm.
- Sizes range from 50 g to 300 kg in weight. Commonly used for castings less than 5 kg.

Quality Issues

- Little porosity and inclusions. Can be minimised by slow die filling to reduce turbulence.
- Redressing of the dies may be required after several thousand castings.
- Collapsible cores improve extraction difficulties on cooling.
- 'Chilling' effect of cold metallic dies on the surface of the solidifying metals needs to be controlled by pre-heating at correct temperature.

- Large castings sometimes require that the die is tilted as molten metal is being poured in to reduce turbulence.
- Mechanical properties are fair to good.
- Surface detail good.
- Surface roughness is in the range 0.8–6.3 μm Ra.
- Process capability charts showing the achievable dimensional tolerances using various materials are provided (Figure 3.3(b)). Allowances of ±0.25 to ±0.75 mm should be added for dimensions across the parting line.

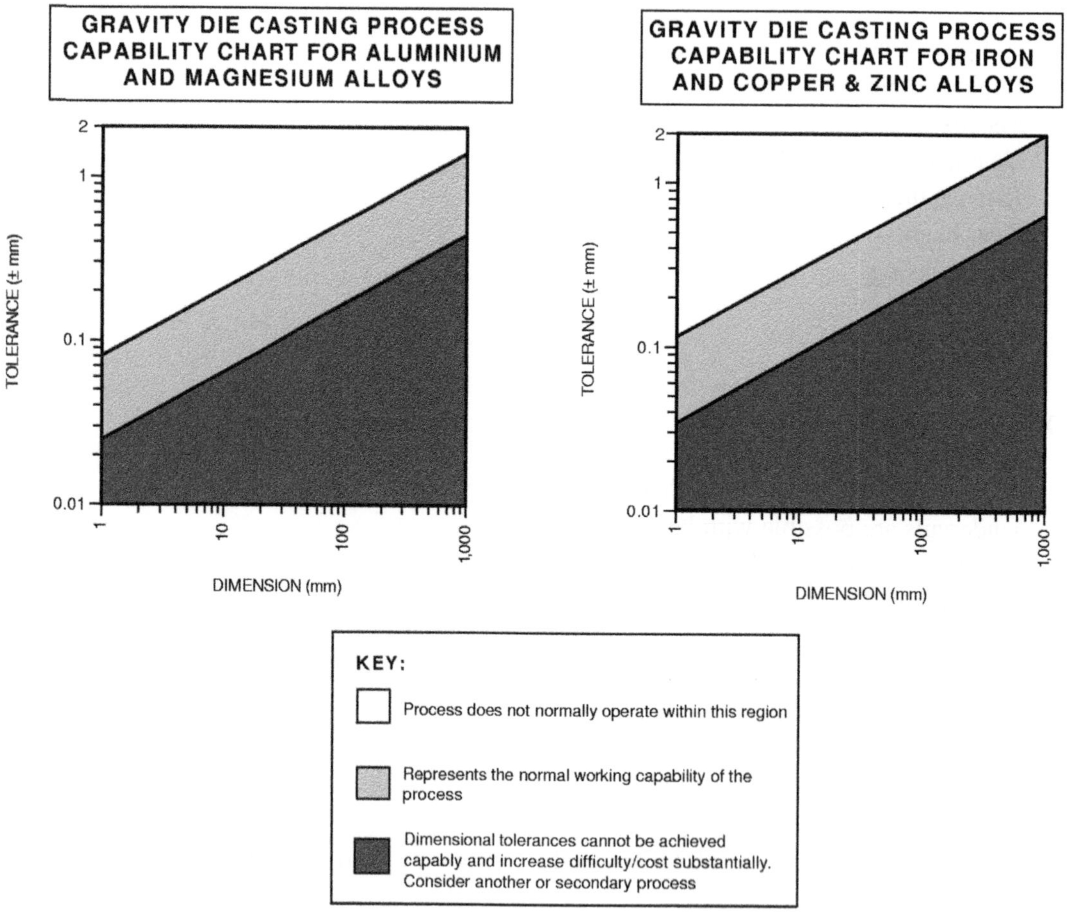

Figure 3.3(b): Gravity Die Casting Process Capability Charts.

3.4 Pressure Die Casting

Process Description

Molten metal is inserted into a metallic mould under very high pressures (100+ bar), where it solidifies. The die is then opened and the casting ejected (Figure 3.4(a)).

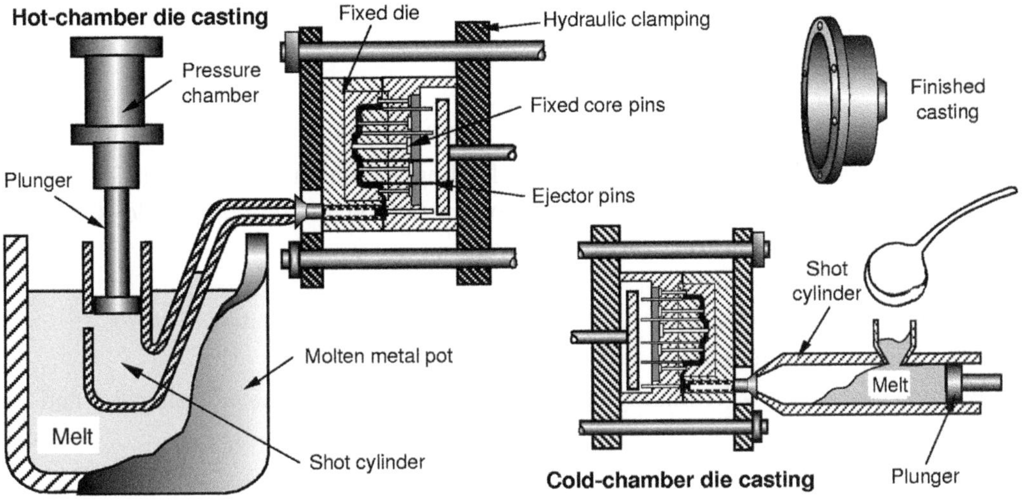

Figure 3.4(a): Pressure Die Casting.

Materials

- Limited to non-ferrous metals, i.e. zinc, aluminium, magnesium, lead, tin and copper alloys.
- Zinc and aluminium alloys tend to be the most popular materials.
- High-temperature metals, such as copper alloys, reduce die life.
- Iron-based materials for casting are under development.

Process Variations

- Dies and cores are made from hardened and tempered alloy steel.
- Cold-chamber die casting: shot cylinder filled with a ladle for each cycle. Used for high melting temperature metals.
- Hot-chamber die casting: shot cylinder immersed in molten metal and then forced using a separate ram. Used for low melting temperature metals due to erosive nature of molten metal. Can be either plunger or goose-neck type.
- Vacuum die casting: overcomes porosity for larger castings.

- Injection metal assembly: variant of hot chamber die casting for the assembly of parts such as tubes and plates, cable terminations and to act as rivets by injecting zinc or lead alloys into a cavity in the assembly.

Economic Considerations

- Very high production rates possible, up to 200/h.
- Lead time very long, months possibly.
- Material utilisation is high.
- Gates, sprues, etc. can be re-melted.
- High initial die costs due to high complexity and difficulty to manufacture.
- Full automation achievable. Robot machine loading and unloading common.
- Production quantities of up to 10,000 are economically viable for copper alloys, but 100,000+ for aluminium, zinc and lead alloys.
- Tooling costs are very high.
- Equipment costs are very high.
- Direct labour costs are low.
- Finishing costs are low. Trimming operations are required to remove flash, gates and sprues.

Typical Applications

- Transmission cases.
- Machine and engine parts.
- Pump components.
- Electrical boxes.
- Domestic appliance components.
- Toy bodies.
- Pump and impeller parts.

Design Aspects

- Shape complexity can be high. Limited by design of movable cores.
- Bosses, large threads, undercuts and inserts are all possible with added cost.
- Moulded-in bearing shells possible.
- Lettering possible.
- Wall thickness should be as uniform as possible; transitions should be gradual.
- Sharp corners should be avoided, but pressure die casting permits smaller radii because metal flow is aided.
- Placing of parting line important, i.e. avoid placement across critical dimensions.

- Holes perpendicular to the parting line can be cast.
- Casting holes for subsequent tapping is generally more economical than drilling.
- Cored holes greater than 0.8 mm diameter.
- Machining allowance is normally in the range 0.25–0.8 mm.
- Draft angle ranges from 0.25° to 3°, depending on section depth.
- Maximum section = 13 mm.
- Minimum section ranges from 0.4 mm for zinc alloys to 1.5 mm for copper alloys.
- Sizes range from 10 g to 50 kg. Castings up to 100 kg have been made in zinc. Copper, tin and lead castings are normally less than 5 kg.

Quality Issues

- Low porosity in small castings typically, but can be a problem in castings with thick or long sections.
- Particularly suited where casting requires high mechanical properties or absence of creep.
- The high melting temperature of some metals can cause significant processing difficulties and die wear.
- Ejector pins may leave small marks and should be positioned at points of strength on the casting.
- Process variables need to be controlled. Variation in temperature, pressure and cycle time especially important for consistency.
- Difficulty is experienced in obtaining sound castings in the larger capacities due to gas entrapment.
- Close control of temperature, pressure and cooling times important in obtaining consistent quality castings.
- Mechanical properties are fair, but poorer than some other casting methods.
- Surface detail excellent.
- Surface roughness is in the range 0.4–3.2 μm Ra.
- Process capability charts showing the achievable dimensional tolerances using various materials are provided (Figure 3.4(b)). Allowances of ±0.05 to ±0.35 mm should be added for dimensions across the parting line.

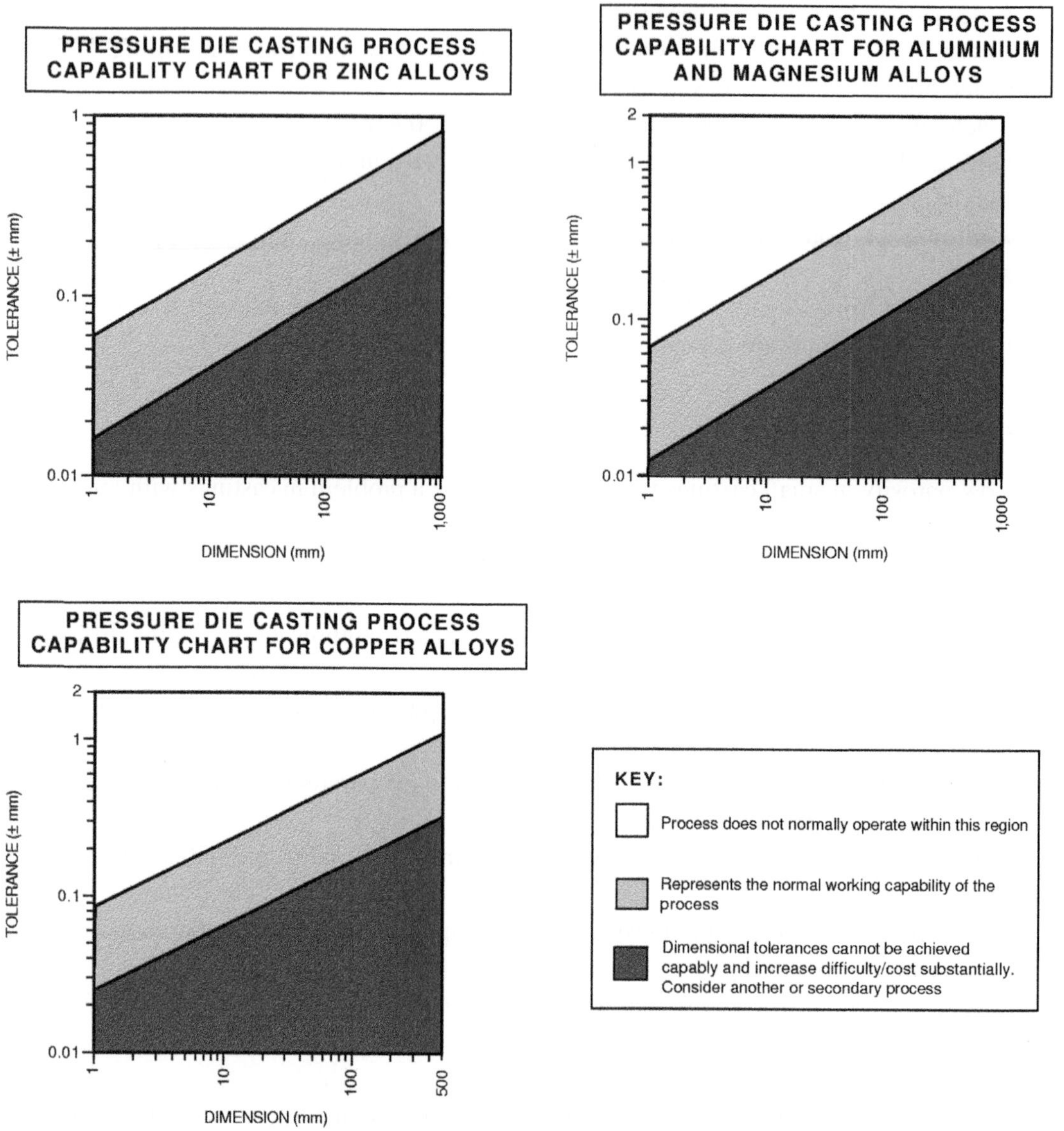

Figure 3.4(b): Pressure Die Casting Process Capability Charts.

3.5 Centrifugal Casting

Process Description

Molten metal is poured into a high-speed rotating mould (300–3000 rpm depending on diameter) until solidification takes place. The axis of rotation is usually horizontal, but may be vertical for short work pieces (Figure 3.5(a)).

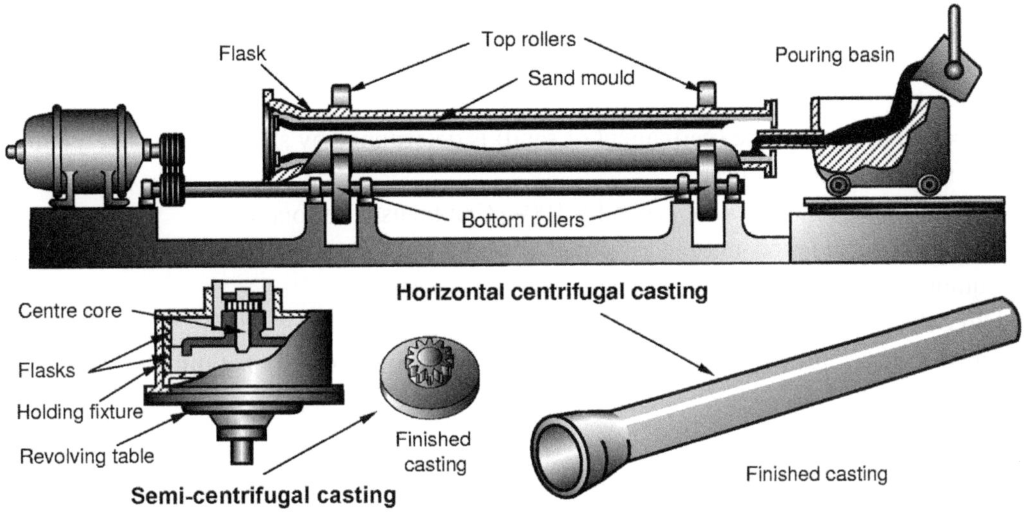

Figure 3.5(a): Centrifugal Casting.

Materials

- Most metals suitable for static casting are suitable for centrifugal casting: all steels, iron, copper, aluminium and nickel alloys.
- Also, glass, thermoplastics, composites and ceramics (metal moulds sprayed with a refractory material) can be moulded by this method.

Process Variations

- Semi-permanent or expendable moulds.
- Semi-centrifugal casting: used to cast parts with radial symmetry in a vertical axis of rotation at low speeds.
- Centrifuge casting: a number of moulds are arranged radially around a central sprue. Molten metal is poured into the sprue and is forced into the mould cavities by centrifugal force due to high-speed rotation. Used for small gears mainly and parts of intricate detail.

Economic Considerations

- Production rates of up to 50/h possible, but dependent on size.
- Lead time may be several weeks.
- Material utilisation high (90–100%). No runners or risers.
- Economic when the mechanical properties of thick-walled tubes are important and high alloy grades of steel are required.
- In large quantities production of other than circular external shapes becomes more economical.
- Small-diameter steel tubes made by this method are not competitive with welded or rolled tubes.
- Selection of mould type (permanent or sand) is determined by shape of casting, quality and number to be produced.
- Production volumes are low, typically 100+. Can be used for one-offs.
- Tooling costs are moderate.
- Equipment costs are low to moderate.
- Direct labour costs are low to moderate.
- Finishing costs are low to moderate. Normally, machining of internal dimension necessary.

Typical Applications

- Pipes.
- Brake drums.
- Pulley wheels.
- Train wheels.
- Flywheels.
- Gun barrels.
- Gear blanks.
- Large bearing liners.
- Engine-cylinder liners.
- Pressure vessels.
- Nozzles.

Design Aspects

- Shape complexity limited by nature of process, i.e. suited to parts with rotational symmetry.
- Contoured surfaces possible.
- Circular bore remains in the finished part.

- Dual metal tubes that combine the properties of two metals in one application are possible.
- Inserts and bosses are possible, but undercuts are not.
- Placing of parting line important, i.e. avoid placement across critical dimensions.
- Cored holes greater than 25 mm diameter.
- Machining allowances range from 0.75 to 6 mm.
- Draft angle approximately 1°.
- Maximum section thickness approximately 125 mm.
- Minimum section ranges from 2.5 to 8 mm, depending on material cast.
- Maximum length = 15 m.
- Sizes range from 25 mm to 2 m diameter.
- Sizes up to 5 t in weight have been cast.

Quality Issues

- Properties of castings vary by distance from the axis of rotation.
- Due to density differences in the molten material, dross, impurities and pieces of the refractory lining tend to collect on the inner surface of the casting. This is usually machined away.
- Tubular castings have higher structural strengths and more distinct cast impressions than gravity die cast or sand cast parts.
- Castings are free of shrinkage due to one-directional cooling.
- The mechanical properties of dense castings are comparable with that of forgings. Fine-grain castings and low porosity are an advantage.
- Good mechanical properties and fine grain structure.
- Surface detail fair to good.
- Surface roughness is in the range 1.6–12.5 μm Ra.
- A process capability chart showing the achievable dimensional tolerances is provided (Figure 3.5(b)). Allowances of approximately ±0.25 to ±0.75 mm should be added for dimensions across the parting line. Note, the chart applies to outside dimensions only. Internal dimensions are approximately 50% greater.

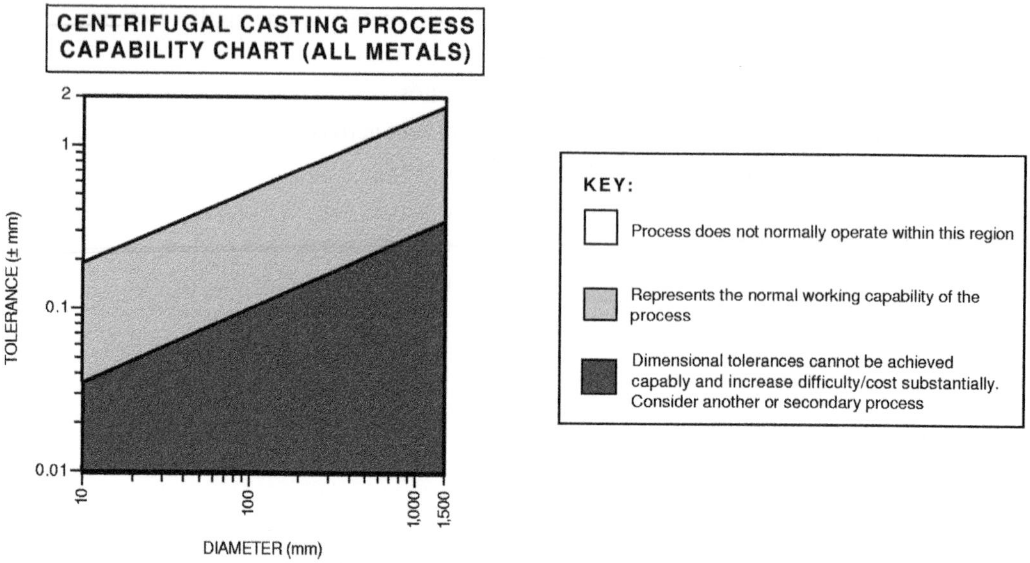

Figure 3.5(b): Centrifugal Casting Process Capability Chart.

3.6 Investment Casting

Process Description

A mould is used to generate a wax pattern of the shape required. A refractory material zircon, then a ceramic slurry and finally a binder is used to coat the pattern, which is slow fired in an oven to cure. The wax is melted out and the metal cast in the ceramic mould. The mould is then destroyed to remove the casting. Process often known as the 'lost wax' process (Figure 3.6(a)).

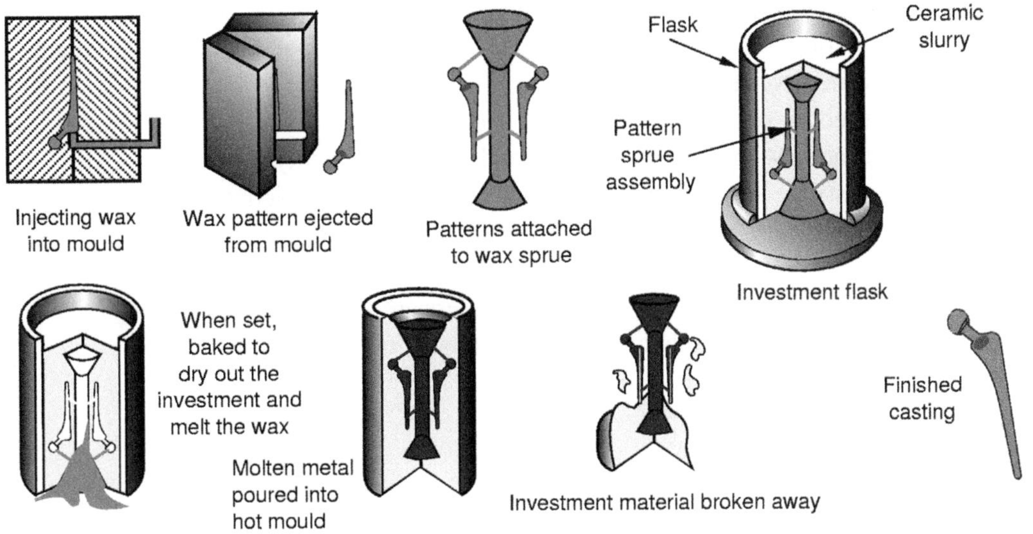

Figure 3.6(a): Investment Casting.

Materials

All metals, including precious, refractory and reactive alloys (cast in vacuum).

Process Variations

- Blends of resin, filler and wax used.
- Use of thermoplastic resin instead of wax.
- Ceramic and water-soluble cores can be used.

Economic Considerations

- Production rates of up to 1,000/h, depending on size.
- Lead times are usually several weeks, but can be shorter.

- Slow process due to many steps in production. Cure time can be as high as 48 hours.
- Wax or plastic patterns can be injection-moulded for high production volumes.
- Best suited to metals having high melting temperatures, and/or which are difficult to machine or are costly.
- Material utilisation is high.
- Some automation possible.
- Pattern costs can be high for low quantities.
- Ceramic and wax cores allow complex internal configurations to be produced, but increase the cost significantly.
- A 'tree' of wax patterns enables many small castings to be handled together.
- Most suitable for small batches (10–1,000) using manual labour, but also high-volume production with automation.
- Sometimes used for one-offs, especially in decorative work.
- Tooling costs are low to moderate, but dependent on complexity.
- Equipment costs are low to moderate (high when processing reactive materials).
- Labour costs are very high. Can be labour intensive as many operations are involved.
- Low to moderate finishing costs. Gates and feeders are removed by machining or grinding. As cast part typically cleaned by shot, bead or sand blasting.

Typical Applications

- Turbine blades.
- Machine tool parts.
- Aerospace components.
- Valve and pump casings.
- Pipe fittings.
- Automotive engine components.
- Decorative work, e.g. figurines.
- Optical instrument parts.
- Small arms parts.
- Gear blanks.
- Levers.
- Jewellery.

Design Aspects

- Very complex castings with unusual internal configurations possible.
- Wax pattern must be easily removable from its mould.
- Complex shapes may be assembled from several simpler shapes.

- Practical way of producing threads in hard-machine materials, or where thread design is unusual.
- Uniform sections are preferred. Abrupt changes should be gradually blended in or designed out.
- Avoid sharp corners.
- Fillets should be as generous as possible.
- Bosses and undercuts are possible with added cost.
- Inserts are not possible, but integral rivets are.
- Lettering possible, either in relief or inset.
- Moulded-in holes, both blind and through, are possible but difficult.
- Length to diameter ratio for blind holes is typically 4:1.
- Minimum hole = 0.5 mm diameter.
- Machining allowance usually between 0.25 and 0.75 mm, depending on size.
- Draft angle usually zero, but 0.5–1° desirable on long extended surfaces, or if mould cavity is deep.
- Minimum section ranges from 1 mm for aluminium alloys and steels, 2 mm for copper alloys, but can be as low as 0.6 mm for some applications.
- Maximum section = 75 mm.
- Maximum dimension = 1 m.
- Sizes range from 0.5 g to 100 kg in weight, but best for parts less than 5 kg.

Quality Issues

- Moderate porosity.
- High-strength castings can be produced.
- Grain growth more pronounced in longer sections, which may limit the toughness and fatigue life of the part.
- Quality of casting depends to a large degree upon the characteristics of wax.
- Very good to excellent surface detail possible.
- Surface roughness is in the range 0.4–6.3 μm Ra.
- Flatness tolerances typically ±0.13 mm per 25 mm, but dependent on surface area.
- Minimum angular tolerance = ±0.5°.
- A process capability chart showing the achievable dimensional tolerances is provided (Figure 3.6(b)).
- No parting line on casting.

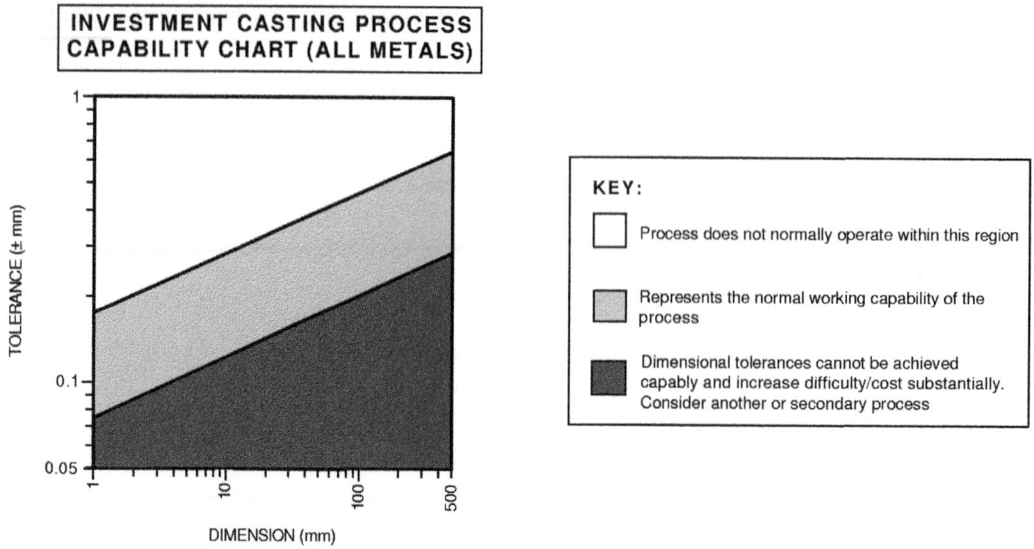

Figure 3.6(b): Investment Casting Process Capability Chart.

3.7 Ceramic Mould Casting

Process Description

A precision pattern generates the mould, which is coated with a ceramic slurry. The mould is dried and baked. The molten metal is then poured into the mould and allowed to solidify. The mould is broken to remove the part (Figure 3.7(a)).

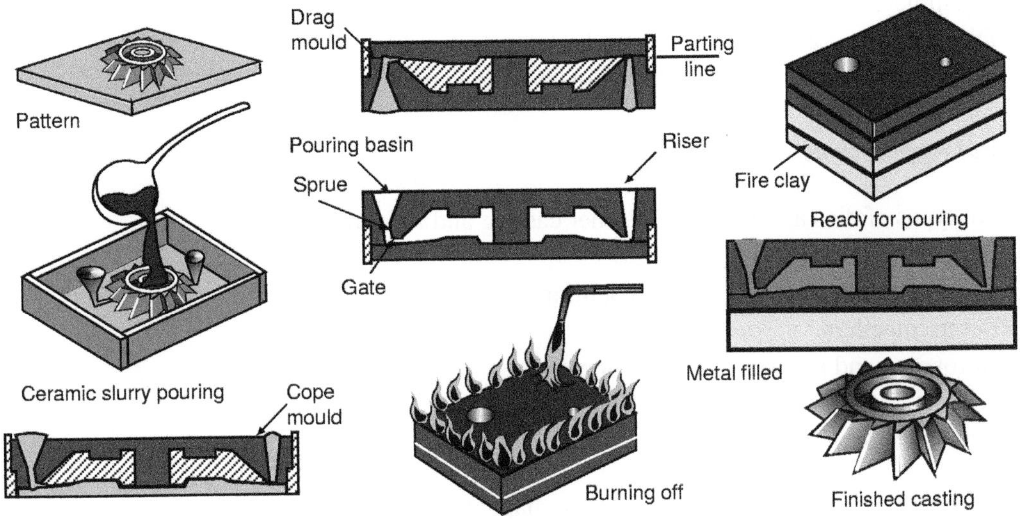

Figure 3.7(a): Ceramic Mould Casting.

Materials

All metals, but to a lesser degree aluminium, magnesium, zinc, tin and copper alloys.

Process Variations

- Variations on the composition of the ceramic slurry and curing mechanism.
- Plaster, wood, metal or rubber are used for patterns.

Economic Considerations

- Production rates of up to 10/h typical.
- Lead times can be several days.
- Material utilisation is high.
- Low scrap losses.

- Best suited to metals having high melting temperatures and/or that are difficult to machine.
- Can be combined with investment casting to produce parts with increased complexity with reduced cost.
- Suitable for small batches and medium-volume production.
- Can be used for one-offs.
- Tooling costs are moderate.
- Equipment costs are moderate to high.
- Direct labour costs are moderate to high.
- Finishing costs are low. Usually no machining is required.

Typical Applications

- All types of dies and moulds for other casting and forming processes.
- Cutting tool blanks.
- Components for food handling machining.
- Pump impellers.
- Aerospace and atomic reactor components.

Design Aspects

- High complexity possible – almost any shape possible.
- Use of cores increase complexity obtainable.
- Inserts, bosses and undercuts are possible.
- Placing of parting line important, i.e. avoid placement across critical dimensions.
- Cored holes greater than 0.5 mm diameter.
- Where machining is required, allowances of up to 0.6 mm should be observed.
- Draft angle usually zero, but 0.1–1° preferred.
- Minimum section ranges from 0.6 to 1.2 mm, depending on material used.
- Sizes range from 100 g to 3 t in weight, but less than 50 kg better.

Quality Issues

- Low porosity.
- Mechanical properties are good.
- Good surface detail possible.
- Surface roughness is in the range 0.8–6.3 μm Ra.
- A process capability chart showing the achievable dimensional tolerances is provided (Figure 3.7(b)). An allowance of ±0.25 mm should be added for dimensions across the parting line.
- Parting lines are sometimes pronounced on finished casting.

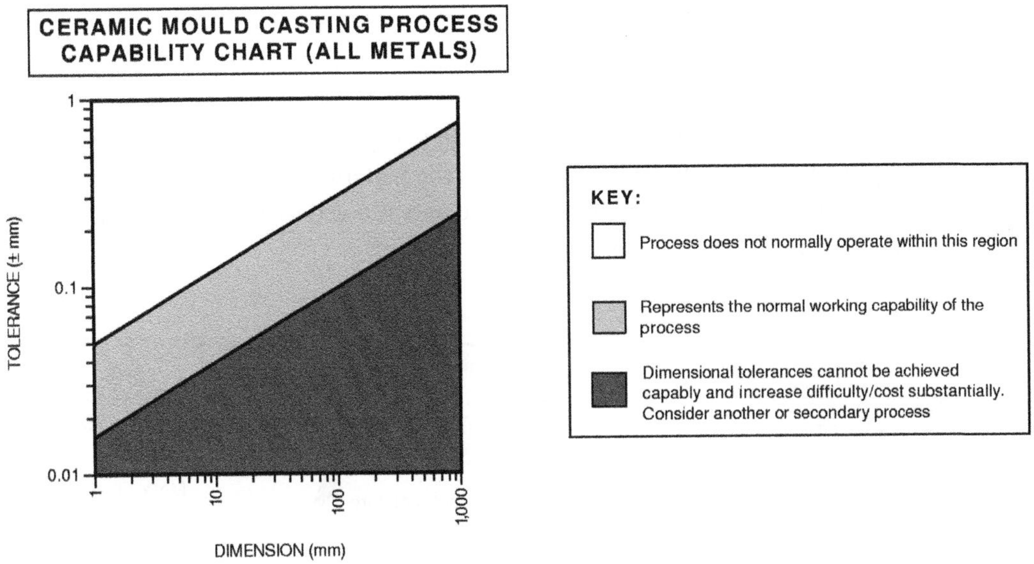

Figure 3.7(b): Ceramic Mould Casting Process Capability Chart.

3.8 Plaster Mould Casting

Process Description

A precision metal pattern (usually brass) generates the two-part mould, which is made of a gypsum slurry material. The mould is removed from the pattern and baked to remove the moisture. The molten metal is poured into the mould and allowed to cool. The mould is broken to remove the part (Figure 3.8(a)).

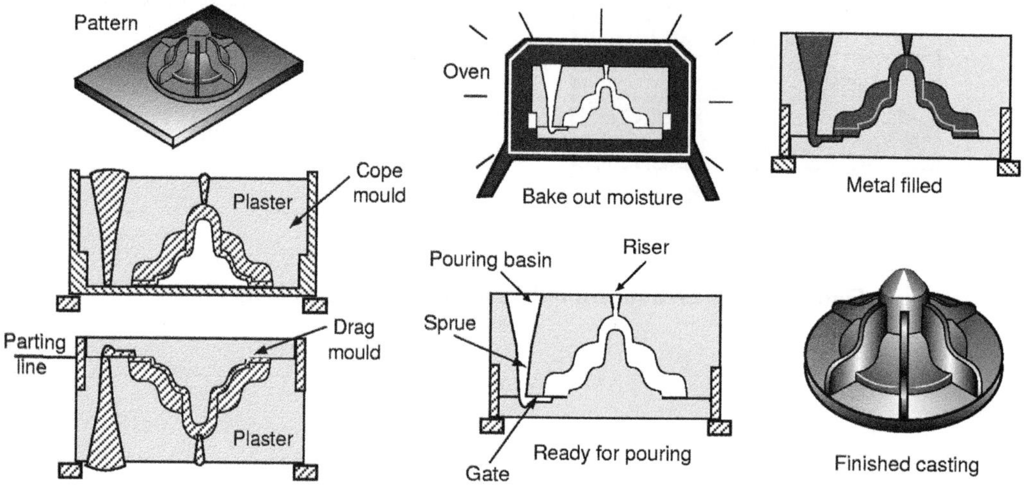

Figure 3.8(a): Plaster Mould Casting.

Materials

- Limited to low melting temperature metals, i.e. aluminium, copper, zinc and magnesium alloys, due to degradation of the plaster mould at elevated temperatures.
- Tin and lead alloys are sometimes processed.

Process Variations

- Patterns can be made from: metal, plaster, wood or thermosetting plastic. Wood has a limited life due to water absorption from the plaster slurry.
- Composition of plaster slurry varies. Additives are sometimes used to control mould expansion and fibres added to improve mould strength.

Economic Considerations

- Production rates of up to 10/h typical.
- Lead times can be several days to weeks.

- Material utilisation is high.
- Low scrap losses. Waste is recycled.
- Mould destroyed in removing casting.
- Easy to change design during production.
- Suitable for small batches of 100 and medium-volume production.
- Tooling costs are low to moderate.
- Equipment costs are moderate.
- Direct labour costs are moderate to high. Some skilled operations necessary.
- Finishing costs are low. Little finishing required except grinding for gate removal and sanding of parting line.

Typical Applications

- Pump impellers.
- Waveguide components (for use in microwave applications).
- Lock components.
- Gear blanks.
- Valve parts.
- Moulds for plastic and rubber processing, i.e. tyre moulds.

Design Aspects

- Moderate to high complexity possible.
- Possible to make mould from several pieces.
- Deep holes are not recommended.
- Sharp corners and features can be cast easily.
- Bosses and undercuts are possible with little added cost.
- Placing of parting line important, i.e. avoid placement across critical dimensions.
- Cored holes greater than 13 mm diameter.
- Where machining is required, allowances of up to 0.8 mm should be observed.
- Draft angles from 0.5° to 2° preferred, but can be zero.
- Minimum section ranges from 0.8 to 1.8 mm, depending on material used.
- Sizes range from 25 g to 50 kg in weight. However, castings up to 100 kg have been made.

Quality Issues

- Little or no distortion on thin sections.
- Plaster mould has low permeability and can create gas evolution problems.
- Moderate to high porosity obtained.
- Mechanical properties are fair.
- Surface detail good.

- Surface roughness is in the range 0.8–3.2 μm Ra.
- A process capability chart showing the achievable dimensional tolerances is provided (Figure 3.8(b)). An allowance of approximately ±0.25 mm should be added for dimensions across the parting line.

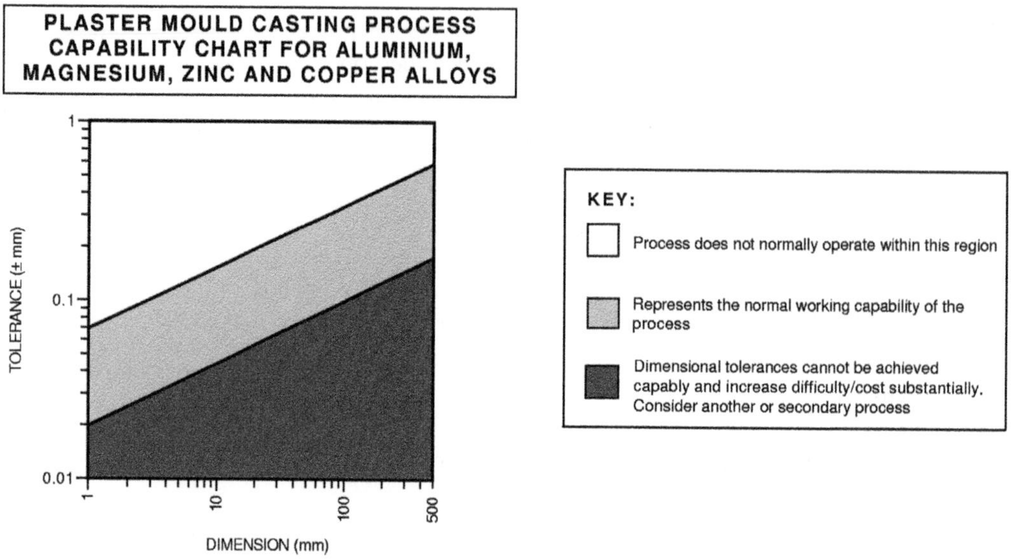

Figure 3.8(b): Plaster Mould Casting Process Capability Chart.

3.9 Squeeze Casting

Process Description

Combination of casting and forging. Molten metal fills a preheated mould from the bottom and during solidification, the top half of the mould applies a high pressure to compress the material into the final desired shape (Figure 3.9). Also known as liquid metal forging and load pressure casting.

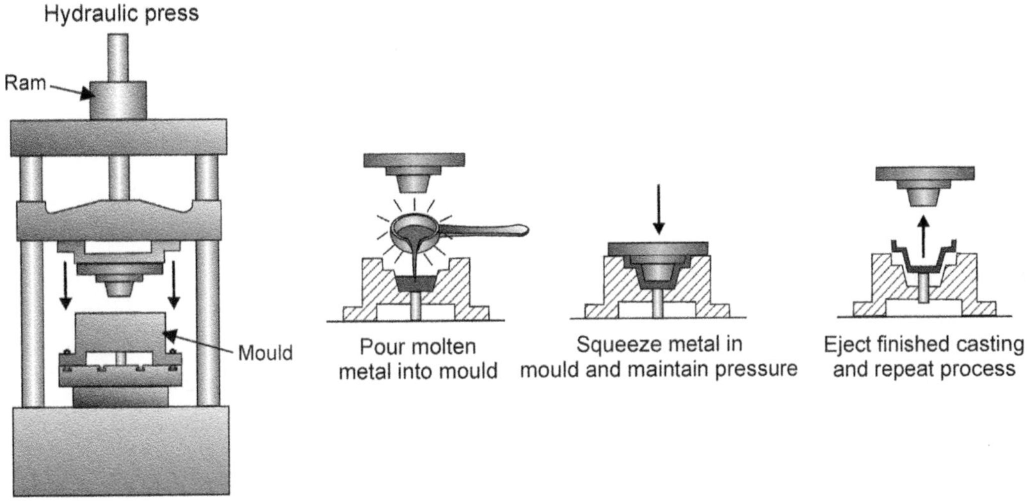

Figure 3.9: Squeeze Casting.

Materials

Typically non-ferrous metals, but occasionally, ferrous alloys.

Process Variations

Pouring can be performed automatically.

Economic Considerations

- Production rates low to medium.
- Cycle times can be of the order of minutes depending on size.
- Lead time is moderate to high.
- Material utilisation is excellent. Near-net shape achieved. No waste.
- High degree of automation possible, but little flexibility.
- Economically viable for medium to high production volumes of 10,000+.

- Tooling costs are high due to complexity.
- Equipment costs are high.
- Direct labour costs are low to moderate.
- Finishing costs are very low.
- Used to minimise or eliminate secondary processing.

Typical Applications

- Aerospace components.
- Suspension parts.
- Steering components.
- Brake rotors.
- Engine pistons.
- Wheels.
- Bearing housings.

Design Aspects

- Complex geometries possible.
- Retractable and disposable cores used to create complex internal features.
- Large variations in cross-section possible, but gradual transition recommended.
- Undercuts, bosses, holes and inserts possible.
- Ribs, pockets and features can improve local sectional properties.
- Placing of parting line important, i.e. avoid placement across critical dimensions.
- Machining allowances are usually in the range 0.6–1.2 mm.
- Draft angle ranges from 0.1° to 3°, depending on section depth.
- Maximum section = 200 mm.
- Minimum section = 6 mm.
- Minimum dimension = 20 mm diameter.
- Sizes range from 25 g to 4.5 kg in weight.

Quality Issues

- Low gas porosity, shrinkage and defect levels compared to other casting processes.
- Adequate process control is important, i.e. metering of molten metal, pressures, solidification times, tooling temperatures, etc. to avoid porosity.
- Defects can be minimised through correct die design, e.g. multiple gate system, by increasing die temperature, or by decreasing delay time before die closure.
- Abrupt changes in section also tend to create shrinkage and porosity problems.
- Low-speed mould filling minimises splashing, but increases cycle times.
- Accurate metering of molten metal required to avoid flashing.

- Excellent mechanical properties can be obtained, similar to forging.
- Graphite releasing agent and ejector pins commonly used to aid removal of finished part.
- Castings can be heat treated.
- Surface detail is good.
- Surface roughness is in the range 1.6–12.5 μm Ra.
- Achievable dimensional tolerances are approximately ±0.15 up to 25 mm, ±0.3 up to 150 mm. Allowances of ±0.25 mm should be added for dimensions across the parting line.

CHAPTER 4

Forming Processes

4.1 Forging

Process Description

Hot metal is formed into the required shape by the application of pressure or impact forces causing plastic deformation using a press or hammer in a single or a series of dies (Figure 4.1(a)).

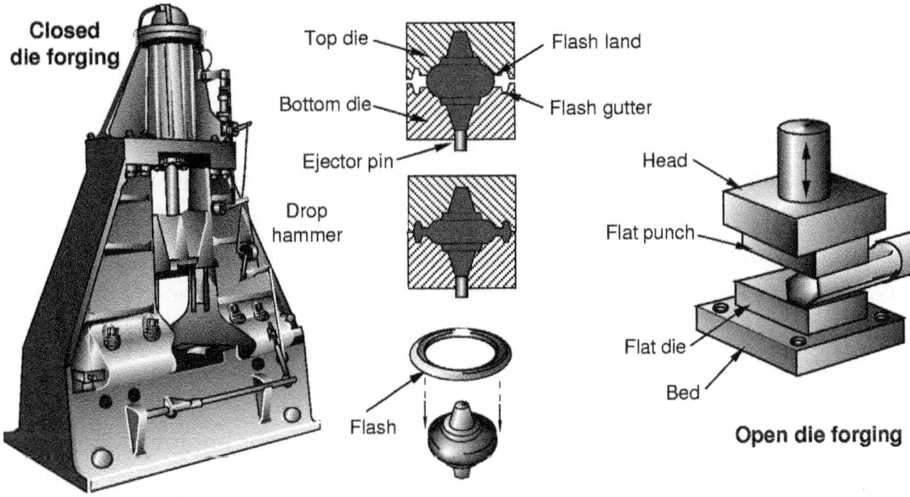

Figure 4.1(a): Forging.

Materials

- Mainly carbon; low alloy and stainless steels; aluminium; copper; and magnesium alloys. Titanium alloys, nickel alloys, high alloy steels and refractory metals can also be forged.
- Forgeability of materials important; must be ductile at forging temperature. Relative forgeability is as follows, with easiest to forge first: aluminium alloys, magnesium alloys, copper alloys, carbon steels, low alloy steels, stainless steels, titanium alloys, high alloy steels, refractory metals and nickel alloys.

Process Variations

- Presses can be mechanical, hydraulic or drop hammer type.
- Closed die forging: series of die impressions used to generate shape.

Manufacturing Process Selection Handbook. http://dx.doi.org/10.1016/B978-0-08-099360-7.00004-5

- Open die forging: hot material deformed between a flat or shaped punch and die. Sections can be flat, square, round or polygonal. Shape and dimensions largely controlled by operator.
- Roll forging: reduction of section thickness of a doughnut-shaped preform to increase its diameter. Similar to ring rolling (see PRIMA 4.2), but uses impact forces from hammers.
- Upset forging: heated metal stock gripped by dies and end pressed into desired shape, i.e. increasing the diameter by reducing height.
- Hand forging: hot material reduced, upset and shaped using hand tools and an anvil. Commonly associated with the blacksmith's trade, used for decorative and architectural work.
- Precision forging: near-net shape generation through the use of precision dies. Reduces waste and minimises or eliminates machining.

Economic Considerations

- Production rates from 1 to 300/h, depending on size.
- Production is most economic in the production of symmetrical rough forged blanks using flat dies. Increased machining is justified by increased die life.
- Lead times are typically weeks.
- Material utilisation moderate (20–25% scrap generated in flash typically).
- Economically viable quantities greater than 10,000, but can be as low as 100 for large parts.
- In the case of open die forging: lower material utilisation, machining of the final shape is necessary, slow production rate, low lead times, commonly used for one-offs and high usage of skilled labour.
- Tooling costs are high.
- Equipment costs generally high.
- Direct labour costs are moderate. Some skilled operations may be required.
- Finishing costs are moderate. Removal of flash, cleaning and fettling important for subsequent operations.

Typical Applications

- Engine components (connecting rods, crankshafts, cam shafts).
- Transmission components (gears, shafts, hubs, axles).
- Aircraft components (landing gear, airframe parts).
- Tool bodies.
- Levers.
- Upset forging: for bolt heads, valve stems.
- Open die forging: for die blocks, large shafts, pressure vessels.

Design Aspects

- Complexity is limited by material flow through dies.
- Deep holes with small diameters are better drilled.
- Drill spots caused by die impressions can be used to aid drill centralisation for subsequent machining operations.

- Locating points for machining should be away from parting line due to die wear.
- Markings are possible at little expense on adequate areas that are not to be subsequently machined.
- Care should be taken with design of die geometry since cracking, mismatch, internal rupture and irregular grain flow can occur.
- Good practice to have approximately equal volumes of material both above and below the parting line.
- Inserts and undercuts are not possible.
- Placing of parting line important, i.e. avoid placement across critical dimensions, keep along simple plane, line of symmetry or follow the part profile.
- Corner radii and fillets should be as large as possible to aid hot metal flow.
- Maximum length to diameter ratio that can be upset is 3:1.
- Avoid abrupt changes in section thickness. Causes stress concentrations on cooling.
- Minimum corner radii = 1.5 mm.
- Machining allowances range from 0.8 to 6 mm, depending on size.
- Drafts must be added to all surfaces perpendicular to the parting line.
- Draft angles range from 0° to 8° depending whether internal or external features, and section depth, but typically 4°. Reduced by mechanical ejectors in dies.
- Minimum section = 3 mm.
- Sizes range from 10 g to 250 kg in weight, but better for parts less than 20 kg.

Quality Issues

- Good strength, fatigue resistance and toughness in forged parts due to grain structure alignment with die impression and principal stresses expected in service.
- Low porosity, defects and voids encountered.
- Forgeability of material important and maintenance of optimum forging temperature during processing.
- Hot material in contact with the die too long will cause excessive wear, softening and breakage.
- Variation in blank mass causes thickness variation. Reduced by allowing for flash generation, but increases waste.
- Residual stresses can be significant. Can be improved with heat treatment.
- Die wear and mismatch may be significant.
- Surface roughness and detail may be adequate, but secondary processing usually employed to improve the surface properties.
- Surface roughness is in the range 1.6–25 μm Ra.
- Process capability charts showing the achievable dimensional tolerances for closed die forging using various materials are provided (Figure 4.1(b)). Note, the total tolerance on charts 1–4 is allocated $+\frac{2}{3}, -\frac{1}{3}$. Allowances of +0.3 to +2.8 mm should be added for dimensions across the parting line and mismatch tolerances range from 0.3 to 2.4 mm, depending on part size.
- Tolerances for open die forging range from ±2 to ±50 mm, depending on size of work and skill of the operator.

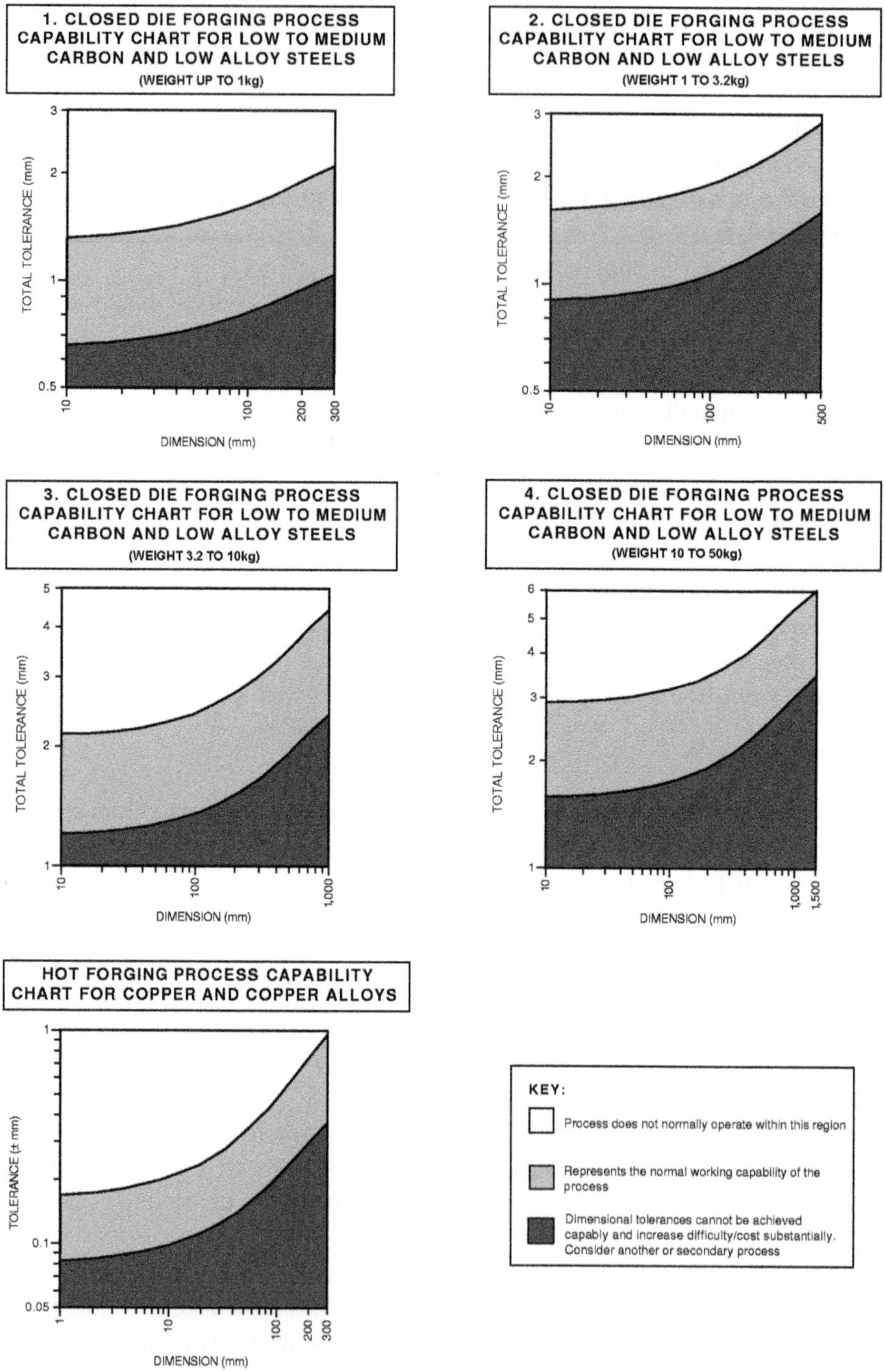

Figure 4.1(b): Forging Process Capability Charts.

4.2 Rolling

Process Description

Continuous forming of metal between a set of rotating rolls whose shape or height is adjusted incrementally to produce desired section through imposing high pressures for plastic deformation. It is the process of reducing thickness, increasing length without increasing the width markedly. Can be performed with the material at a high temperature (hot) or initially at ambient temperature (cold) (Figure 4.2(a)).

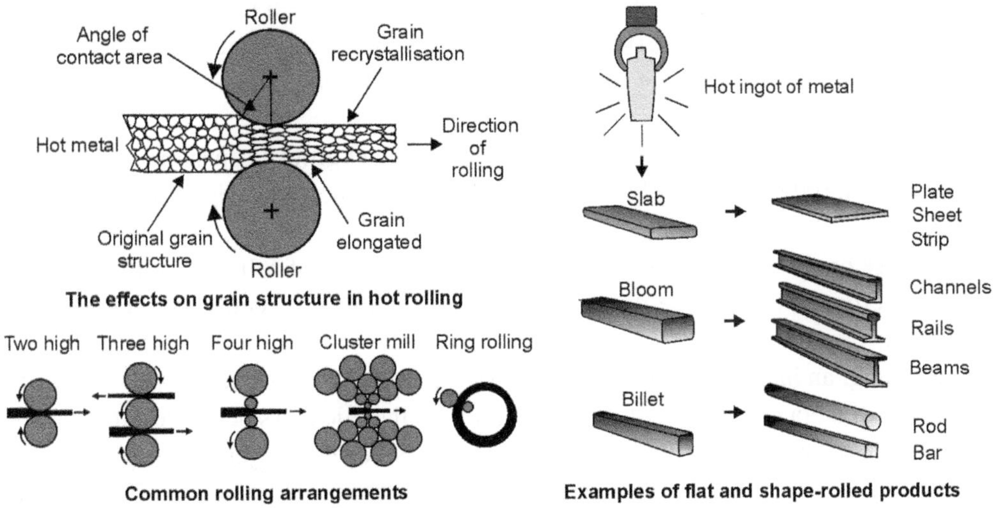

Figure 4.2(a): Rolling.

Materials

- Most ductile metals such as: low carbon alloy and stainless steels; aluminium alloys; copper alloys and magnesium alloys.
- Metal ingots called blooms, slabs or billets, used to load the mill. Blooms are used to produce structural sections (beams, channels, rail sections); slabs are used to produce flat products such as sheets and plate; and billets are rolled into rods and bars using shaped rolls.
- Continuous casting is also used for higher efficiency and lower cost.

Process Variations

- Variety of roll combinations exist (called mills):
  - Two high: commonly used for hot rolling of plate and flat product, either reversing or non-reversing type.
  - Two high with vertical rolls: commonly used for hot rolling of structural sections. Vertical rolls maintain uniform deformation of section and prevent cracking.

- Three high: for reversing one length above the other simultaneously.
- Four high (tandem): backing rolls give more support to the rolls in contact with product for initial reduction of ingots.
- Cluster mills: very low roll deflection obtained due to many supporting rolls above the driven rolls that are in contact with product. For cold rolling thin sheets and foil to close dimensional tolerances.

- Levelling rolls: used to improve flatness of strip product after main rolling operations.
- Flat rolling: for long continuous lengths (long discontinuous lengths in reality) of flat product. The height between the rolls is adjusted lower on each reversing cycle, or the product is passed through a series of tandem rollers with decreasing roller gap and increasing speed, to reduce the product to its final thickness. Tandem roll system has higher production rates.
- Shape rolling: billet is passed through a series of shaped grooves on same roll or a set of rolls in order to gradually form the final shape; typically used for structural sections.
- Transverse or cross rolling: wedge-shaped forms in a pair of rolls create the final shape on short-cropped bars in one revolution. For parts with axial symmetry such as spanners.
- Ring rolling: an internal roller (idler) and external roller (driven) impart pressure on to the thickness of a doughnut-shaped metal preform. As the thickness decreases, the diameter increases. For creating seamless rings used for pressure vessels, jet engine parts and bearing races. Rectangular cross-sections and contours are also possible. Can be readily automated.
- Pack rolling: operation where two or more layers of metal are rolled together.
- Thread rolling: wire or rod is passed between two flat plates, one moving and the other stationary, with a thread form engraved on surfaces. Used to produce threaded fasteners with excellent strength and surface integrity at high production rates and no waste.
- Roll forming: forming of long lengths of sheet metal into complex profiles using a series of rolls (see PRIMA 4.9).
- Calandering: thermoplastic raw material is passed between a series of heated rollers in order to produce sheet product.

Economic Considerations

- Production rates are high. Continuous process with speeds ranging from 20 to 500 m/min.
- Production rates for related processes: transverse rolling up to 100/h and thread rolling up to 30,000/h.

- Lead times are typically months due to number of mills required and complexity of profile.
- Long set-up times for shaped rolls.
- Hot rolling requires less energy than cold rolling.
- Material utilisation very good (rolling is a constant volume process). Less than 1% scrap generated, commonly through line stoppages or when cutting to lengths. Can be recycled.
- High degree of automation possible.
- Plane rolls are flexible in the range of flat products they can produce. Shaped rolls are dedicated and therefore not flexible.
- Economical for very high production runs. Minimum quantity is 50,000 m of rolled product (equivalent to 100,000+).
- Tooling costs are high.
- Equipment costs are high.
- Direct labour costs are low to moderate.
- Very low finishing costs.

Typical Applications

- Rolling is an important process for producing the stock material for many other processes, e.g. machining, cold forming and sheet-metal work. Around 90% of all stock product used is produced by rolling for many industries:
  - Flat, square, rectangular and polygonal sections.
  - Structural sections, e.g. I-beams, H-beams, T-sections, channels, rails, angles and plate.
  - Strip, foil and sheet.
- Sheet for shipbuilding.
- Structural fabrication.
- Sheet metal for shearing and forming operations.
- Tube forming.
- Automotive trim.

Design Aspects

- Simple shapes using flat rolling, fairly complex two-dimensional profiles using shape rolling and three-dimensional shapes for transverse rolling.
- Re-entrant angles possible on profile.
- No draft angles required, except in transverse rolling.
- Hot rolling:
  - Minimum section = 1.6 mm.
  - Maximum section = 1 m.

- Cold rolling:
  - Minimum section = 0.0025 mm.
  - Maximum section = 200 mm.
- Maximum width = 5 m.

Quality Issues

- Coarse grain structure and porosity of hot ingot or continuous casting is gradually improved and finer grain structure produced with few or no voids.
- Hot rolling takes place above recrystallisation temperature and therefore sections are free from residual stresses. No working hardening of material.
- Anisotropy in cold-rolled sections due to directionality of grains during rolling and work hardening. Can be used to advantage, but does mean high compressive residual stresses that exist in surface are balanced by high tensile residual stresses in section bulk. Can lead to surface delamination.
- High sulphur contents in steels can cause cracking and flaring of rolled section ends. Possibility of jamming when introduced to a subsequent set of rolls. High scrap rates and downtime can be experienced if this occurs.
- Hot-rolled material is more difficult to handle than cold-rolled. Cold-rolled strip product can be coiled for subsequent processing, hot-rolled cannot.
- Rough surface finish of rolls is used in hot rolling to aid traction of metal through the rolls. Cold rolling rolls have a high surface finish.
- Lubrication can be used for ferrous alloys (graphite) and non-ferrous alloys (oil emulsion) to minimise friction during rolling.
- Cold rolling can be performed with low-viscosity lubricants such as paraffin or oil emulsion.
- Hot rolling requires the preparation of stock material to remove surface oxides before processing.
- Maintenance of rolling temperature dictates quality. Too low and it becomes difficult to deform. Too high and surface quality is reduced.
- Roll material must be highly wear resistant. Made to withstand 5,000,000 m of rolled section production. Can be re-coated and ground back to size.
- Surface defects may result from inclusions and impurities in the material (scale, rust, dirt, roll marks and other causes related to prior treatment of ingots).
- Surface detail is poor in hot-rolled product (oxide layer called mill scale is always present). Oxide layer can be removed by pickling in acid.
- Surface detail is excellent for cold rolling.
- Surface roughness values are in the range 6.3–50 μm Ra for hot rolling and 0.2–6.3 μm Ra for cold rolling.
- Process capability charts showing the achievable dimensional tolerances for cold rolling various materials are provided (Figure 4.2(b)).

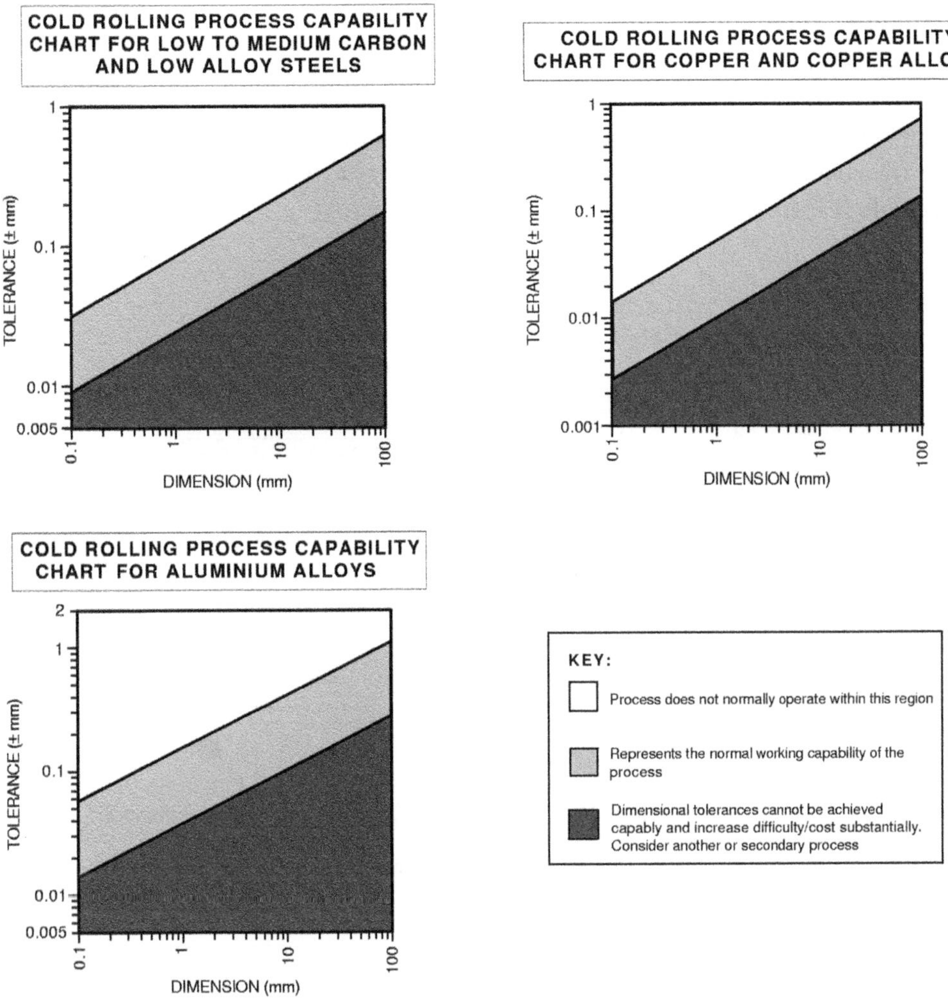

Figure 4.2(b): Rolling Process Capability Charts.

- Achievable tolerances range from ±1% to ±2.5% of the dimension for hot rolling. Dimensional variations are greater than for cold rolling due to non-uniformities in material properties such as hardness, roll deflection and surface conditions.

4.3 Drawing

Process Description

A number of processes where long lengths of rod, tube or wire are pulled through dies to progressively reduce the original cross-section through plastic deformation. The process is performed cold (Figure 4.3(a)).

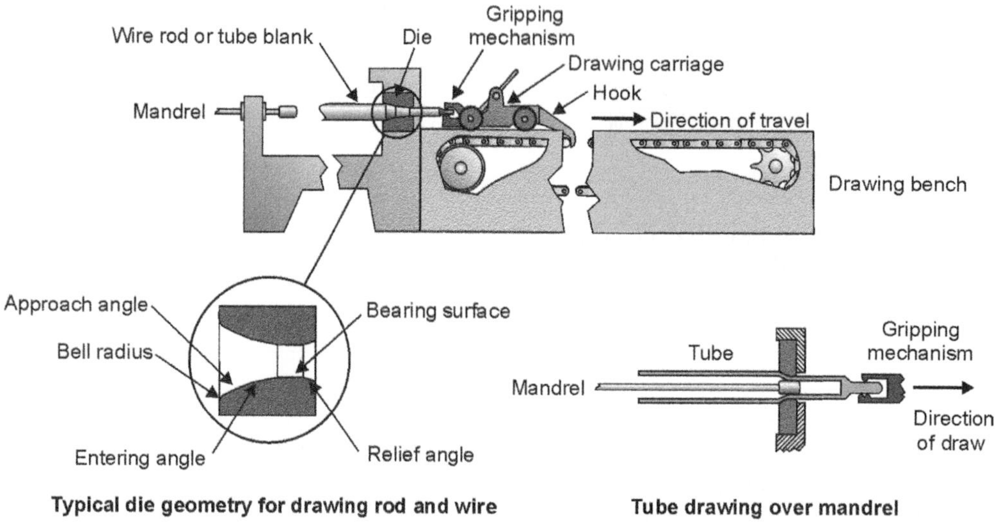

Figure 4.3(a): Drawing.

Materials

Any ductile metal at ambient temperatures.

Process Variations

- Rod or bar drawing: reduction of the diameter of rod or bar through a single die or progressive reduction through a number of dies.
- Wire drawing: performed on multiple wire drawing machine where the wire is wrapped around blocks before being pulled through the next die to successively reduce diameter. Wire diameters that cannot be wrapped around blocks are drawn out on long benches at low speeds, but give lower production rates.
- Tube drawing: reduction of either the diameter of a tube or simultaneous reduction of diameter and thickness using mandrel.

- Can use rollers in place of dies for plastic deformation.
- Sizing is a low deformation operation sometimes used to finish the drawn section giving closer dimensional accuracy and improved surface roughness.

Economic Considerations

- Production rates vary from 10 (rod, tube) to 2000 m/min (wire).
- Lead time typically days.
- Material utilisation is excellent. Some scrap may be generated when cutting to length.
- High degree of automation possible.
- Economical for high production runs (1,000+).
- Tooling costs are low.
- Equipment costs are moderate.
- Direct labour costs are low to moderate.
- Very low finishing costs. Cutting long lengths of rod, bar and tube to length only.

Typical Applications

- Drawing is an important process for producing the stock material for many other processes, e.g. machining and cold heading.
- Rod, bar, wire, tubes.
- Fabrication and machine construction.
- Spring wire, musical instrument wire.

Design Aspects

- Simple shapes with rotational symmetry only.
- No draft angles required.
- Rod drawing:
  - Minimum diameter = 0.1 mm
  - Maximum diameter = 50 mm
- Wire drawing:
  - Minimum diameter = 0.1 mm
  - Maximum diameter = 20 mm
- Tube drawing:
  - Minimum diameter = 6 mm
  - Maximum diameter = 600 m
  - Minimum section = 0.1 mm
  - Maximum section = 25 mm

Quality Issues

- Strain hardening occurs in material during cold working, giving high strength.
- High directionality (anisotropy) due to nature of plastic deformation and grain orientation in direction of drawing.
- High friction between work and die causes high temperatures, which must be reduced through external cooling.
- Surface detail is excellent.
- Surface roughness is in the range 0.2–6.3 μm Ra.
- Finer surface roughness values are obtained with finer grit grades.
- Process capability charts showing the achievable dimensional tolerances for cold drawing various materials are provided (Figure 4.3(b)).

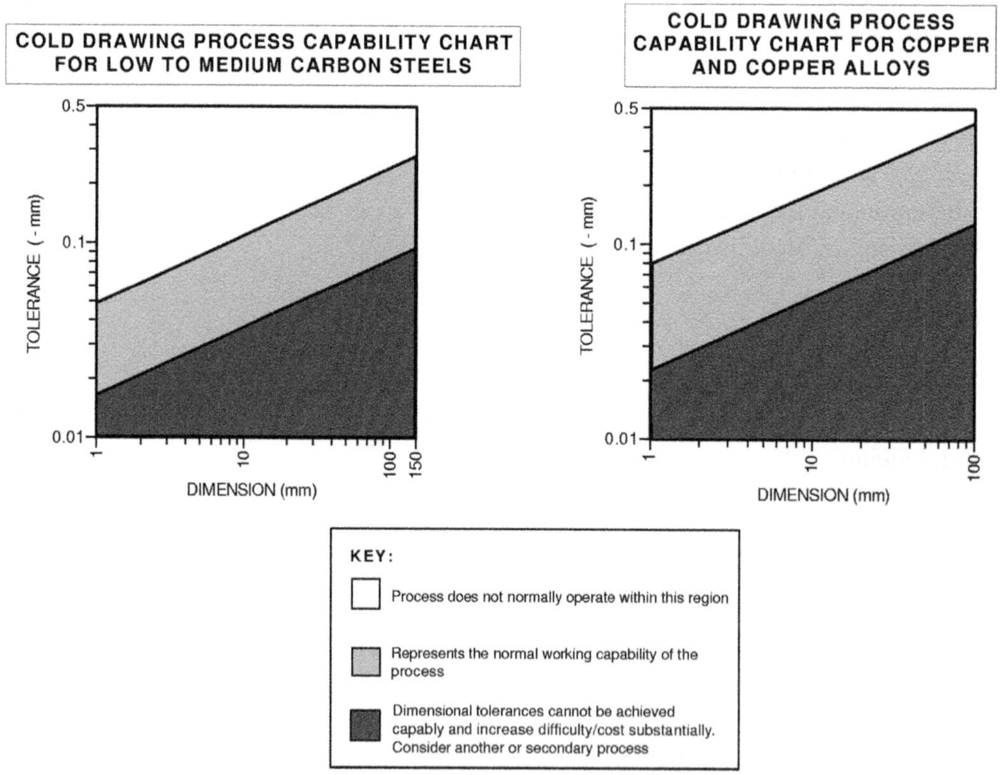

Figure 4.3(b): Drawing Process Capability Charts.

4.4 Cold Forming

Process Description

Various processes under the heading of cold forming tend to combine forward and backward extrusion to produce near-net-shaped components by the application of high pressures and forces (Figure 4.4(a)).

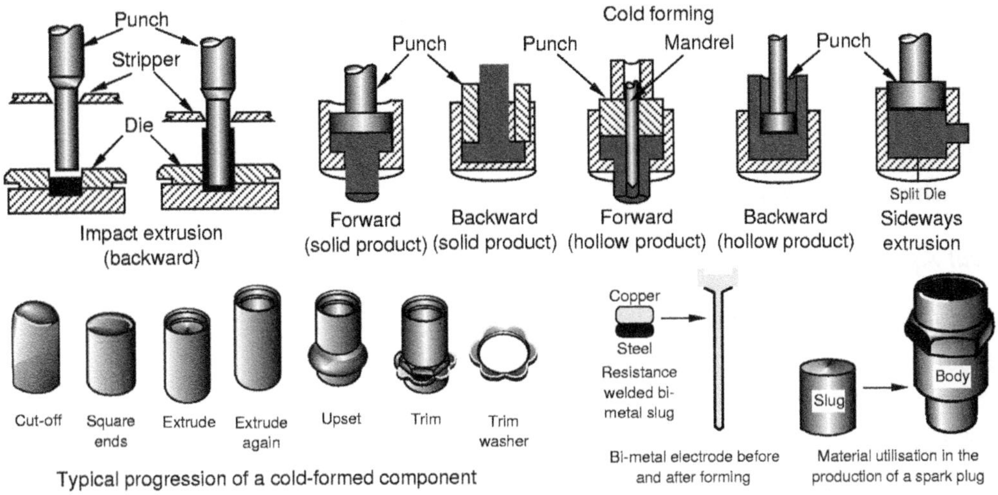

Figure 4.4(a): Cold Forming.

Materials

Any ductile material at ambient temperature, including: aluminium, copper, zinc, lead and tin alloys, and low carbon steels. Also, alloy and stainless steels, nickel and titanium alloys are processed on a more limited basis.

Process Variations

- Impact extrusion: similar to cold extrusion, but cold billet is plastically deformed by a single blow of the tool. Can be forward or backward extrusion (Hooker process).
- Cold forming: can be forward, backward or a combination of both.
- Hydrostatic extrusion: metal forced through die by high fluid pressure. Used for high-strength, brittle and refractory alloys.
- Can incorporate other processes such as: cold heading, drawing, swaging, nosing, sizing and coining to produce complex parts at one station.

Economic Considerations

- Production rates up to 2000/h.
- Lead times are usually weeks.
- High utilisation of material (95%). Possible material cost savings over machining can be high. Near elimination of heat treatment and machining requirements.
- Can be economical for quantities down to 10,000, depending on complexity of part. More suited to high production volumes (100,000+).
- Most applications are in the formation of symmetrical parts with solid or hollow cross-sections.
- Tooling costs are high.
- Equipment costs are high.
- Direct labour costs are low.
- Finishing costs are very low.

Typical Applications

- Fasteners.
- Tool sockets.
- Spark plug bodies.
- Gear blanks.
- Collapsible tubes.
- Bearing races.
- Valve seats.

Design Aspects

- Complexity limited. Symmetry of the part is important: concentric, round or square cross-sections typical. Limited asymmetry possible.
- To avoid mismatch of dies, every effort should be made to balance the forces, especially on unsymmetrical parts.
- Length to diameter ratios of secondary formed back extruded parts may approach 10:1; forward extrusion unlimited.
- Any parting lines should be kept in one plane and placement across critical dimensions should be avoided.
- Can be used to process two materials simultaneously to produce parts such as steel coated copper electrodes.
- Inserts are not recommended.
- Undercuts are not possible.
- Draft angles not required.

- Maximum section ranges from 0.25 to 22 mm depending on material for impact extrusion. No limit for cold forming.
- Minimum section ranges from 0.09 to 0.25 mm depending on material.
- Sizes range from 1.3 to 150 mm diameter depending on cold formability of material being processed.

Quality Issues

- Inside shoulders require secondary processing to ensure flatness.
- Cold working offers valuable increase in mechanical properties, including extended fatigue life.
- Concentricity of blank and punch important in providing uniform section thickness.
- Supply of lubrication (commonly phosphate based) to the die surfaces is important in providing uniform material flow and to reduce friction.
- Small quantities of sulphur, lead, phosphorus, silicon, etc. reduce the ability of ferrous metals to withstand cold working.
- Surface cracking: tearing of the surface of the part, especially with high-temperature alloys, aluminium, zinc, magnesium. Control of the billet temperature, extrusion speed and friction are important.
- Pipe or fishtailing: metal flow tends to draw surface oxides and impurities towards centre of part. Governing factors are friction, temperature gradients and amount of surface impurities in billets.
- Internal cracking or chevron cracking: similar to the necked region in a tensile test specimen. Governing factors are the die angle and amount of impurities in the billet.
- Surface detail excellent.
- Surface roughness is in the range 0.1–1.6 μm Ra.
- Process capability charts showing the achievable dimensional tolerances for impact extrusion and cold forming are provided (Figure 4.4(b)).
- Dimensional tolerances for non-circular components are at least 50% greater than those shown on the charts.

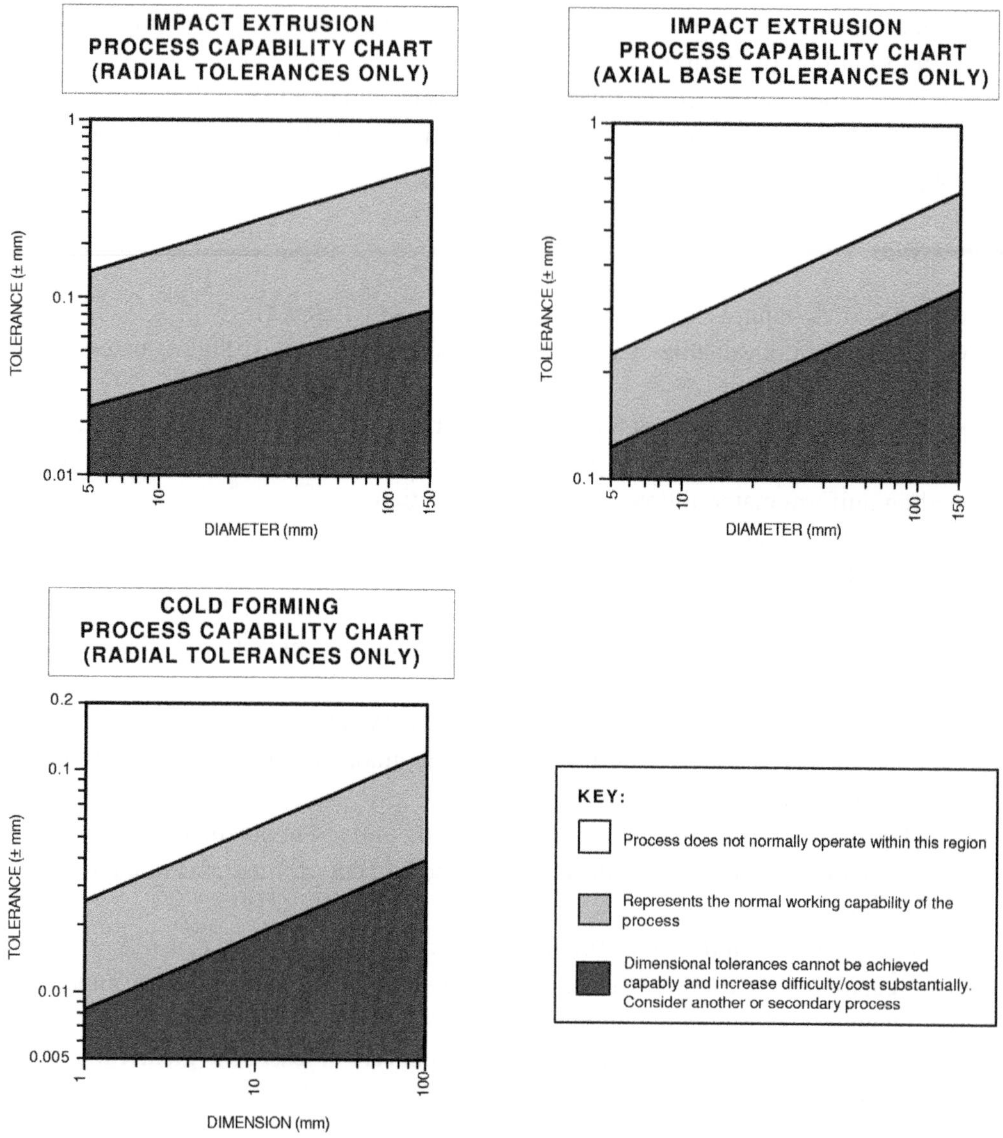

Figure 4.4(b): Cold Forming Process Capability Charts.

4.5 Cold Heading

Process Description

Wire form stock material is gripped in a die with usually one end protruding. The material is subsequently formed (effectively upset) by successive blows into the desired shape by a punch or a number of progressive punches. Shaping of the shank can be achieved simultaneously (Figure 4.5(a)).

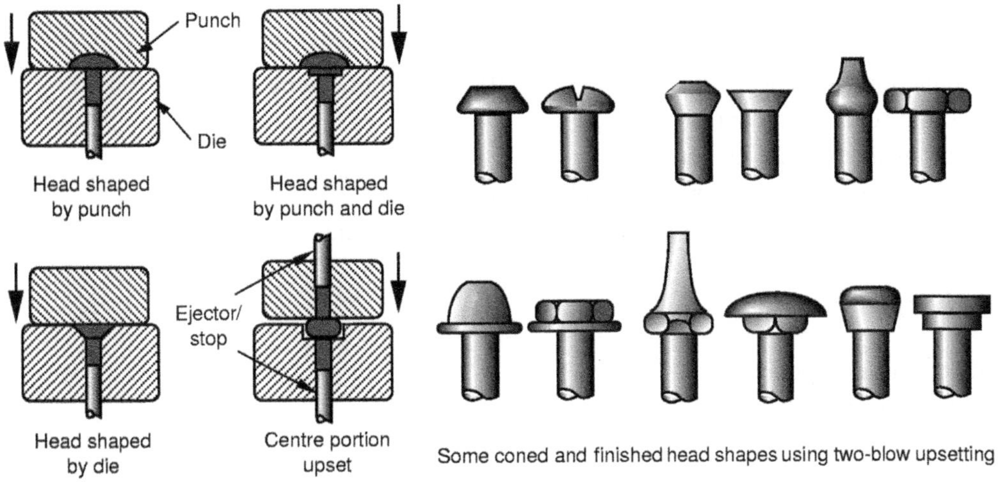

Figure 4.5(a): Cold Heading.

Materials

- Suitable for all ductile metals: principally carbon steels, aluminium, copper and lead alloys.
- Alloy and stainless steels, zinc, magnesium, nickel alloys and precious metals are also processed.

Process Variations

- Usually performed with stock material at ambient temperature (cold), but also with stock material warm or hot.
- Solid die: single stroke, double stroke, three blow, two die, progressive bolt makers, cold or hot formers – the choice is determined by the length to diameter ratio of the raw material.
- Open die: parts made by this process have wide limits and are too long for solid dies.
- Continuous rod or cut lengths of material can be supplied to the dies.

- Can incorporate other forming processes, for example knurling, thread rolling and bending, to produce complex parts at one machine.
- Upset forging: heated metal stock gripped by dies and end pressed into desired shape, i.e. increasing the diameter by reducing height.

Economic Considerations

- Production rates between 35 and 120/min are common.
- Lead times are relatively short due to simple dies.
- High material utilisation. Virtually no waste.
- Flexibility is moderate. Tooling tends to be dedicated.
- Production quantities typically very high, 100,000+, but can be as low as 10,000.
- Tooling costs are moderate.
- Equipment costs are moderate.
- Direct labour costs are low. Process highly automated.
- Finishing costs are low: normally no finishing is required.

Typical Applications

- Electronic components.
- Electrical contacts.
- Nails.
- Bolts and screws.
- Pins.
- Small shafts.

Design Aspects

- Complexity limited to simple cylindrical forms with high degree of symmetry.
- Significant asymmetry difficult.
- Minimisation of shank diameter and upset volume important.
- Radii should be as generous as possible.
- Threads on fasteners should be rolled wherever possible.
- Head volumes are limited due to amount of deformation possible.
- Inserts are possible at added cost.
- Undercuts are produced via secondary operations.
- Machining usually not required.
- Draft angles not required.
- Minimum diameter = 0.8 mm.
- Maximum diameter = 50 mm.
- Minimum length = 1.5 mm.
- Maximum length = 250 mm.

Quality Issues

- Cold working process gives improved mechanical properties.
- Fatigue, impact and surface strength are increased, giving a tough, ductile, crack-resistant structure.
- Small quantities of sulphur, lead, phosphorus, silicon, etc. reduces the ability of ferrous metals to withstand cold working.
- Length to diameter ratio of protruding shank to be formed should be below 2:1 to avoid buckling.
- Residual stresses may be left at critical points.
- Sharp corners reduce tool life.
- Surface detail is good to excellent.
- Surface roughness is in the range 0.8–6.3 μm Ra.
- Process capability charts showing the achievable dimensional tolerances for cold heading are provided (Figure 4.5(b)).

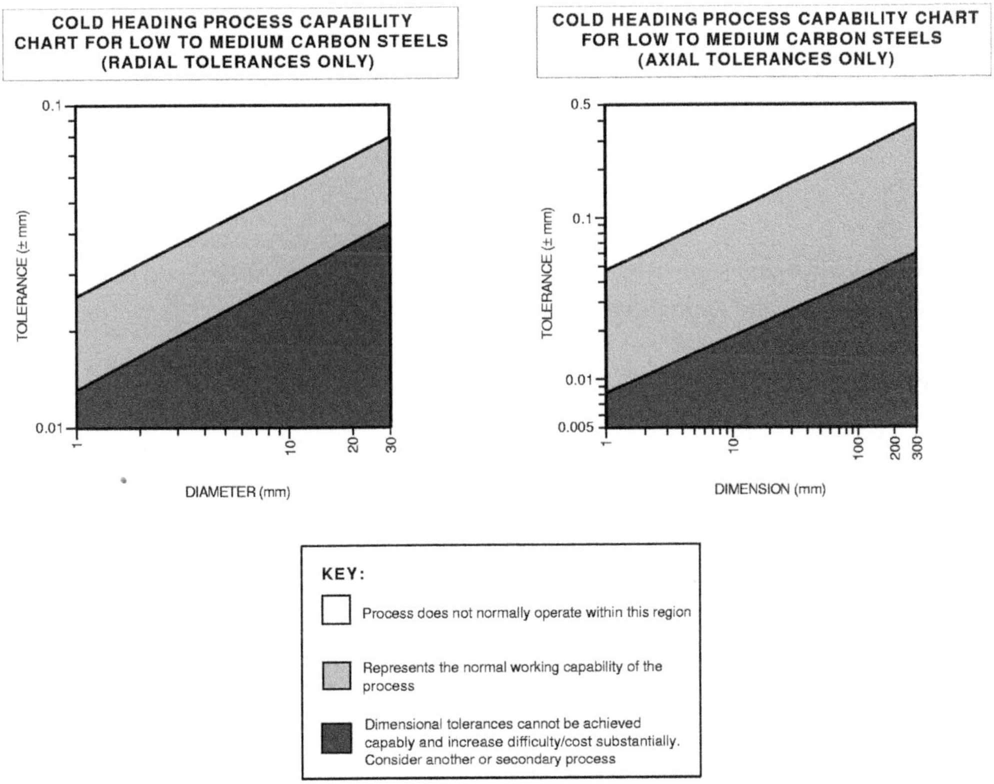

Figure 4.5(b): Cold Heading Process Capability Charts.

4.6 Swaging

Process Description

Process of gradually shaping and reducing the cross-section of tubes, rods and wire using successive blows from hard dies rotating around the material (on a mandrel if necessary for tubular sections). Operation performed at ambient temperature (Figure 4.6(a)).

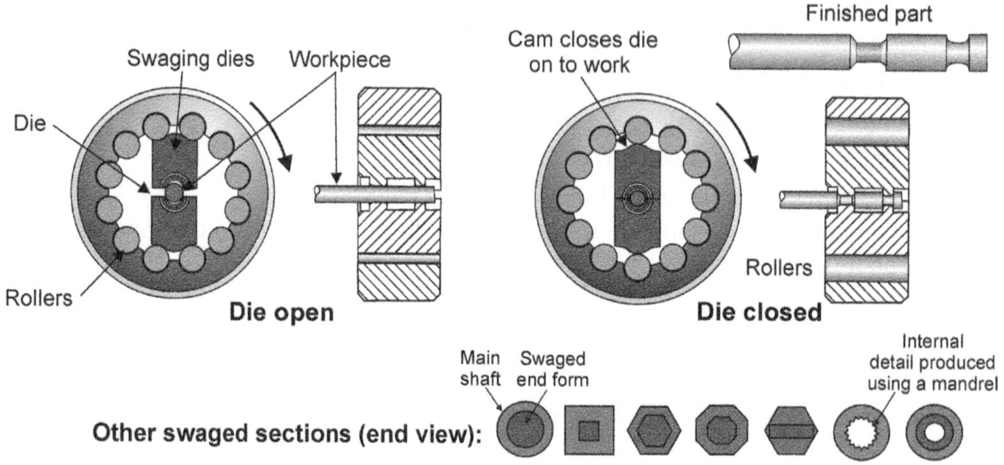

Figure 4.6(a): Swaging.

Materials

- Carbon; low alloy and stainless steels; aluminium; magnesium; nickel and their alloys.
- Copper, zinc, lead and their alloys, less commonly.

Process Variations

- Using a shaped mandrel can generate inner section profiles different to outer.
- Hand forging: hot material reduced, upset and shaped using hand tools and an anvil, commonly associated with the blacksmith's trade. For decorative and architectural work.

Economic Considerations

- Production rates moderate to high (100–300/h).
- Lead time typically days, depending on complexity of tool.
- Special tooling not necessarily required for each job.
- Material utilisation is excellent. No scrap generated.
- Some automation possible.
- Economical for low production runs. Can be used for one-offs.

- Tooling costs are high.
- Equipment costs are generally moderate.
- Direct labour costs are low to moderate.

Typical Applications

- Used to close tubes, produce tapering, clamping and steps in sections.
- Many section types possible, either parallel or tapered.
- Tool shafts and handles.
- Punches.
- Chisels.
- Exhaust pipes.
- Cable assemblies.
- Architectural work.

Design Aspects

- Complexity fairly high.
- Round, square, rectangular and polygonal sections possible, either parallel or tapered.
- Splines and contoured surfaces.
- Holes are possible, but only through the length of the part.
- No undercuts or inserts possible.
- Draft angles range from 0° to 3.5°.
- Minimum section = 2.5 mm.
- Maximum section = 50 mm.
- Minimum solid diameter = 2.5 mm.
- Maximum solid diameter = 150 mm.
- Maximum tube diameter = 350 mm.
- Minimum length = 1.5 mm.
- Maximum length = 250 mm.

Quality Issues

- Cold working of material gives good mechanical properties and compressive surface stresses improve fatigue life.
- Surface finish of stock material is markedly improved.
- Surface detail is good to excellent.
- Surface roughness values are in the range 0.8–3.2 μm Ra.
- A process capability chart showing the achievable dimensional tolerances for swaging is provided (Figure 4.6(b)).

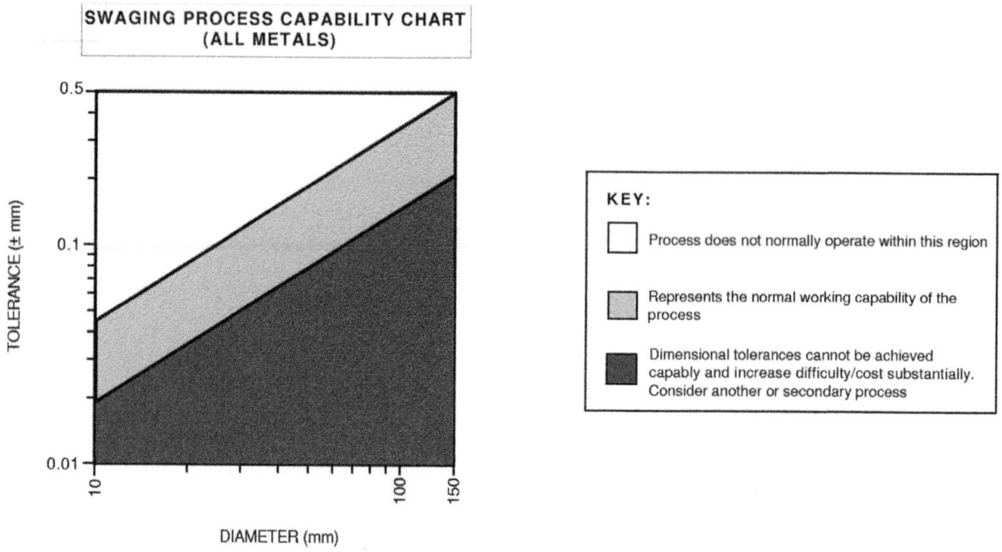

Figure 4.6(b): Swaging Process Capability Chart.

4.7 Superplastic Forming

Process Description

Sheet metal is clamped over a male or female form tool and heated to a high enough temperature to give the material high ductility at low strain rates. Pressurised gas (typically argon) on the back face of the sheet forms the material into a cavity or over the surface of the tool (Figure 4.7).

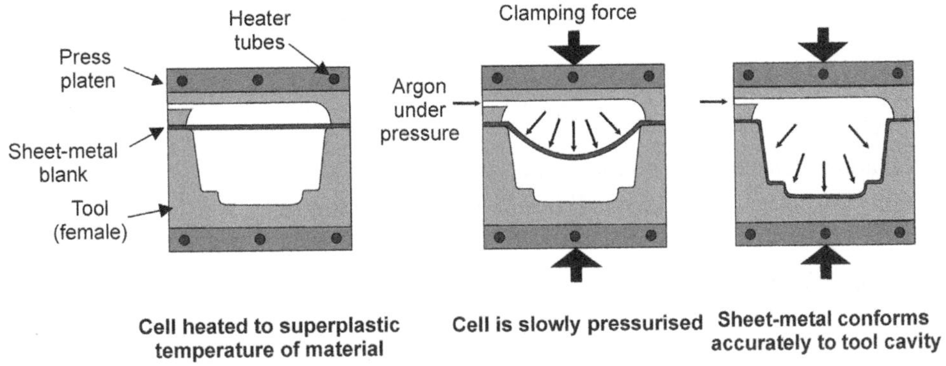

Figure 4.7: Superplastic Forming.

Materials

- For stainless steels, aluminium and titanium alloys typically.
- Material must be able to deform at low strain rates and high temperatures and possess a stable microstructure.

Process Variations

- Either male or female tool: male forming more complex, but offers greater design freedom and more uniform material distribution.
- Additional use of tool movement with gas pressure gives deeper parts with more uniform wall thickness.
- Diaphragm forming: uses additional diaphragm sheet behind material to give better control of thickness distribution.
- Can be used in conjunction with diffusion bonding to create complex parts (see PRIMA 11.9).

Economic Considerations

- Production rates are low. Long cycle times.
- Slower than conventional deep drawing.
- Lead times are moderate, typically weeks, depending on complexity of mould.
- Material utilisation is good. Some waste may be generated during subsequent trimming operations.
- Scrap not recyclable directly.
- Some aspects can be automated.
- Economically viable for low to moderate production volumes, 10–10,000. Can be used for one-offs.
- Tooling costs are high.
- Equipment costs are high.
- Direct labour costs are low to moderate.
- Finishing costs low to moderate. Trimming typically required.

Typical Applications

- Used to generate deep and intricate forms in sheet metal.
- Aerospace fuselage panels.
- Automotive body panels.
- Containers.
- Casings.
- Architectural and decorative work.
- Can also be used to clad other materials.

Design Aspects

- Complexity limited to shape of female or male tool and constant thickness parts.
- Ribs, bosses and recesses possible.
- Re-entrant features not possible.
- Radii should be greater than five times the wall thickness.
- Sharp radii at extra cost can be produced.
- Draft angles range from 2° to 3°.
- Maximum drawing ratio (height to width) = 0.6.
- Maximum dimension = 2.5 m.
- Maximum thickness = 4 mm.
- Minimum thickness = 0.8 mm.

Quality Issues

- No spring back exhibited after processing.
- No residual stresses.
- Creep performance poor due to small grain sizes produced.
- Cavitation and porosity can occur in some alloys at high temperatures and low strain rates.
- Graphite coating on the blank sheet is used to reduce friction.
- Surface detail is good.
- Secondary operations such as heat treatment, paint, powder coating and anodising (see PRIMA 9.4) are commonly used to improve finish.
- Surface roughness is in the range 0.4–6.3 μm Ra.
- Achievable dimensional tolerances range from ±0.13 to ±0.25 mm up to 25 mm, ±0.45 to ±0.78 mm up to 150 mm. Wall thickness tolerances are typically ±0.25 mm.

4.8 Sheet-metal Shearing

Process Description

Various shearing processes used to cut cold-rolled, sheet metal with hardened punch and die sets. The most common shearing processes are: cutting, piercing, blanking and fine blanking (Figure 4.8(a)).

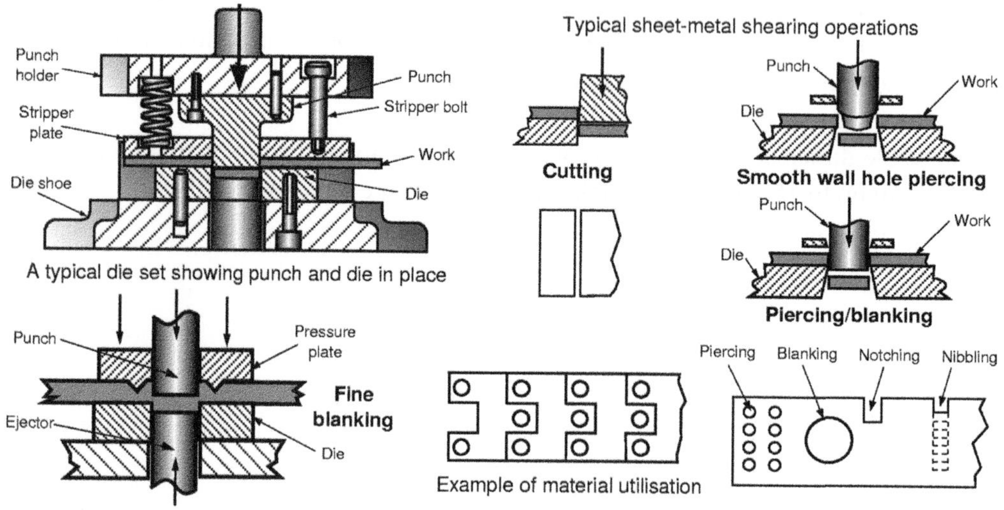

Figure 4.8(a): Sheet-metal Shearing.

Materials

- All ductile metals available in cold-rolled sheet form, supplied flat or coiled.
- Most commonly used metals are: carbon steels, low alloy steels, stainless steels, aluminium alloys and copper alloys.
- Also, nickel, titanium, zinc and magnesium alloys are processed to a lesser degree.

Process Variations

- Mechanical drives: faster action and more positive displacement control.
- Hydraulic drives: greater forces and more flexibility.
- Cutting: large sheets of metal are clamped and cut along a straight line.
- Piercing: removal of material from a blank, for example, a hole.
- Blanking: parts are blanked to obtain the final outside shape.
- Fine blanking: uses special clamping tooling to produce a smooth and square-edged contoured blank or hole.

- Smooth wall hole piercing: special punch profiles are used to produce crack-free holes.
- Other operations include: nibbling, notching, trimming and shaving.
- Computer numerical control (CNC) common on piercing and blanking machines.

Economic Considerations

- Production rates are high, 10,000+/h for small components.
- High degree of automation possible.
- Cycle time is usually determined by loading and unloading times for stock material.
- Progressive dies can incorporate shearing and forming processes.
- Lead times can be several weeks depending on complexity and degree of automation, but more typically several days.
- Material utilisation is moderate to high; however, substantial amounts of scrap can be produced in piercing and blanking.
- Production quantities should be high for dedicated tooling, 10,000+.
- Economical quantities can range from one for blanking and piercing to 2000 for fine blanking.
- Tooling cost moderate to high, depending on process and degree of automation.
- Equipment costs vary greatly. Low for simple guillotines to high for high-speed, precision CNC presses.
- Labour costs are low to moderate depending on degree of automation.
- Finishing costs are low to moderate. Deburring and cleaning usually required.

Typical Applications

- Blanks for forming work.
- Cabinet panels.
- Domestic appliance components.
- Machine parts.
- Gears and levers.
- Washers.

Design Aspects

- Complex patterns of contours and holes possible in two dimensions.
- Material used dictates press forces and die clearances.
- Blanked parts should be designed to make the most use of the stock material.
- Pierced holes with their diameter greater than the material thickness should be drilled.
- Fine blanked holes with diameters 60% of the material thickness are possible.
- Holes should be spaced at least 1.5 times the thickness of the material away from each other.
- Maximum sheet thickness = 13 mm.

- Minimum sheet thickness = 0.1 mm.
- Maximum sheet dimension for cutting is 3 m and for fine blanking 1 m.

Quality Issues

- Conventional hole piercing, blanking and cutting does not result in a perfectly smooth and parallel cut.
- Acceptable hole wall and blank edge quality may be achieved with fine blanking and piercing processes.
- Holes placed too close to a bend line can be distorted subsequently in forming operations.
- Inspection and maintenance of die wear and breakage is important.
- Variations in stock material thickness and flatness should be controlled.
- Surface detail is good.
- Surface roughness values range from 0.1 to 12.5 μm Ra.
- Process capability charts showing the achievable dimensional tolerances for several sheet-metal shearing processes are provided (Figure 4.8(b)).

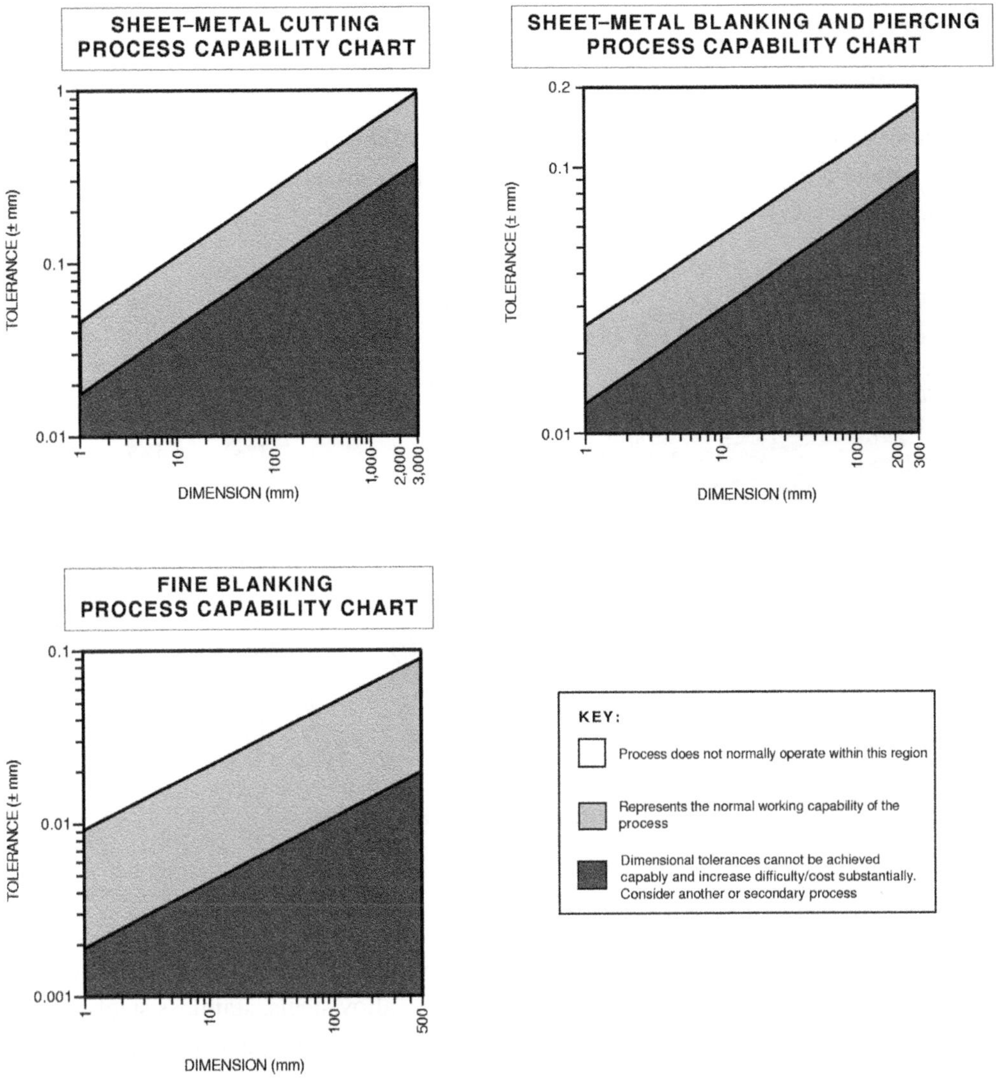

Figure 4.8(b): Sheet-metal Shearing Process Capability Charts.

4.9 Sheet-metal Forming

Process Description

Various processes are used to form cold-rolled sheet metal using die sets, formers, rollers, etc. The most common processes are: deep drawing, bending, stretch forming and roll forming (Figure 4.9(a)).

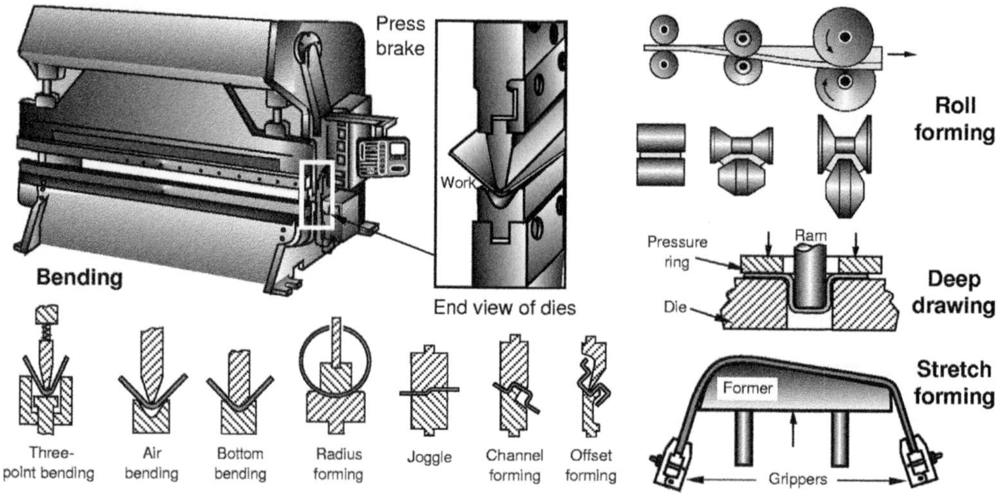

Figure 4.9(a): Sheet-metal Forming.

Materials

- All ductile metals available in cold-rolled sheet form, supplied as blanks, flat or coiled.
- Most commonly used metals are: carbon steels, low alloy steels, stainless steels, aluminium alloys and copper alloys. Also, nickel, titanium, zinc and magnesium alloys are processed to a lesser degree.
- Also coated materials, such as galvanised sheet steel.

Process Variations

- Mechanical drives: faster action and more positive displacement control.
- Hydraulic drives: greater forces and more flexibility.
- Deep drawing: forming of a blank into a closed cylindrical or rectangular shaped die. Incorporating an ironing operation improves dimensional tolerances.
- Bending: deformation about a linear axis to form an angled or contoured profile.
- Stretch forming: sheet metal is clamped and stretched over a simple form tool.
- Roll forming: forming of long lengths of sheet metal into complex profiles using a series of rolls.

- Beading: edge of sheet bent into cavity of die. May be used to remove sharp edges.
- Hemming: edge of sheet folded over. May be used to remove sharp edges.
- Can incorporate initial sheet-metal shearing operations.

Economic Considerations

- Production rates vary, up to 3000/h for small components using automated processes.
- Deep drawing punch speeds a function of material: high to low – brass, aluminium, copper, zinc, steel, stainless steel (typically 800/h).
- High degree of automation possible.
- Cycle time is usually determined by loading and unloading times for the stock material.
- Lead times vary, up to several weeks for deep drawing and stretch forming; could be less than an hour for bending.
- Material utilisation is moderate to high (10–25% scrap generated). Bending and roll forming do not produce scrap directly. Deep drawing and stretch forming may require a trimming operation.
- Production quantities should be high for dedicated tooling, 10,000+. Minimum economical quantities range from one for bending to 1,000 for deep drawing.
- Tooling cost moderate to high, depending on component complexity.
- Equipment costs vary greatly: low for simple bending machines, moderate for roll forming machines, and high for automated deep drawing, sheet-metal presses and stretch forming.
- Labour costs are low to moderate depending on degree of automation.
- Finishing costs are low. Trimming and cleaning may be required.

Typical Applications

- Cabinets.
- Mounting brackets.
- Electrical fittings.
- Cans.
- Machine frames.
- Automotive body panels.
- Aircraft fuselage panels.
- Light structural sections.
- Domestic appliances.
- Kitchen utensils.

Design Aspects

- Complex forms possible: several processes may be combined to produce one component, or a series of operations used to progressively form the part.
- Working envelope of machine and uniform thickness of sheet can restrict design options.

- No inserts or re-entrant angles.
- Draft angles of 0.25° may be required.
- Minimum bend radii are a function of material and sheet thickness, but typically four times the sheet thickness.
- Keep radii as large as possible, particularly if parallel with grain of material.
- Square or rectangular boxes limited by sharpness of corner detail required.
- Minimum sheet thickness = 0.1 mm.
- Maximum sheet thickness: deep drawing = 12 mm, bending = 25 mm, roll/stretch forming = 6 mm.
- Sizes range from 2 to 600 mm diameter for deep drawing; 10 mm to 1.5 m width for roll forming; 2 mm to 3.6 m width for bending.

Quality Issues

- Bending and stretch forming are limited by the onset of necking.
- The limiting drawing ratio (blank diameter/punch diameter) is between 1.6 and 2.2 for most materials. This should be observed where drawing takes place without progressive dies, otherwise excessive thinning and tearing could occur.
- Variations in stock material thickness and flatness should be controlled.
- Other problems include: spring-back (metal returns to original form) and wrinkling during drawing (comparable with forcing a circular piece of paper into a drinking glass), eliminated by adjustment of blank holder force.
- Spring-back can also be compensated for by over-bending, coining and stretch-bending operations.
- High residual stresses can be generated. Subsequent heat treatment may be necessary.
- Surface detail is good.
- Surface roughness is approximately that of the sheet material used.
- Process capability charts showing the achievable dimensional tolerances are provided (Figure 4.9(b)).

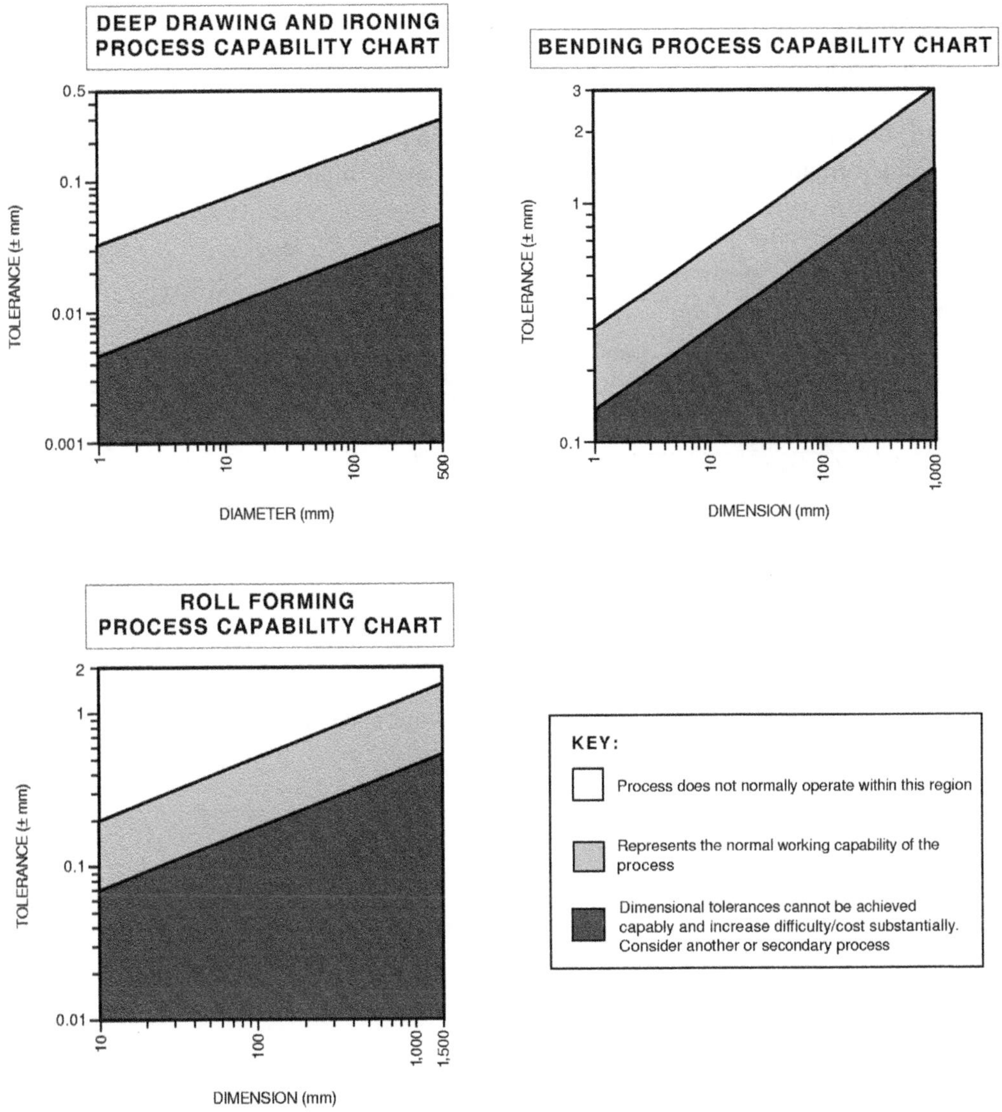

Figure 4.9(b): Sheet-metal Forming Process Capability Charts.

4.10 Spinning

Process Description

Forming of sheet-metal or thin tubular sections using a roller or tool to impart sufficient pressure for deformation against a mandrel while the work rotates (Figure 4.10(a)).

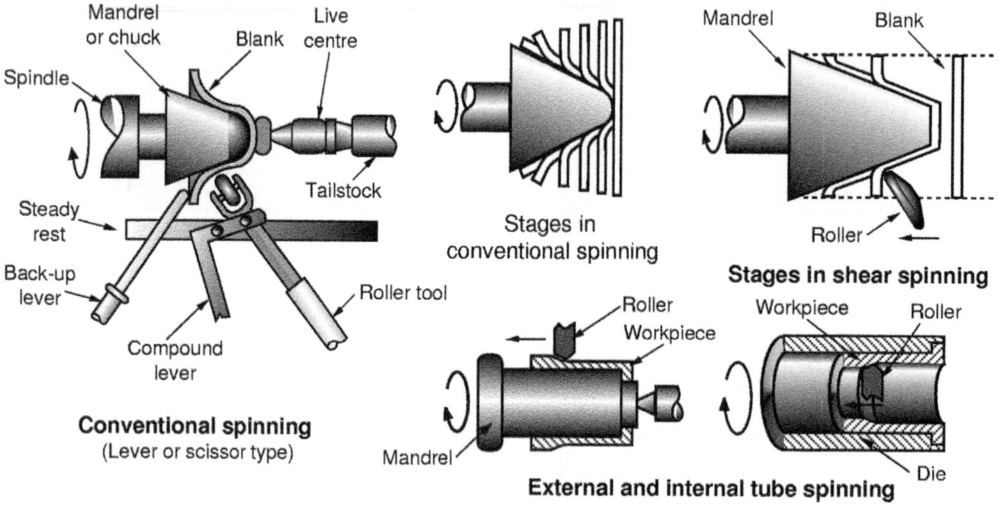

Figure 4.10(a): Spinning.

Materials

- All ductile metals that are available in sheet form. The most common metals used are carbon steels, stainless steels, aluminium alloys, copper alloys and zinc alloys.
- Used on a more limited basis are: magnesium, tin, lead, titanium and nickel alloys.

Process Variations

- Tube spinning: thickness reduction of a cylindrical section on a mandrel, either internally or externally.
- Flame spinning: oxyacetylene flame heats material prior to forming. Permits rapid forming of parts with thick sections.
- Shear spinning: point extrusion process reduces thickness of starting blank or shape to produce the final form.

Economic Considerations

- Production rates are low, typically 10–30/h.
- Lead times are short. Simple mandrels made quickly.

- Material utilisation is moderate. The main losses occur in cutting blanks.
- Flexibility is high: formers are changed quickly and set-up times are low.
- Production volumes are viable from 10 to 10,000. Can be used for one-offs.
- Tooling costs are low.
- Equipment costs are moderate.
- Labour costs are high. Skilled labour is needed.
- Finishing costs are low. Cleaning and trimming required.

Typical Applications

- Flanged, dished, spherical and conical shapes.
- Nose cones.
- Missile heads.
- Bells.
- Light shades.
- Cooking utensils.
- Funnels.
- Reflectors.

Design Aspects

- Complexity limited to thin-walled, conical, concentric shapes. Typically, the diameter is twice the depth.
- Cylindrical or cup-shaped pieces are the most difficult of the simple shapes.
- Oval or elliptical parts are possible, but expensive.
- Material thickness, bend radii, depth of spinning, diameter, steps in diameter and workability of the material are important issues in spinning.
- Radii should be at least 1.5 times the material thickness.
- Stiffening beads should be formed externally rather than internally.
- No draft angle is required.
- Undercuts are possible, but at added cost.
- Maximum section is 75 mm for automated spinning, but approximately 6 mm for hand spinning.
- Minimum section = 0.1 mm.
- Sizes range from 6 to 7.5 m diameter.

Quality Issues

- Skill and experience are required to cause the metal to flow at the proper rate, avoiding wrinkles and tears.
- Streamlined or smooth curves and large radii are an aid both to manufacture and improved appearance.
- Associated problems are blank development and proper feed pressure.
- Grain flow and cold working give good mechanical properties.
- Surface detail is good.
- Surface roughness is in the range 0.4–3.2 μm Ra.
- A process capability chart showing the achievable dimensional tolerances is provided (Figure 4.10(b)).

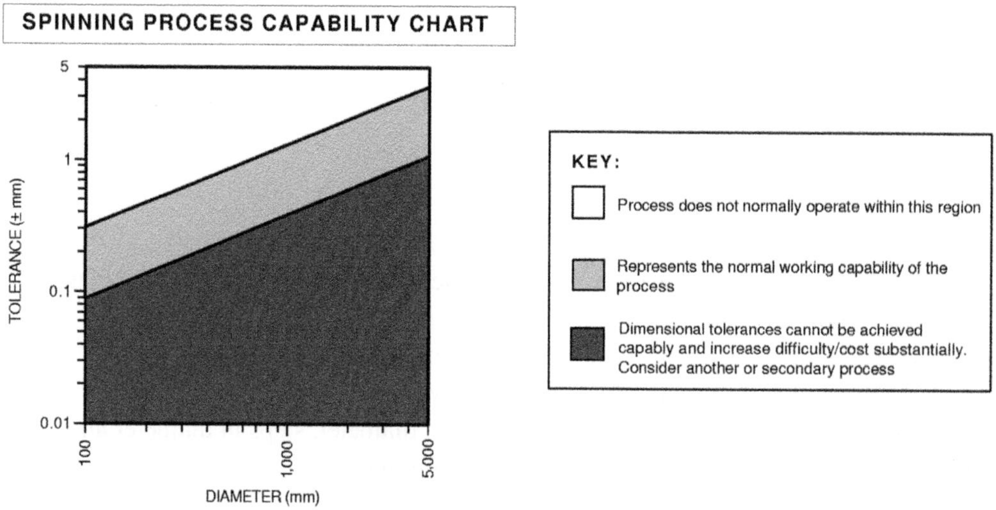

Figure 4.10(b): Spinning Process Capability Chart.

4.11 Powder Metallurgy

Process Description

Die compaction of a blended powdered material into a 'green' compact, which is then sintered with heat to increase the bond strength. Usually secondary operations are performed to improve dimensional accuracy, surface roughness, strength and/or porosity (Figure 4.11(a)).

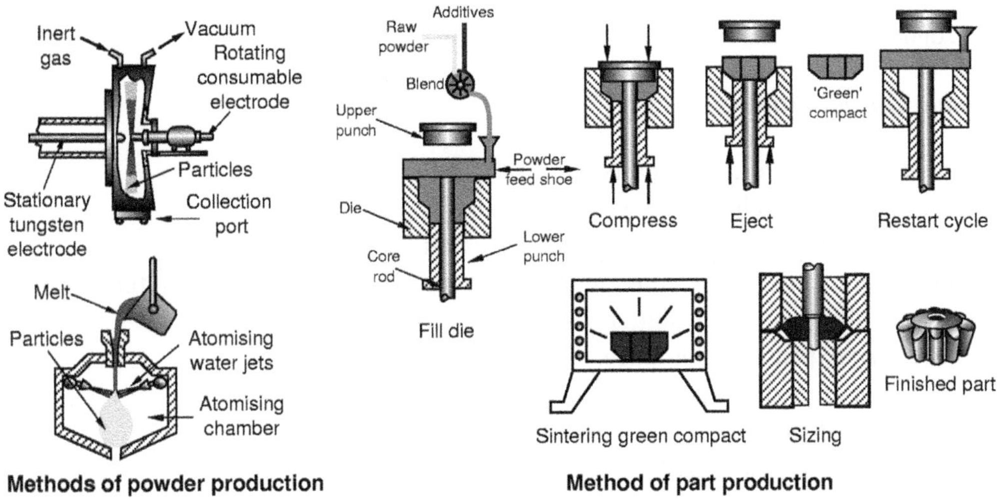

Figure 4.11(a): Powder Metallurgy.

Materials

- All materials, typically metals and ceramics. Iron, copper alloys and refractory metals most common.
- Can process materials not formable by other methods, as long as the material can be powdered.
- Powder production: gas and water atomisation, electrolysis and chemical reduction methods.

Process Variations

- Cold die compaction: performed at room temperature, producing high-porosity and low-strength parts.
- Hot forging: deformation of reheated, sintered compact to final density and shape.
- Continuous compaction: for strip or sheet product. Slower than conventional rolling.

- Isostatic compaction (hot or cold): compaction of powder in a membrane using pressurised fluid (oil, water) or gas. Permits more uniform compaction and near-net shapes. Undercuts and reverse tapers possible, but not transverse holes. Used for ceramics mainly.
- Extrusion: a high-pressure ram forces powder through an orifice determining the section profile.
- Spark sintering: gives magnetic and electrical properties.
- Pressureless compaction: for porous components.
- Metal injection moulding: metal or ceramics powders with binder are injection-moulded, the binder is then stripped from the part and sintered (see PRIMA 4.12).
- Secondary operations include: repressing, sizing and machining.

Economic Considerations

- High production rates, small parts up to 1800/h.
- Cycle times dictated by sintering mechanisms.
- Lead times are several weeks. Dies must be carefully designed and made.
- Production quantities of 20,000+ preferred, but may be economic for 5000 for simple parts.
- Material utilisation is very high. Less than 5% lost in scrap.
- Powders are expensive to produce.
- Automation of process common.
- Each new product requires a new set of die and punches, i.e. flexibility low.
- Tooling costs are very high. Dedicated tooling.
- Equipment costs are high. Sintering equipment not dedicated though.
- Labour costs are low to moderate. Some skilled labour may be required.
- Finishing costs are generally low.
- Final grinding may be more economical than sizing for very close tolerances.

Typical Applications

- Cutting tools.
- Small arms parts.
- Bearings.
- Filters (porous).
- Lock components (keys, barrels).
- Machine parts (ratchets, pawls, cams, gears).

Design Aspects

- Complexity and part size limited by powder flow through die space (powders do not follow hydrodynamic laws) and pressing action.
- Near-net shapes generated.

- Concentric, cylindrical shapes with uniform, parallel walls preferred.
- Multiple-action tooling can be used to create complex parts.
- Complex profiles on one side only.
- Parts can be quenched, annealed and surface-treated like wrought products to alter mechanical properties.
- Inert plastics can be impregnated for pressure sealing or a low-melting-point metal for powder forging.
- Densities typically between 90% and 95% of original material.
- Density can be controlled for special functional properties, e.g. porosity for filters.
- Spheres approximated. Complicated radial contours possible.
- Avoid marked changes in section thickness.
- Narrow slots, splines, long thin section, knife-edges and sharp corners should be avoided. Use secondary processing operations.
- Threads not possible.
- Tapered, blind and non-circular holes, vertical knurls possible in direction of powder compaction.
- Grooves, cut-outs and off-axis holes perpendicular to the pressing direction cannot be produced directly.
- Undercuts perpendicular to compaction direction not possible. Better to secondary process.
- Radii should be as generous as possible.
- Chamfers preferred to radii on part edges.
- Maximum length to diameter ratio is 4:1.
- Maximum length to wall thickness ratio is 8:1.
- Inserts are possible at extra cost.
- Draft angles can be zero.
- Minimum section can be as low as 0.4 mm, but 1.5 mm typically.
- Sizes range from 10 g to 15 kg in weight or 4 mm^2 to 0.016 m^2 in projected area.

Quality Issues

- Density and strength variations in product can occur with asymmetric shapes. Can be minimised by die design.
- High densities required for subsequent welding of sintered parts.
- Porosity in sintered parts means excessive absorption of braze and solder fillers.
- Product strength determinable by powder size, compacting pressure, sintering time and temperature, but generally lower mechanical properties than wrought materials.
- Can give a highly porous structure, but can be controlled and used to advantage, e.g. filters and bearing lubricant impregnation (10–30% oil by volume). Also, resin impregnation can greatly improve machinability of sintered products.

- Generally, lower mechanical properties than wrought materials.
- Avoid sharp edges on tools. Causes excessive tool wear.
- Remnants of contaminants at grain boundaries may act as crack initiators.
- Oxide film may impair properties of finished part, for example chromium and high-temperature superalloys.
- Surface detail good.
- Surface roughness is in the range 0.2–3.2 μm Ra.
- Process capability charts showing the achievable dimensional tolerances are provided (Figure 4.11(b)).
- Repressing, coining and sizing improves surface finish, density and dimensional accuracy. Also for embossing.

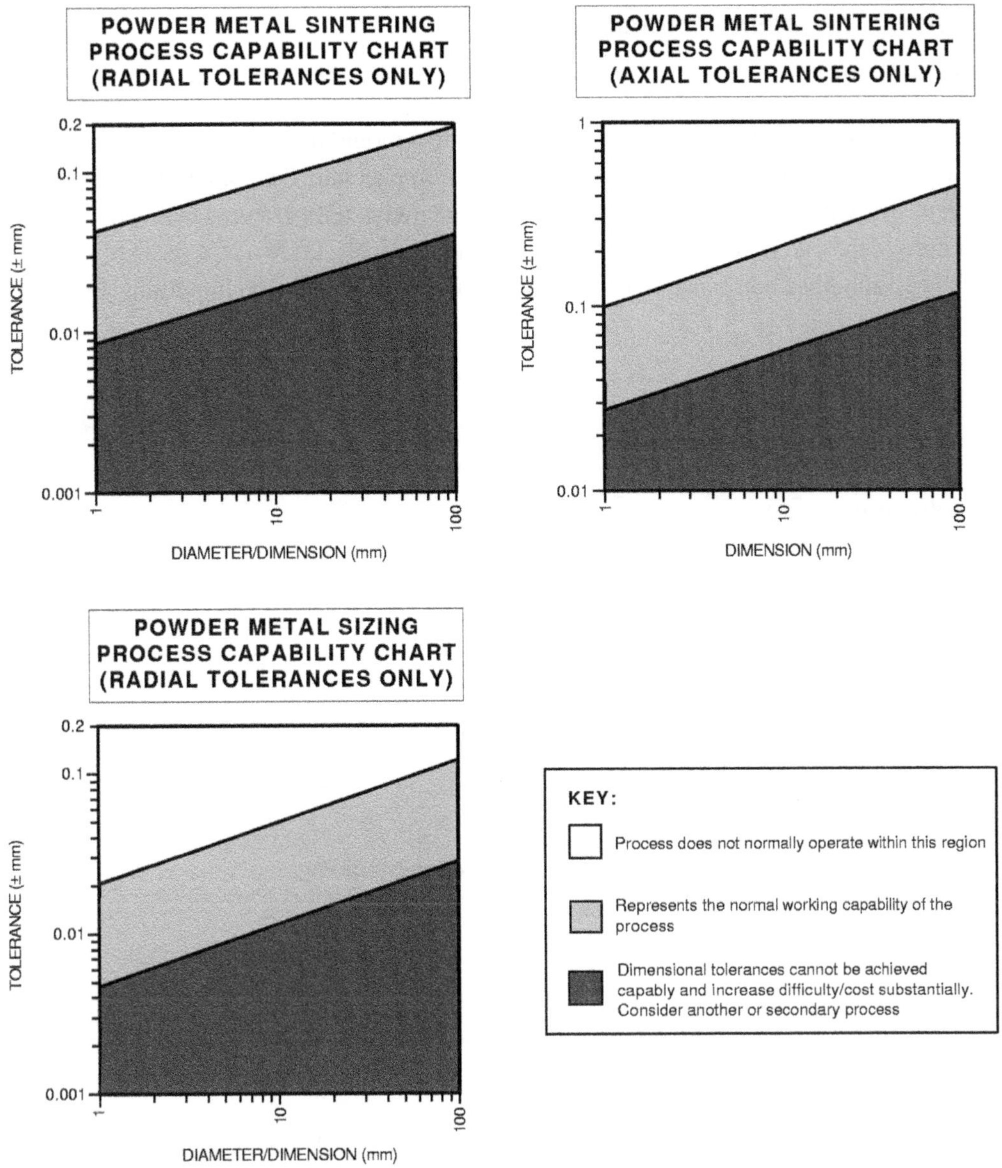

Figure 4.11(b): Powder Metallurgy Process Capability Charts.

4.12 Metal Injection Moulding

Process Description

Fine metal powder mixed with a binder is injected into a mould under high pressure, similar to plastic injection moulding, to create a brittle 'green' compact part. The binder is stripped from the 'green' part with solvents and/or heat and then sintered at temperatures below the melting temperature of the parent material to bond the powder particles, known as a 'brown' part. Process is essentially a combination of powder metallurgy and injection moulding (Figure 4.12).

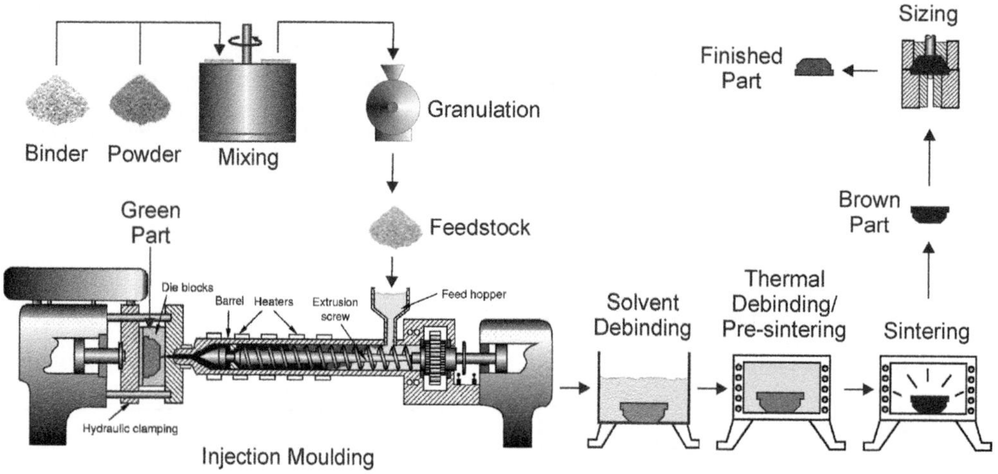

Figure 4.12: Metal Injection Moulding.

Materials

- Powder form (micron-sized) of plain and low alloy steels, stainless steels, high-speed steels, copper base alloys, nickel superalloys, titanium, ceramics, refractory and precious metals.
- Binder choices of natural waxes, thermoplastics (polyacetals) and water/agar.
- Powder production: gas and water atomisation, electrolysis and chemical reduction methods (see PRIMA 4.11).

Process Variations

- Also known as powder injection moulding.
- Choice of large continuous furnaces and smaller batch furnaces for sintering.
- Removal of binder (debinding) achieved through different mechanisms depending on binder material: catalytic debinding with nitric acid for thermoplastic binder; evaporation

for water/agar; water for water-soluble polymers; and thermal or solvent for wax-based binders.
- When required, secondary operations such as sizing and coining can be used to improve form and tolerances.

Economic Considerations

- Production rates 10–60/min, depending on part complexity.
- Lead times are 4–6 weeks typically, depending on mould complexity.
- Near-net or net shape process, therefore material utilisation is very high.
- Waste material from runners, sprues, etc. can be reused.
- Medium to high volumes are economical, 10,000+ per annum.
- Tooling costs are high.
- Equipment costs are high.
- Direct labour costs are low.
- Finishing costs are low.
- Tumbling, burnishing and polishing can be used to improve finish, particularly for the removal of gate marks.

Typical Applications

- Valves and pumps.
- Cogs and gears.
- Rotors.
- Watch and lock components.
- Surgical and dental instruments.
- Hydraulic fittings.
- Gas manifolds.
- Fuel nozzles.
- Power and hand tools.
- Firearm parts.
- Electronics enclosures.
- Connectors.
- Heat sinks.

Design Aspects

- Complex shapes with fine details, but limited to thin-walled parts with no cavities.
- Can incorporate knurls, studs, protrusions, logos and lettering in mould design.
- Machining should only be used to add precise holes and thread forms.

- Ribs and webs are an efficient way to increase your part strength.
- External undercuts/threads/projections possible; however, internal undercuts/threads are difficult and costly.
- Section changes from thin to thick should be gradual.
- Largest dimension <100 mm.
- Draft angle = 0.25–2°, but dependent on length of taper.
- Radii should be as large as possible, typically >0.4 mm.
- Maximum section = 12 mm.
- Minimum section = 0.2 mm (1 mm more practical).
- Smallest hole diameter = 0.1 mm (0.4 mm more practical).
- Geometric aspect ratio and hole length to diameter ratio <10:1.
- Size range = <1 g to 250 g.

Quality Issues

- Fine powders sinter more readily than coarse powders.
- Selection, preparation and mixing of binder material important for successful processing of components.
- Typically, component densities are 95–99% of parent material.
- Up to 75% of fatigue strength of the parent material can be achieved.
- Selection of parting line needs consideration to allow removal of the component from mould and to minimise failure due to fatigue.
- Care should be taken in transferring 'green' compact parts due to their brittle nature.
- Shrinkage of 20% in volume can occur after sintering stage of process.
- Stress concentrations can occur on internal and external radii of less than 1 mm.
- Sink marks occur at the junctions to thicker sections due to poor mould design.
- Removal of gate marks requires secondary processing, e.g. burnishing, tumbling.
- Surface/bulk treatments and coatings can be applied, e.g. black oxide, electroless nickel, chromating and PTFE coatings, to improve finish and properties.
- Surface roughness is in the range 0.4–1.6 μm Ra, dependent on material and powder particle size.
- Typical tolerances are ±0.3% to ±0.5% of the nominal dimension, but dependent on size, shape, material and mould configuration.
- Flatness and straightness geometrical tolerance approximately 0.1% of the nominal dimension.
- Angular tolerances >±0.5° should be specified.

4.13 Continuous Extrusion (Metals)

Process Description

A billet of the metal, either hot or cold, is placed into a chamber and forced through a die of the required profile with a ram to produce long sections (Figure 4.13(a)).

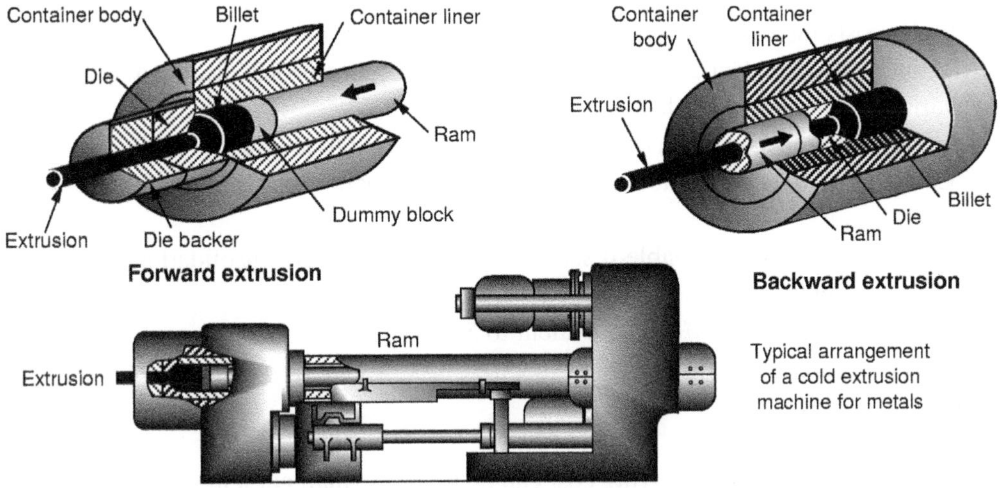

Figure 4.13(a): Continuous Extrusion (Metals).

Materials

Most ductile metals, for example aluminium, copper and magnesium alloys. To a lesser degree zinc; lead; tin; nickel and titanium alloys; refractory metals; and carbon, low alloy and stainless steels are processed.

Process Variations

- Forward extrusion: billet extruded by ram from behind.
- Backward extrusion: billet displaced by advancing ram with die attached to front and extrudate travels through the centre of the ram. Limited to short lengths only. Similar to Hooker process.
- Cold extrusion: increases friction and therefore processing energy/forces, but increases dimensional accuracy. May be performed warm. Viable for materials possessing adequate cold-working ability.

Economic Considerations

- Production rates are high but are dependent on size and complexity. Continuous lengths up to 12 m/min.
- Cut extruded length up to 1,000/h possible.
- Extruders are often run below their maximum speed for trouble-free production.
- Can have multiple holes in die for increased production rates and lower wear rates.
- Extruder costs increase steeply at the higher range of output.
- Lead times are dependent on the complexity of the two-dimensional die, but normally weeks.
- Material utilisation is high (less than 1.5% scrap).
- Waste is only produced when cutting continuous section to length.
- Process flexibility is moderate: tooling is dedicated, but changeover and set-up times are short.
- Short production runs are viable if section designed with part consolidation and integral fastening in mind.
- Minimum billet size is 250 kg (equivalent to 500 m).
- Tooling costs are moderate.
- Equipment costs are high.
- Direct labour costs are low.
- Finishing costs are low. Deburring cost can be high for small cut lengths.

Typical Applications

- Profiles cut to required length, e.g. gear blanks.
- Wrought bar and sections for other processing methods.
- Window frames.
- Structural sections, corner and edge members.
- Decorative trim.
- Railings.

Design Aspects

- Dedicated to long products with uniform, symmetrical or varying cross-section.
- Cross-section may be fairly complex (round, tube, square, T-section, L-section).
- Difficult to control internal dimensions of hollow sections that use complex mandrels or spiders held in the die.
- Can eliminate or minimise secondary processing operations. Section profile designed to decrease amount of machining required and/or increase assembly efficiency by integrating part consolidation features.
- Solid forms including re-entrant angles.

- Grooves and holes not parallel to the axis of extrusion must be produced by a secondary operation.
- Avoid too great a variation in adjacent section thicknesses (less than 2:1 if possible).
- Avoid very small holes through the profile.
- Use of materials other than aluminium and copper alloys can cause shape restrictions.
- Radii should be as generous as possible. Concave radii greater than 0.5 mm.
- No draft angle required.
- Maximum extrusion ratios: 40:1 for aluminium alloys, 5:1 for carbon steel.
- Section thickness should be greater than 1.5% of the maximum dimension.
- Maximum diameter = 250 mm.
- Minimum section = 1 mm for aluminium and magnesium alloys, 3 mm for steel.
- Sizes range from 8- to 500-mm sections, but dependent on complexity and material used.

Quality Issues

- Cold extrusion eliminates oxidation and gives a better surface finish.
- Avoid knife-edges and long, thin sections.
- Warp and twist of sections can occur.
- Plastic working in cold extrusion produces favourable grain structure and directional properties (anisotropy).
- The rate and uniformity of cooling are important for dimension control because of shrinkage and distortion in hot extrusion.
- Die swell, where the extruded product increases in size as it leaves the die, may be compensated for by:
  - Increasing haul-off rate compared with extrusion rate
  - Decreasing extrusion rate
  - Increasing the length of the die land
  - Decreasing the melt temperature
- High die wear reduced by selection of appropriate lubricant to reduce friction.
- Stretching post extrusion eliminates bowing and twisting.
- Surface detail is good to excellent.
- Surface roughness is in the range 0.4–12.5 μm Ra.
- Process capability charts showing the achievable dimensional tolerances are provided (Figure 4.13(b)).
- Wall thickness tolerances are between ±0.15 and ±0.25 mm.
- Straightness tolerance is 0.3 mm/m, typically.
- Twist tolerance up to 2°/m, not untypical.

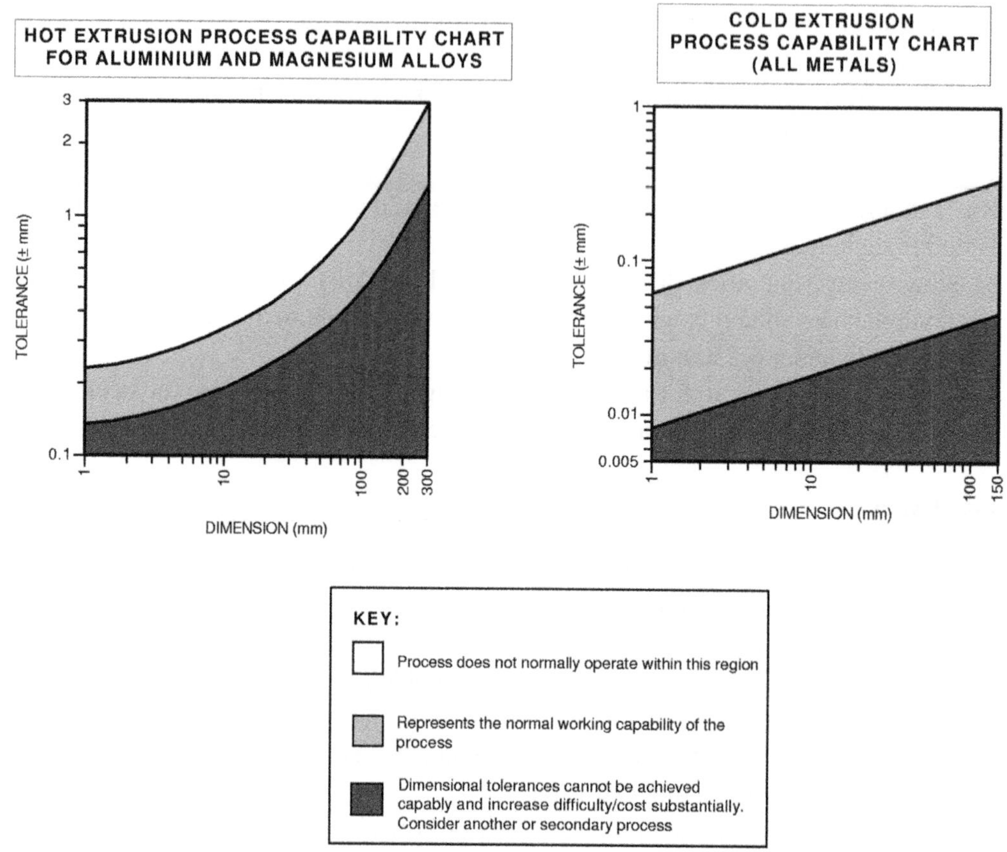

Figure 4.13(b): Continuous Extrusion Process Capability Charts.

CHAPTER 5

Plastics and Composites Processing

5.1 Injection Moulding

Process Description

Granules of polymer material are heated and then forced under pressure using a screw into the die cavity. On cooling, a rigid part or tree of parts is produced (Figure 5.1(a)).

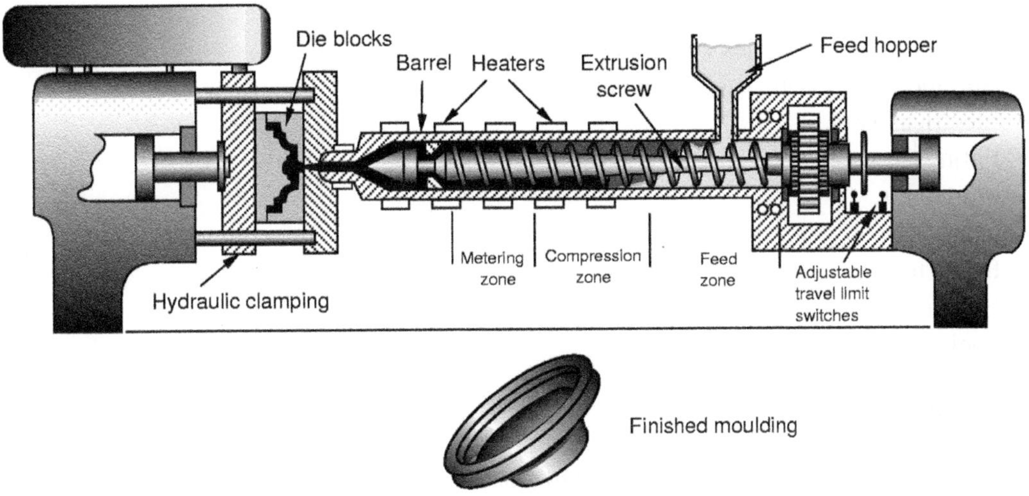

Figure 5.1(a): Injection Moulding.

Materials

Mostly thermoplastics, but thermosets, composites and elastomers can be processed.

Process Variations

- Injection blow moulding: allows small hollow parts with intricate neck detail to be produced.
- Co-injection: for products with rigid cores pre-placed in the die before injection or simultaneous injection of different materials into the same die.

Economic Considerations

- Production rates are high, 1–60/min, depending on size.
- Thermoset parts usually have a longer cycle time.

Manufacturing Process Selection Handbook. http://dx.doi.org/10.1016/B978-0-08-099360-7.00005-7

- Lead times can be many weeks due to manufacturing of complex dies.
- Material utilisation is good. Scrap generated in sprues and risers.
- If material permits, gates and runners can be reused resulting in lower material losses.
- Flexibility limited by dedicated dies, die changeover and machine set-up times.
- Economical only for high production runs, typically 20,000+.
- Full automation achievable. Robot machine loading and unloading common.
- Tooling costs are very high. Dies are usually made from hardened tool steel.
- Equipment costs are very high.
- Direct labour costs are low to moderate.
- Finishing costs are low. Trimming is required to remove gates and runners.

Typical Applications

- High-precision, complex components.
- Automotive and aerospace components.
- Electrical parts.
- Fittings.
- Containers.
- Cups.
- Bottle tops.
- Housings.
- Tool handles.

Design Aspects

- Very complex shapes and intricate detail possible.
- Holes, inserts, threads, lettering, colour, bosses and minor undercuts possible.
- Uniform section thickness should be maintained.
- Unsuitable for the production of narrow-necked containers
- Variation in thickness should not exceed 2:1.
- Marked section changes should be tapered sufficiently.
- Living hinges and snap features allow part consolidation.
- Placing of parting line important, i.e. avoid placement across critical dimensions.
- The clamping force required is proportional to the projected area of the moulded part.
- Radii should be as generous as possible. Minimum inside radii = 1.5 mm.
- Draft angle ranges from less than 0.25° to 4°, depending on section depth.
- Maximum section, typically = 13 mm.
- Minimum section = 0.4 mm for thermoplastics, 0.9 mm for thermosets.
- Sizes range from 10 g to 25 kg in weight for thermoplastics, 6 kg maximum for thermosets.

Quality Issues

- Thick sections can be problematic.
- Care must be taken in the design of the running and gating system, where multiple cavities are used to ensure complete die fill.
- Control of material and mould temperature is critical, also injection pressure and speed, condition of resin, dwell and cooling times.
- Adequate clamping force is necessary to prevent the mould creating flash.
- Thermoplastic moulded parts usually require no de-flashing: thermoset parts often require this operation.
- Excellent surface detail obtainable.
- Surface roughness is a function of the die condition. Typically, 0.2–0.8 μm Ra is obtainable.
- Process capability charts showing the achievable dimensional tolerances using various materials are provided (Figure 5.1(b)). Allowances of approximately ±0.1 mm should be added for dimensions across the parting line. Note, that charts 1–3 are to be used for components that have a major dimension greater than 50 mm and typically large production volumes. The chart titled 'Light Engineering' is used for components with a major dimension less than 150 mm and for small production volumes.

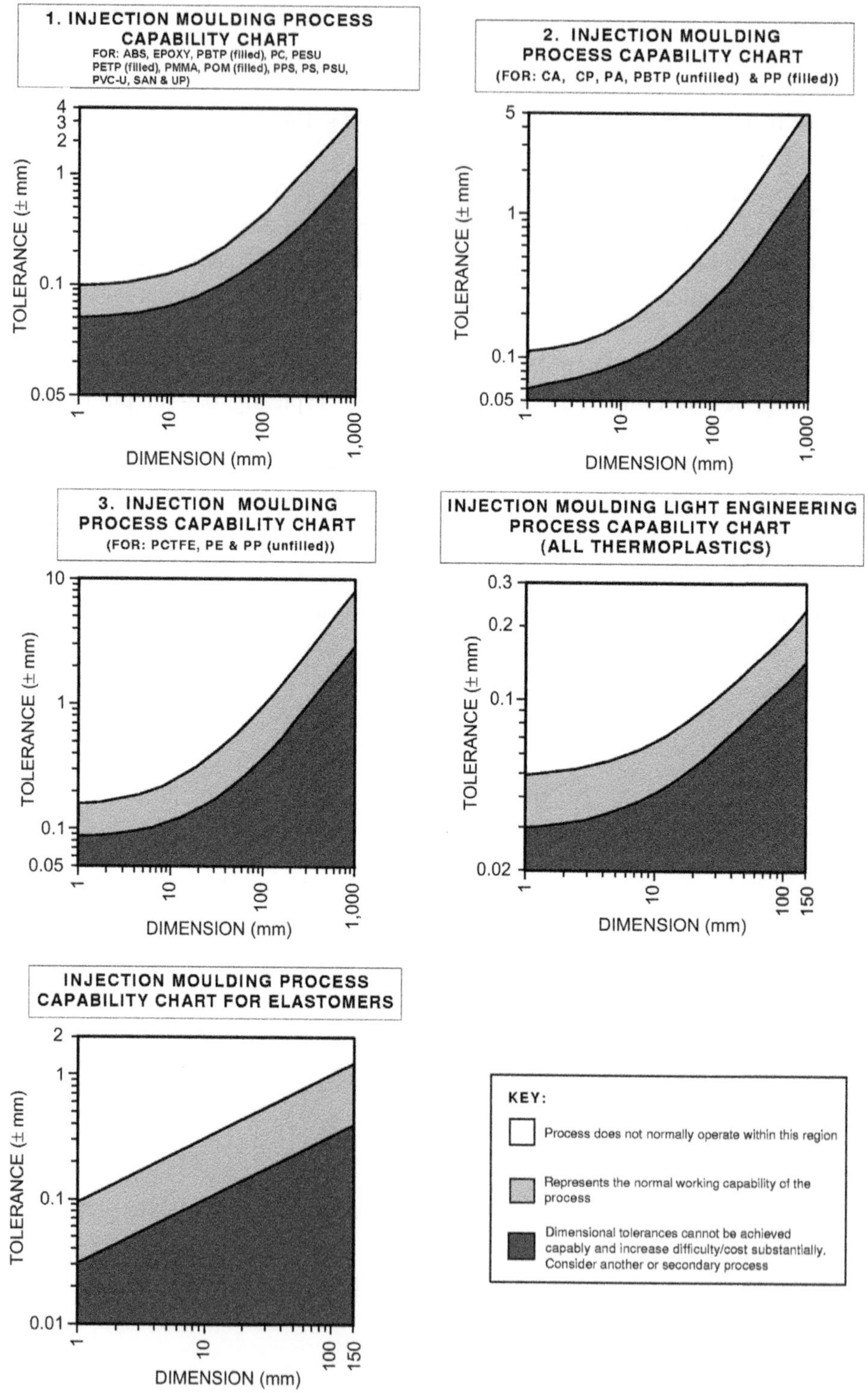

Figure 5.1(b): Injection Moulding Process Capability Charts.

5.2 Reaction Injection Moulding

Process Description

Two components of a thermosetting resin are injected into a mixing chamber and then injected into the mould at high speed, where polymerisation and subsequent solidification takes place (Figure 5.2).

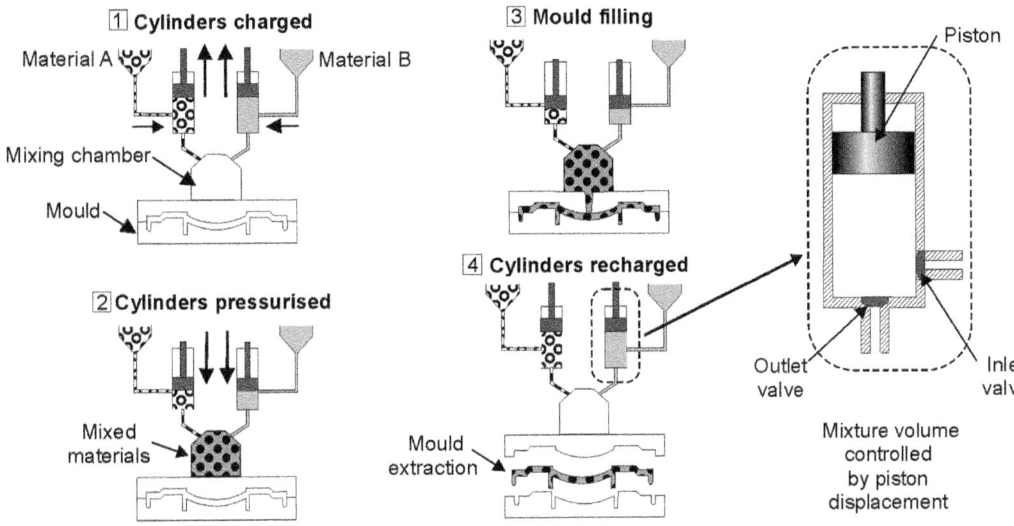

Figure 5.2: Reaction Injection Moulding.

Materials

- Mostly thermosets.
- Foamed materials possessing a solid skin can be created by setting up a pressure differential between mixing chamber and mould.
- Can add chopped fibre material (glass, carbon) for added stiffness to mixing, to produce composites.

Process Variations

- Mould material is usually aluminium. Can also use resin for low production runs or hardened tool steel for very high volumes.
- Further heating of resin components before mixing is dependent on material used.

Economic Considerations

- Production rates are from 1 to 10/h.
- Lead times can be several weeks.
- Material utilisation is good. Less than 1% lost in scrap.
- Scrap cannot be recycled.
- Flexibility limited by dedicated dies, die changeover and machine set-up times.
- Economical for low to medium production volumes (10–10,000).
- Can be used for one-offs, e.g. prototyping.
- Tooling costs are low.
- Equipment costs are high.
- Direct labour costs are moderate to high.
- Finishing costs are low. A little trimming required.

Typical Applications

- Car bumpers.
- Cups.
- Containers.
- Panels.
- Housings.
- Footware.
- Garden furniture.

Design Aspects

- Very complex shapes and intricate detail possible.
- Ribs, holes, bosses and inserts possible.
- Small re-entrant features possible.
- Radii should be as generous as possible.
- Uniform section thickness should be maintained.
- Marked section changes should be tapered sufficiently.
- Placing of parting line important, i.e. avoid placement across critical dimensions.
- Draft angles range from 0.5° to 3°, depending on section depth.
- Maximum section = 10 mm.
- Minimum section = 1.5 mm; foamed material = 3 mm.
- Maximum dimension = 1.5 m.
- Sizes range from 100 g to 10 kg in weight.

Quality Issues

- Thick sections can be problematic.
- Care must be taken in the design of the running and gating system, where multiple cavities are used to ensure complete die fill.
- Problems can be created by premature reaction before complete filling of mould.
- Excellent surface detail obtainable.
- Surface roughness is variable, but mainly dependent on mould finish.
- Achievable dimensional tolerances are approximately ±0.05 up to 25 mm, ±0.3 up to 150 mm. Allowances of approximately ±0.2 mm should be added for dimensions across the parting line.

5.3 Compression Moulding

Process Description

A measured quantity of raw, unpolymerised plastic material is introduced into a heated mould, which is subsequently closed under pressure, forcing the material into all areas of the cavity as it melts (Figure 5.3(a)). Analogous to closed die forging of metals.

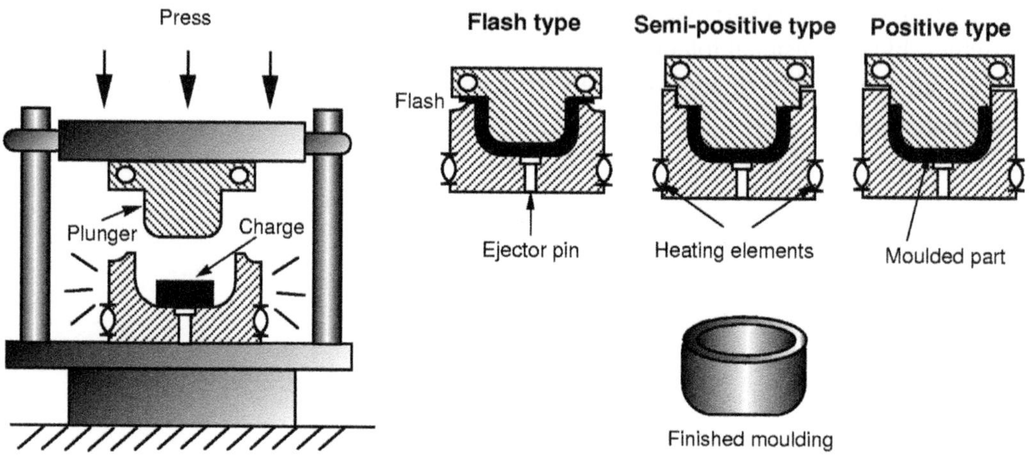

Figure 5.3(a): Compression Moulding.

Materials

- Mainly thermosets, but also some composites, elastomers and a limited number of thermoplastics.
- Raw material supplied in either powder or liquid resin form.

Process Variations

- Flash type: for shallow parts, but more material lost.
- Semi-positive (partly positive, partly flash): used for closer tolerance work or when the design involves marked changes in section thickness.
- Positive: high-density parts involving composite Sheet Moulding Compounds (SMC), Bulk Moulding Compounds (BMC) or impact-thermosetting materials.
- Cold moulding: powder or filler is mixed with a binder, compressed in a cold die and cured in an oven. Strictly for thermosets.

Economic Considerations

- Production rates from 20 to 120/h depending on size and compounds used.
- Cycle time is restricted by material handling. Each cavity must be loaded individually.

- The greater the thickness of the part, the longer the curing time.
- Multiple cavity mould increases production rate.
- Mould maintenance is minimal.
- Certain amount of automation possible.
- Time required for polymerisation (curing) depends mainly on the largest cross-section of the product and the type of moulding compound.
- Lead times may be several weeks according to die complexity.
- Material utilisation is high. No sprues or runners.
- Flexibility is low. Difference in shrinkage properties reduces the capability to change from one material to another.
- Production volumes are typically 10,000+, but can be lower than 1,000 per annum depending on size.
- Tooling costs are moderate to high.
- Equipment costs are moderate.
- Direct labour costs are low to moderate.
- Finishing costs are generally low. Flash removal required.

Typical Applications

- Dishes.
- Housings.
- Automotive parts.
- Panels.
- Handles.
- Container caps.
- Electrical components and fittings.

Design Aspects

- Shape complexity is limited to relatively simple forms. Moulding in one plane only.
- Threads, ribs, inserts, lettering, holes and bosses all possible.
- When moulding materials with reinforcing fibres, directionality is maintained enabling high strength to be achieved.
- Thin-walled parts with minimum warping and dimensional deviation may be moulded.
- Placing of parting line important, i.e. avoid placement across critical dimensions.
- A draft angle of greater than 1° required.
- Maximum section typically = 13 mm.
- Minimum section = 0.8 mm.
- Maximum dimension typically = 450 mm.
- Minimum area = 3 mm^2.
- Maximum area = 1.5 m^2.
- Sizes range from several grammes to 16 kg in weight.

Quality Issues

- Variation in raw material charge weight results in variation of part thickness and scrap.
- Air entrapment possible.
- Internal stresses are minimal.
- Dimensions in the direction of the mould opening and the product density will tend to vary more than those perpendicular to the mould opening.
- Flash moulds do not require that the quantity of material is controlled.
- Tumbling may be required as a finishing process to remove flash.
- Surface detail is good.
- Surface roughness is a function of the die condition. Typically, 0.8 μm Ra is obtained.
- Process capability charts showing the achievable dimensional tolerances using various materials are provided (Figure 5.3(b)). Allowances of approximately ±0.1 mm should be added for dimensions across the parting line.

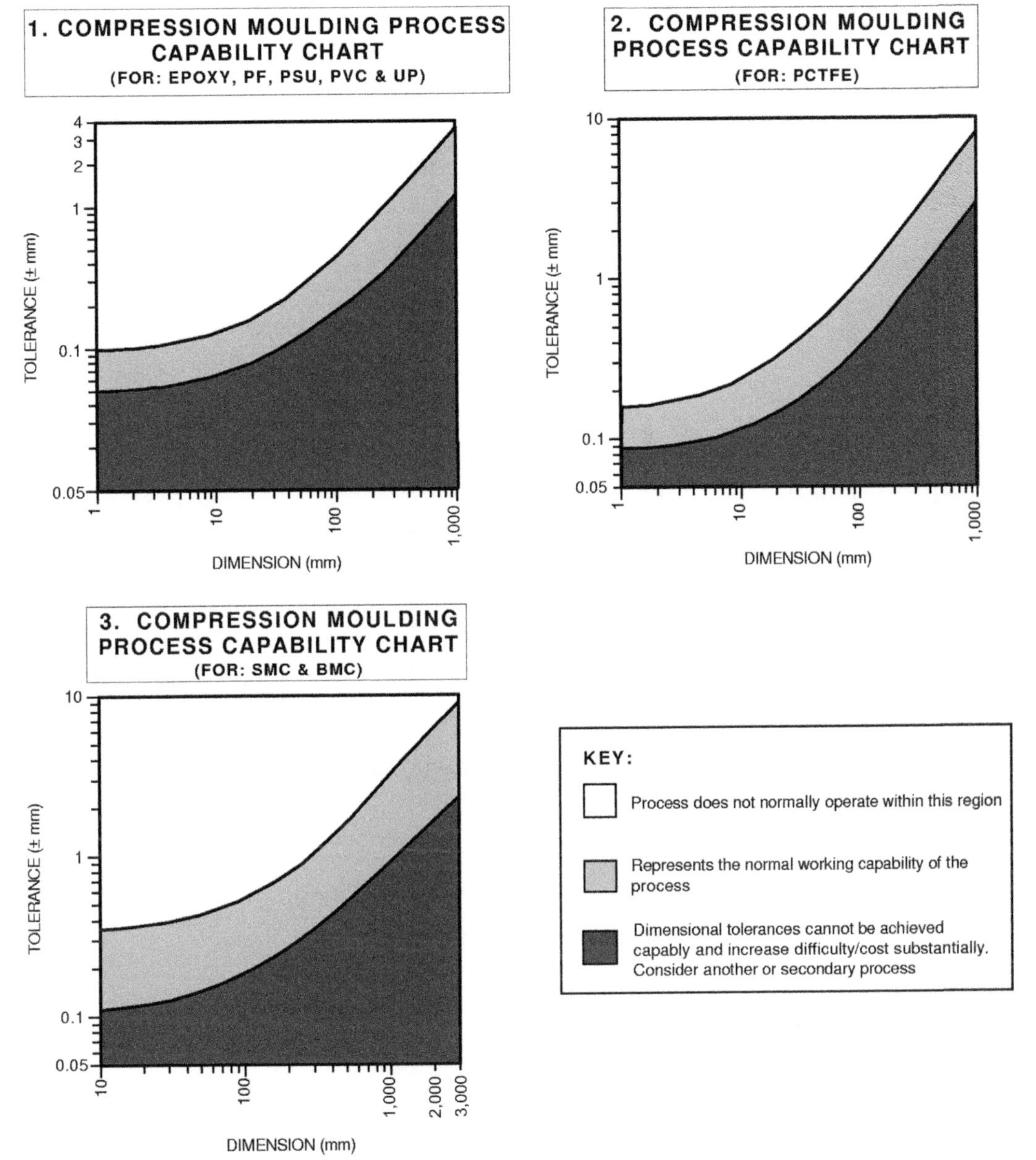

Figure 5.3(b): Compression Moulding Process Capability Charts.

5.4 Resin Transfer Moulding

Process Description

A heated mould is closed under low pressure and then a liquid resin and catalyst is loaded into an adjacent mixing head and forced via a plunger into the cavity, where curing takes place (Figure 5.4).

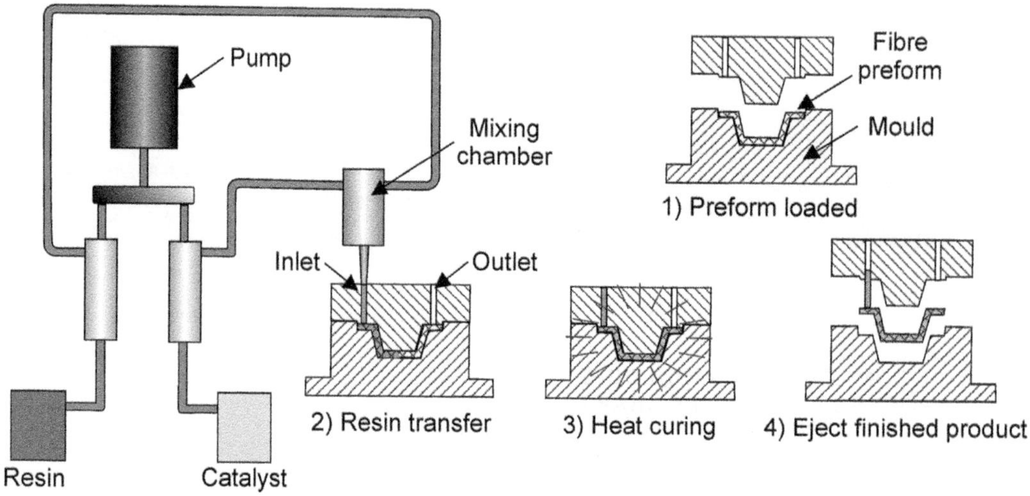

Figure 5.4: Resin Transfer Moulding.

Materials

- Limited to only several thermosetting plastics and elastomers, with or without fillers.
- Can use pre-pressed fibre packs to fit the mould, called preforms.
- Fibres can be glass or carbon.

Process Variations

- Powder material placed in a heated melting pot and forced under pressure into a heated mould.
- Vacuum-assisted resin injection: additional vacuum can be used in mould cavity to assist resin filling of fibre preforms.

Economic Considerations

- Production rates 2–300/h. Fast curing speed.
- Lead time typically days depending on complexity of tool.

- Material utilisation very good. Less than 3% scrap, typically.
- Scrap material cannot be recycled directly.
- High degree of automation possible.
- Economical for production runs of 1,000–10,000.
- Can be as low as 200 depending on size and complexity.
- Tooling costs are moderate to high.
- Equipment costs are generally moderate.
- Direct labour costs are low to moderate.
- Some skilled labour required, but easily reduced with automation.
- Finishing costs are low, but no opportunity for in-mould trimming.

Typical Applications

- Electrical cabinets.
- Housings and panels.
- Car body panels.
- Wind turbine blades.
- Seating.
- Yacht hulls and decks.
- Plant growing trays.
- Garden ponds.

Design Aspects

- Complex geometries possible and hollow shapes.
- Cores possible for increased complexity.
- Can mould around inserts and delicate cores easily.
- Lettering, ribs, holes, inserts and threads possible.
- Undercuts possible, but at added cost.
- Thickness variation should be less than 2:1.
- Draft angles from 2° to 3° preferred, but can be as low as 0.5°.
- Minimum inside radius = 6 mm.
- Minimum section ranges from 0.8 to 1.5 mm, depending on material used.
- Maximum section = 90 mm.
- Maximum dimension = 450 mm.
- Minimum area = 3 mm^2.
- Maximum size 16 kg in weight, but suited to smaller parts.

Quality Issues

- Differential stress distribution may occur due to flow characteristics of mould, resulting in minor distortion.
- High temperatures above resin-melting temperatures must be maintained prior to and during mould filling.
- Improperly placed fibre preforms can cause dry spots or pools of resin on surface of finished part.
- Fibre preforms can also move during injection mould filling without proper fixing arrangements within mould.
- Variation in resin/fibre concentration is difficult to control in sharp corners.
- Not recommended for parts subjected to high loads in service.
- Surface detail is excellent.
- Surface roughness is a function of the die condition, with 0.8 μm Ra readily obtainable.
- Achievable dimensional tolerances are ±0.05 up to 25 mm, ±0.15 up to 150 mm. Wall thickness tolerances are typically ±0.25 mm.

5.5 Vacuum Forming

Process Description

A plastic sheet is softened by heating elements and pulled under vacuum on to the surface form of a cold mould and allowed to cool. The part is then removed (Figure 5.5).

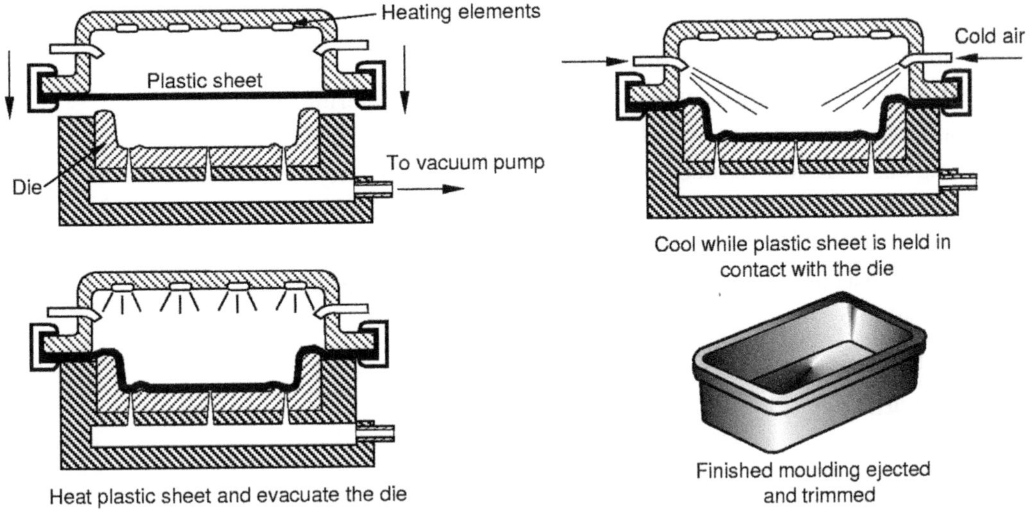

Figure 5.5: Vacuum Forming.

Materials

- Several thermoplastics that can be produced in sheet form.
- The material to be processed should exhibit high uniform elongation.
- Can also introduce some fibre-reinforcing material to improve strength and rigidity.

Process Variations

- Moulds are usually made of cast aluminium or aluminium-filled epoxy.
- Sheets can be heated by infrared heaters or in ovens.
- Can have top and bottom heating elements, or top heating element only.
- For thick sheets, a top enclosure and compressed air are used.
- Sheet is drawn over mould with additional force, other than that provided by the vacuum, until cooled.
- Thermoforming: for thin-walled parts such as packaging and cups in large volumes, although greater sheet thicknesses are possible for prototypes and shorter runs.

Economic Considerations

- Production rates from 60 to 360/h, commonly. Cups can be produced at 3600/h.
- Lead times of a few days, typically.
- Material utilisation is moderate to low. Unformed parts of the sheet are lost and cannot be directly recycled.
- Full automation achievable.
- Multiple moulds may be used.
- Set-up times and changeover times are low.
- Sheet material much more expensive than raw pellet material.
- Production volumes economical in small batches of 10–1,000.
- Tooling costs are low to moderate, depending on complexity.
- Equipment costs are low to moderate, but can be high if automated.
- Labour costs are low to moderate.
- Finishing costs are low. Some trimming of unformed material after moulding.

Typical Applications

- Open plastic containers and panels.
- Pages of Braille text.
- Food packaging and containers.
- Automotive interior parts.
- Electrical cabinets and enclosures.
- Bath tubs, sink units and shower panels.
- Dinghy hulls.
- Signs.

Design Aspects

- Shape complexity limited to mouldings in one plane.
- Open forms of constant thickness.
- Undercuts possible with a split mould.
- Cannot produce parts with large surface areas.
- Bosses, ribs and lettering possible, but at large added cost.
- Parts with moulded-in holes not possible.
- Corner radii should be large compared to thickness of material.
- Sharp corners should be avoided.
- No parting lines.
- Draft angles of 1° or greater recommended.

- Maximum section = 3 mm.
- Minimum section = 0.05–0.5 mm, depending on material used.
- Sizes range from 25 mm^2 to 7.5 m × 2.5 m in area.

Quality Issues

- Control of temperature, clamping force and vacuum pressure important if variability is to be minimised.
- Thermoplastic material must possess a high uniform elongation otherwise tearing at critical points in the mould may occur.
- Sheet material will have a plastic memory and so at high temperatures the formed part will revert back to original sheet profile. Operating temperature therefore important.
- Uniform temperature control of sheet important.
- If multiple moulds are used it is necessary that there is sufficient distance between cavities to avoid flow interference.
- Excessive thinning can occur, particularly at sharp corners.
- Surface detail fair.
- Surface finish is good and is related to the condition of mould surface.
- Achievable tolerances range from ±0.25 to ±2 mm, and are largely mould dependent. Wall thickness tolerances are typically ±20% of the nominal.

5.6 Blow Moulding

Process Description

A hot hollow tube of plastic, called a parison, is extruded or injection-moulded downwards and then caught between two halves of a shaped mould that closes the top and bottom of the parison. Hot air is blown into the parison, expanding it until it uniformly contacts the inside contours of the cold mould. The part is allowed to cool and is then ejected (Figure 5.6(a)).

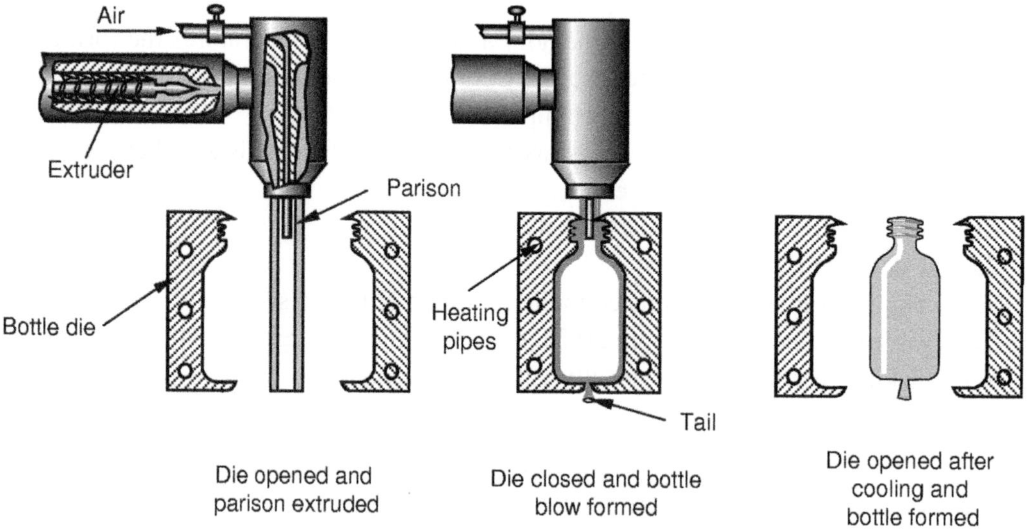

Figure 5.6(a): Blow Moulding.

Materials

Most thermoplastics.

Process Variations

- Extrusion blow moulding: more applicable to asymmetrical parts, integrated handles possible.
- Injection blow moulding: parison is injection-moulded (see PRIMA 5.1) and then transferred to blow moulding machine. For small parts with intricate neck detail.
- Multiple parisons: can create multilayered parts. This requires close control since uneven parisons produce waste.
- Parisonless blowing: similar to dip coating followed by expansion into the mould.
- Stretch blow moulding: the simultaneous axial and radial expansion of a parison, yielding a biaxially orientated container.

Economic Considerations

- Production rates between 100 and 2500/h, depending on size.
- Lead times are a few days.
- Integration with extrusion process to produce parison provides continuous operation.
- There is generally little material waste, but can increase with some complex geometries using extrusion blow moulding.
- Full automation readily achievable.
- Flexibility is limited since moulds are dedicated.
- Set-up times and changeover times are relatively short.
- Production volumes of 1,000, but better suited to very high volumes.
- Tooling costs are moderate to high.
- Equipment costs are moderate to high, especially for full automation.
- Direct labour costs are low. One operator can manage several machines.
- Finishing costs are low. Some trimming required.

Typical Applications

- Hollow plastic parts with relatively thin walls.
- Bottles.
- Bumpers.
- Ducting.

Design Aspects

- Complexity limited to hollow, well-rounded, thin-walled parts with low degree of asymmetry.
- Asymmetrical mouldings, e.g. offset necks, are possible with movable blowing spigots.
- Undercuts, bosses, ribs, lettering, inserts and threads possible.
- Corner radii should be as generous as possible (>3 mm).
- Placing of parting line important, i.e. avoid placement across critical dimensions.
- Holes cannot be moulded in.
- Draft angles are not required.
- Maximum section = 6 mm. Thick sections may need cooling aids (carbon dioxide or nitrogen gas).
- Minimum section = 0.25 mm.
- Sizes range from 12 mm in length to volumes up to 3 m^3.

Quality Issues

- Poor control of wall thickness, typically ±50% of nominal.
- Creep and chemical stability of product are important considerations.
- Residual stresses, e.g. non-uniform deformation, may relax in time, causing distortion of the part.
- Good surface detail and finish possible.
- The higher the pressure, the better the surface finish of the product.
- A process capability chart showing the achievable dimensional tolerances is provided (Figure 5.6(b)). Allowances of approximately ±0.1 mm should be added for dimensions across the parting line.

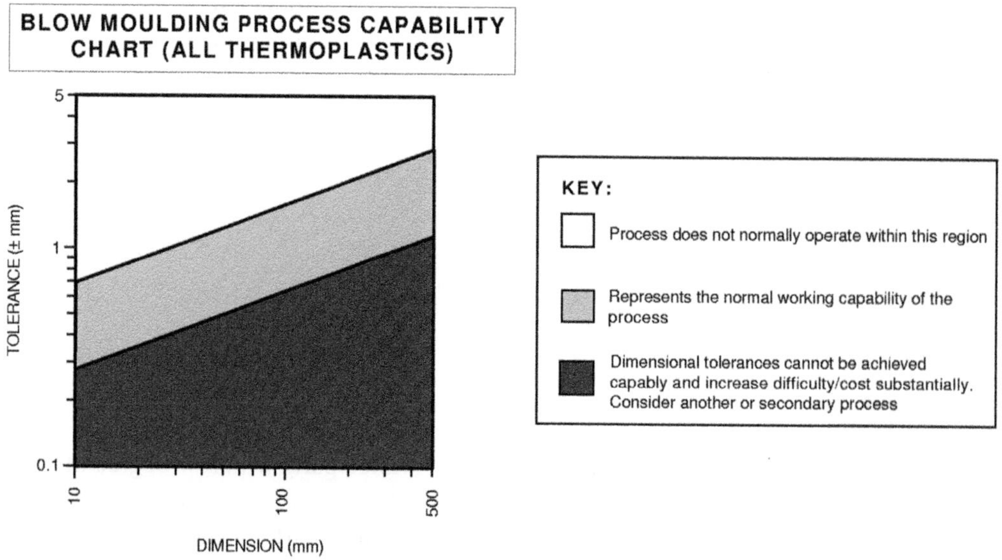

Figure 5.6(b): Blow Moulding Process Capability Chart.

5.7 Rotational Moulding

Process Description

Raw material is placed in the mould and simultaneously heated and rotated, forcing the particles to deform and melt on the walls of a female mould without the application of external pressure or centrifugal forces. The part is cooled whilst rotating. The mould is designed to rotate about two perpendicular axes (Figure 5.7(a)).

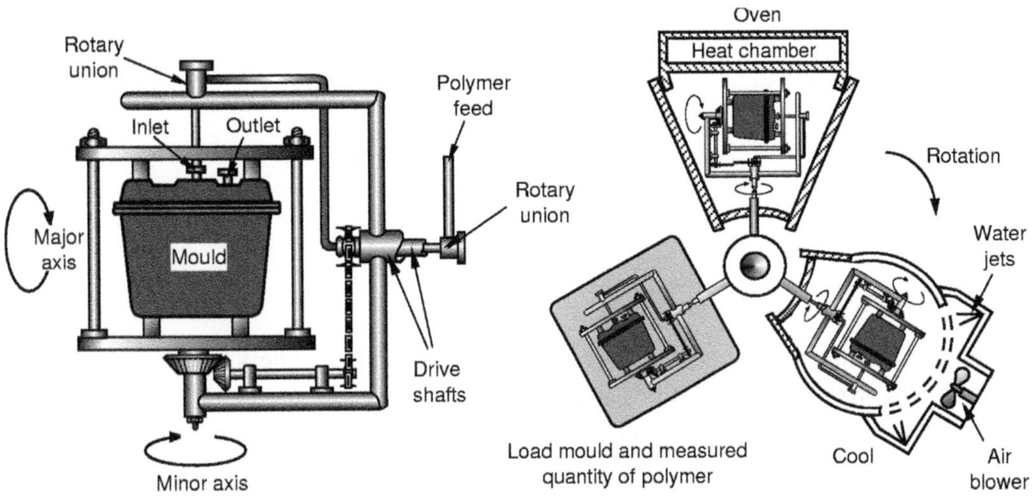

Figure 5.7(a): Rotational Moulding.

Materials

- Several common thermoplastics, including difficult fluoropolymers.
- Raw material supplied as finely ground powder.

Process Variations

- Slush moulding: uses liquid polymers (plastisols) for small hollow parts.
- Air, water or mist can be used for mould cooling.

Economic Considerations

- Production rates of 3–50/h, but dependent on size.
- To increase production rates, three-arm carousels are often used with one mould each in the load–unload, heat and cool positions.
- Lead time several days.
- Material utilisation very high. Little waste material.

- Production volumes are typically in the range of 100–1,000.
- Tooling costs are low.
- Equipment costs are low to moderate.
- Labour costs are moderate.
- Finishing costs are low. Little finishing required.

Typical Applications

- Water tanks.
- Storage vessels.
- Dustbins.
- Buckets.
- Housings.
- Drums.
- Prototypes.

Design Aspects

- Complexity limited to large, hollow parts of uniform wall thickness.
- Long, thin projections are not possible.
- Large flat surfaces should be avoided due to distortion and difficulty to form. Use stiffening ribs.
- Internal walls need to be well spaced.
- Moulded-in holes, bosses, finishes and lettering are all possible at added cost and limited accuracy.
- With rotation speed variation, can build up thicker layers at key points in the mould.
- Integral handles possible.
- Large threads can be moulded in.
- Undercuts should be avoided.
- Sharp corners difficult to fill in the mould. Radii should be as generous as possible (greater than five times the wall thickness) and tend to become thicker than the wall thickness on moulding.
- Metal or higher-melting-point plastic inserts can be moulded in.
- Can clad the inside of the finished part using another polymer.
- Placing of parting line important, i.e. avoid placement across critical dimensions.
- Holes cannot be moulded, although open-ended articles are possible.
- Thickness variation should be less than 2:1.
- Draft angles are generally greater than 1°, typically 3°.
- Maximum section = 13 mm.
- Minimum section is typically 2 mm, but can be as low as 0.5 mm for certain applications.
- Sizes up to 4 m^3.

Quality Issues

- The part is practically free from residual stresses.
- Surface detail is fair.
- Outer surface finish of the part is a replica of the inside finish of the mould walls.
- Control of inside surface finish is not possible.
- Wall thickness is determined by the close control of the amount of raw material used.
- Dimensional variations can be large if sufficient setting time is not allowed before removal of the part.
- A process capability chart showing the achievable dimensional tolerances is provided (Figure 5.7(b)). Allowances of approximately ±0.5 mm should be added for dimensions across the parting line.
- Wall thickness tolerances are generally between ±5% and ±20% of the nominal.

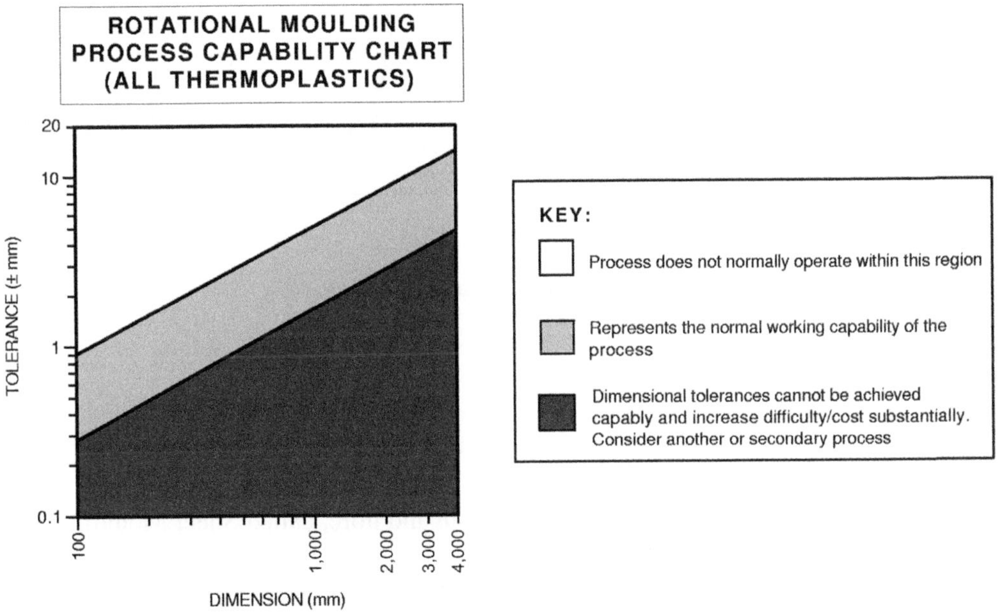

Figure 5.7(b): Rotational Moulding Process Capability Chart.

5.8 Contact Moulding

Process Description

Glass fibre-reinforced material (30–45% by volume) and a liquid thermosetting resin are simultaneously applied to a male or female mould using a hand lay-up or a spray lay-up process. The resulting part is subsequently cured at room temperature or with the application of heat to accelerate the process (Figure 5.8(a)).

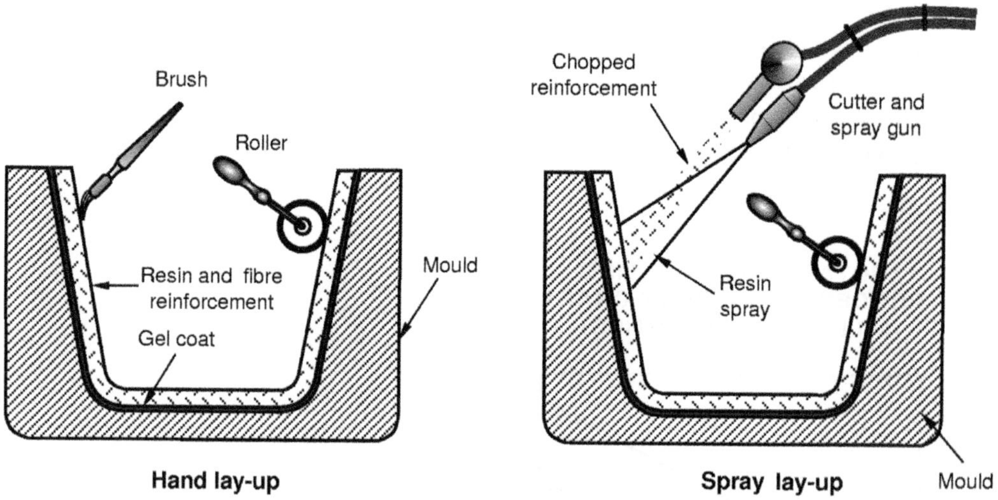

Figure 5.8(a): Contact Moulding.

Materials

- Glass-reinforced fibre in woven, continuous and chopped roving, mat and cloth forms.
- Can use pre-impregnated sheets of uncured resin and fibre, called Sheet Moulding Compound (SMC).
- Thermosetting liquid resin: commonly catalysed polyester or epoxy.

Process Variations

- Hand lay-up: manual laying of fibre-reinforced material and application of resin to mould to build up the thickness. Hand or roller pressure removes any trapped air. Variations on hand lay-up are:
  - Vacuum bag moulding: uses a rubber bag clamped over the mould. A vacuum is applied between the mould and the bag to squeeze the resin/reinforcement together, removing any trapped air. Curing performed in an oven.

  - Pressure bag moulding: as vacuum bag moulding, but pressure is applied above the bag. Can be used for thicker section parts.
  - Hand lay-up using SMC: cured by heat and clamped if necessary to further reduce air pockets.
- Spray lay-up: use of an air spray gun incorporating a cutter that chops continuous rovings to a controlled length before being blown into the mould simultaneously with the resin.
- Moulds can be made of wood, plaster, concrete, metal or glass fibre-reinforced plastic.
- Cutting of composites can be performed using knives, disc cutters, lasers and water jets.

Economic Considerations

- Production rates are low at 100–500 per annum using hand lay-up. Faster using spray lay-up method.
- Long curing cycle typically.
- Production rates increased using SMC materials.
- Lead times usually short, depending on size and material used for the mould.
- Mould life is approximately 1,000 parts.
- Multiple moulds incorporating heating elements should be used for higher production rates.
- Material utilisation is moderate. Scrap material cannot be recycled.
- Limited amount of automation possible.
- Economical for low production runs, 10–1000. Can be used for one-offs, mock-ups, etc.
- Tooling costs are low to medium.
- Equipment costs generally low to medium.
- Direct labour costs are high. Can be very labour intensive, but not skilled.
- Finishing costs are moderate. Some part trim is required.

Typical Applications

- Hulls for boats and dinghies.
- Large containers.
- Swimming pools and garden pond mouldings.
- Bath tubs.
- Small cabins and buildings.
- Machine covers.
- Car body panels.
- Sports equipment.
- Wind turbine blades.
- Prototypes and mock-ups.
- Architectural work.

Design Aspects

- High degree of shape complexity possible, limited only by ability to produce mould.
- Produces only one finished surface.
- Fibres should be placed in the expected direction of loading, if any. Random layering gives less strength.
- Avoid compressive stresses and buckling loads.
- Used for parts with a high surface area to thickness ratio.
- Moulded-in inserts, ribs, holes, lettering and bosses are possible.
- Draft angles are not required.
- Undercuts are possible with flexible moulds.
- Minimum inside radius = 6 mm.
- Minimum section = 1.5 mm.
- Maximum economic section = 30 mm, but can be unlimited.
- Sizes range from 0.01 to 500 m^2 in area.
- Maximum size depends on ability to produce the mould and the transport difficulties of finished part.

Quality Issues

- Air entrapment and gas evolution can create a weak matrix and low-strength parts.
- Non-reinforcing gel coat helps to create smoother mould surface and protects the moulding from moisture.
- Resin and catalyst should be accurately metered and thoroughly mixed for correct cure times.
- Excessive thickness variation can be eliminated by sufficient clamping and adequate lay-up procedures.
- Toxicity and flammability of resin is an important safety issue, especially because of high degree of manual handling and application.
- Surface roughness and surface detail can be good on moulded surface, but very poor on opposite surface.
- Shrinkage increases with higher resin volume fraction.
- A process capability chart showing the achievable dimensional tolerances for hand/spray lay-up is provided (Figure 5.8(b)). Wall thickness tolerances are typically ±0.5 mm.

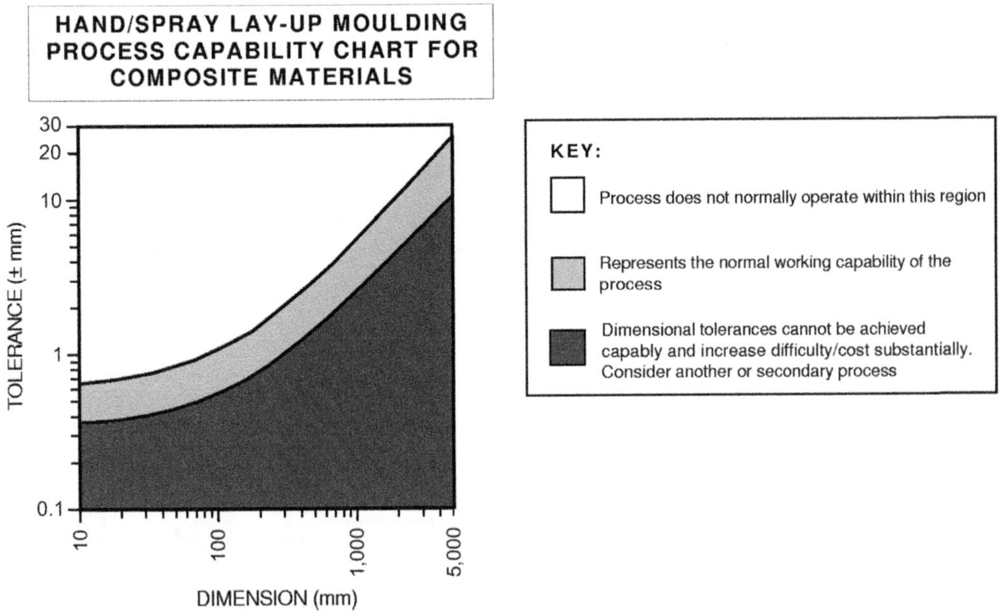

Figure 5.8(b): Contact Moulding Process Capability Chart.

5.9 Pultrusion

Process Description

Continuous process where long strands of reinforcing fibre mats and rovings supplied on creels are impregnated with a bonding resin by pulling them through a bath, and then through a heated die (150–250°C) of the desired cross-sectional profile, where the section subsequently cures, is air-cooled and is cut to length (Figure 5.9).

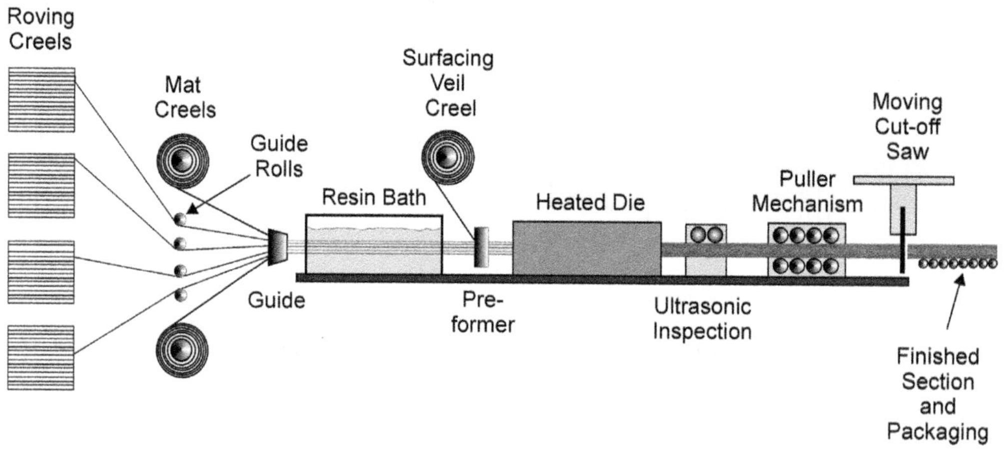

Figure 5.9: Pultrusion.

Materials

- Bonding matrix is typically a polyester resin, with epoxy, vinyl ester, acrylic and phenolic resins also used.
- Commonly glass fibre, but also carbon and aramid fibres to a lesser extent in roving, woven mat or cloth veil form.

Process Variations

- Hollow sections require a cantilevered mandrel at the entrance of the heated die, which the impregnated fibre mat/roving passes around.
- Different mechanisms for pulling are used, e.g. caterpillar or reciprocating puller.
- Pre-preg fibre system: uses resin bath as the mat/roving is already supplied impregnated with resin. Requires an additional softening pre-heater before the main heated die.
- Reaction injection moulded pultrusion: the two components of a thermoset resin are injected into the heating die, rather than using a bath (see PRIMA 5.2).
- Pulwinding/pulbraiding: additional fibre filaments are wound on to the core of the pultrusion before it enters the heated die.

- Pulforming: partly cured pultruded component shaped by split dies for varying cross-section components.
- Filament winding: typically resin-coated glass fibre is wound under tension on to a mandrel to create cylindrical composite structures.

Economic Considerations

- Production speeds medium to high (typical speeds 0.5–3 m/min), but dependent on section thickness.
- Lead times are relatively short.
- Material utilisation is very high.
- Production volumes are high. Minimum order quantity = 3000 m.
- Tooling costs are low to moderate.
- Equipment costs are low to moderate.
- Direct labour costs are low to moderate. Some aspects of process are highly automated.
- Finishing costs are low.
- The pultruded section is cut to length by an automatic flying cut-off saw, and cleaned using solvents, sanding or sand blasting in some cases.

Typical Applications

- Tubes, rods, and structural sections (beam, column, angle, channel).
- Flooring, piping, bridge, walkway and platform support sections.
- Cable trays and gutters.
- Ladders and hand rails.
- Sign, lamp and fence posts.
- Window and door frames, panel supports.
- Fishing, kite, tent and ski poles.
- Shafts for golf clubs, umbrellas, paddles and hockey sticks.
- Antennas and flag poles.
- Tool handles.
- High-performance springs.
- Drive shafts.
- Car bumpers.

Design Aspects

- Typically for production of long parts of constant cross-section or profile, hollow or solid, and ideally of uniform wall thickness.
- Wire inserts, undercuts and ribs are possible.

- High degree of functionality: strong, lightweight, corrosion resistant, electrically and thermal insulating.
- Conventional machining processes, e.g. drilling, milling and tapping, need to be used for features such as holes, threads and cut-outs.
- Strength of structural profile can be optimised by choice and direction of fibre. Mat gives a degree of transverse strength; roving provides the high longitudinal strength.
- Corner radii should be as large as possible to allow material flow and improved strength distribution.
- Structures requiring complex loading cannot be produced using this process because the properties are mostly limited to the axial direction.
- Uniform colour is obtained throughout the section.
- Maximum profile width = 1.8 m.
- Maximum profile depth = 500 mm.
- Maximum length = 10 m (6 m is a typical maximum length for transporting).
- Minimum radii = 1 mm for roving fibre, 2.5 mm for mat.
- No draft angle required.
- Maximum thickness = 75 mm (practical maximum thicknesses is 25 mm).
- Minimum section = 0.4 mm (practical minimum thicknesses is 1 mm).
- Sizes from 5 g/m to 50 kg/m.

Quality Issues

- Pultruded parts could lose up to 50% strength at service temperatures of 100°C from those at ambient conditions.
- Fibre content varies between 50% and 80% by weight depending on strength requirements.
- Vinyl ester and epoxy resins provide up to 30% more strength than polyester resins.
- Residual stresses and distortions can be minimised by specifying constant wall thicknesses, which cool more uniformly.
- Small cracks in the pultrusion surface can be repaired using pure resin.
- Rolling contact ultrasonic inspection devices can be incorporated into the process to detect large cracks and abnormalities.
- Approximately 2–3% shrinkage can occur on the cross-section when fully cured.
- Polyesters and vinyl esters are fire-retardant resins.
- Vinyl esters and epoxy resins are aggressive and reduce die life compared to polyesters.
- The more uniform the cross-section of the profile, the better the fibre/resin distribution and fibre alignment will be. Voids in the section will also be reduced.
- Dies generally require refurbishment after 50,000 m of pultruded product.

- Calcium carbonate added to resin to improve opacity and is also a low-cost filler.
- Alumina trihydrate/antimony trioxide are fire retardancy additives to the resin.
- Aluminum silicate (kaolin clay) is a resin additive that provides improved insulation, opacity, surface finish and chemical resistance.
- Ultraviolet inhibitors can also be added to the resin to avoid weathering damage, discoloration and fibre exposure (blooming).
- Urethane coatings provide superior weather protection for pultruded parts, providing up to 20 years protection.
- Polyester surfacing veils can improve corrosion resistance and profile finish.
- Temperature control for curing is required as thermoset resins can burn and decompose at too high temperatures.
- Tension control required to provide correct strength in section without fibre breakage.
- Typical surface roughness = 0.8 μm Ra, on all sides.
- Cutting tolerances: length <6 m = 0/+6 mm; length >6 m = 0/+10 mm; cut angle = ±2°.
- Dimensional tolerances = ±0.2 to ±1 mm, depending on profile size.
- Wall thickness tolerance = ±0.25 mm.
- Straightness tolerance = 0.04% of length.
- Twist tolerance = 1°/m.
- Angular tolerances = ±1.5°.

5.10 Continuous Extrusion (Plastics)

Process Description

The raw material is fed from a hopper into a heated barrel and pushed along a screw-type feeder, where it is compressed and melts. The melt is then forced through a die of the required profile, where it cools on exiting the die (Figure 5.10(a)).

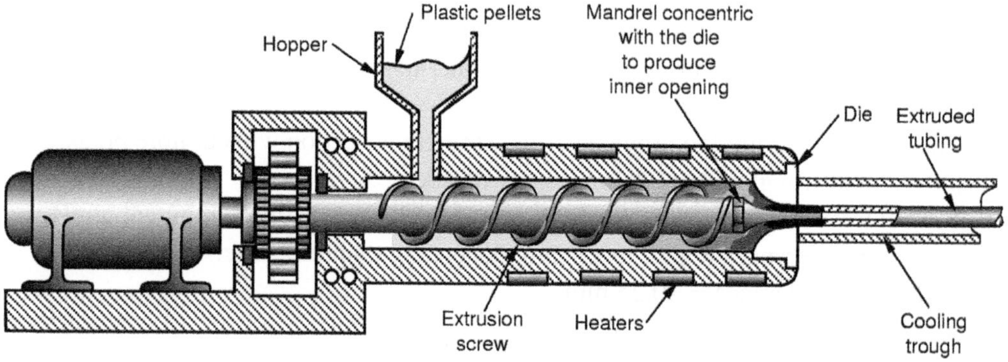

Figure 5.10(a): Continuous Extrusion (Plastics).

Materials

- Most plastics, especially thermoplastics, but also some thermosets and elastomers.
- Raw material in pellet, granular or powder form.

Process Variations

- Most extruders are equipped with a single screw, but two-screw or more extruders are available. These are able to produce coaxial fibres or tubes and multi-component sheets.
- Metal wire, strips and sections can be combined with the extrusion process using an offset die to produce plastic coatings.

Economic Considerations

- Production rates are high but are dependent on size. Continuous lengths up to 60 m/min for some tube sections and profiles, up to 5 m/min for sheet and rod sections.
- Extruders are often run below their maximum speed for trouble-free production.
- Can have multiple holes in die for increased production rates.
- Extruder costs increase steeply at the higher range of output.
- Lead times are dependent on the complexity of the two-dimensional die, but normally weeks.

- Material utilisation is good. Waste is only produced when cutting continuous section to length.
- Process flexibility is moderate. Tooling is dedicated, but changeover and set-up times are short.
- Production of 1,000 kg of profile extrusion is economical, 5000 kg for sheet extrusions (equates to about 10,000 items).
- Tooling costs are generally moderate.
- Equipment costs are high.
- Some materials give off toxic or volatile gases during extrusion. Possible need for air extraction and washing plant, which adds to equipment cost.
- Direct labour costs are low.
- Finishing costs are low. Cutting to length only real cost.

Typical Applications

- Complex profiles; all types of thin-walled, open or closed profiles.
- Rods, bar, tubing and sheet.
- Small-diameter extruded bar that is cut into pellets and used for other plastic processing methods.
- Fibres for carpets, tyre reinforcement, clothes and ropes.
- Cling film.
- Plastic pipe for plumbing.
- Plastic coated wire, cable or strips for electrical applications.
- Window frames.
- Trim and sections for decorative work.

Design Aspects

- Dedicated to long products with uniform cross-sections.
- Cross-sections may be extremely intricate.
- Solid forms including re-entrant angles, closed or open sections.
- Section profile designed to increase assembly efficiency by integrating part consolidation features.
- Grooves, holes and inserts not parallel to the axis of extrusion must be produced by secondary operations.
- No draft angle required.
- Maximum section = 150 mm.
- Minimum section = 0.4 mm for profiles (0.02 mm for sheet).
- Sizes range from 6 mm^2 to 1800-mm-wide sheet, and from 1 to 150 mm diameter for tubes and rods.

Quality Issues

- The rate and uniformity of cooling are important for dimensional control because of shrinkage and distortion.
- Extrusion causes the alignment of molecules in solids.
- Die swell, where the extruded product increases in size as it leaves the die, may be compensated for by:
  - Increasing haul-off rate compared with extrusion rate.
  - Decreasing extrusion rate.
  - Increasing the length of the die land.
  - Decreasing the melt temperature.
- There is a tendency for powdered materials to carry air into the extruder barrel: trapped gases have a detrimental effect on both the output and the quality of the extrusion.
- Surface roughness is good to excellent.
- Process capability charts showing the achievable dimensional tolerances for various materials are provided (Figure 5.10(b)).

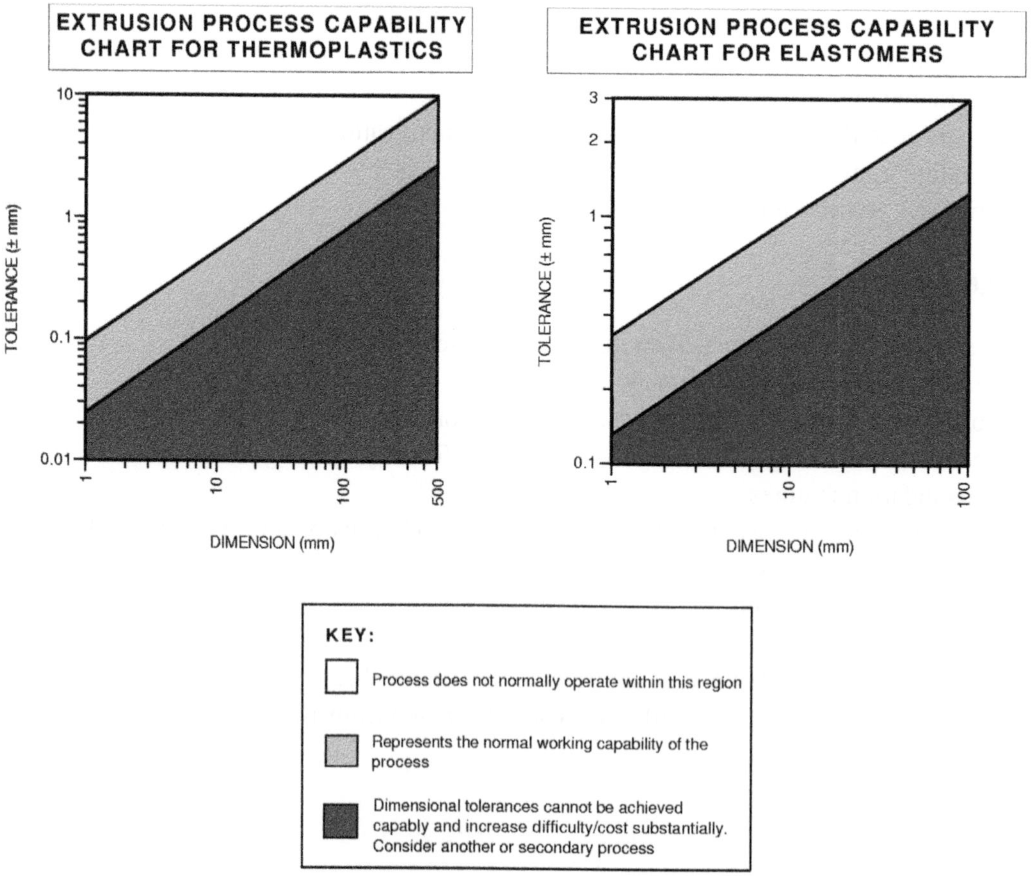

Figure 5.10(b): Continuous Extrusion Process Capability Charts.

CHAPTER 6

Machining Processes

6.1 Turning and Boring

Process Description

The removal of material by chip processes using sequenced or simultaneous machining operations on cut-to-length bar or coiled bar stock. The stock can be automatically or manually fed into the machine (Figure 6.1(a)).

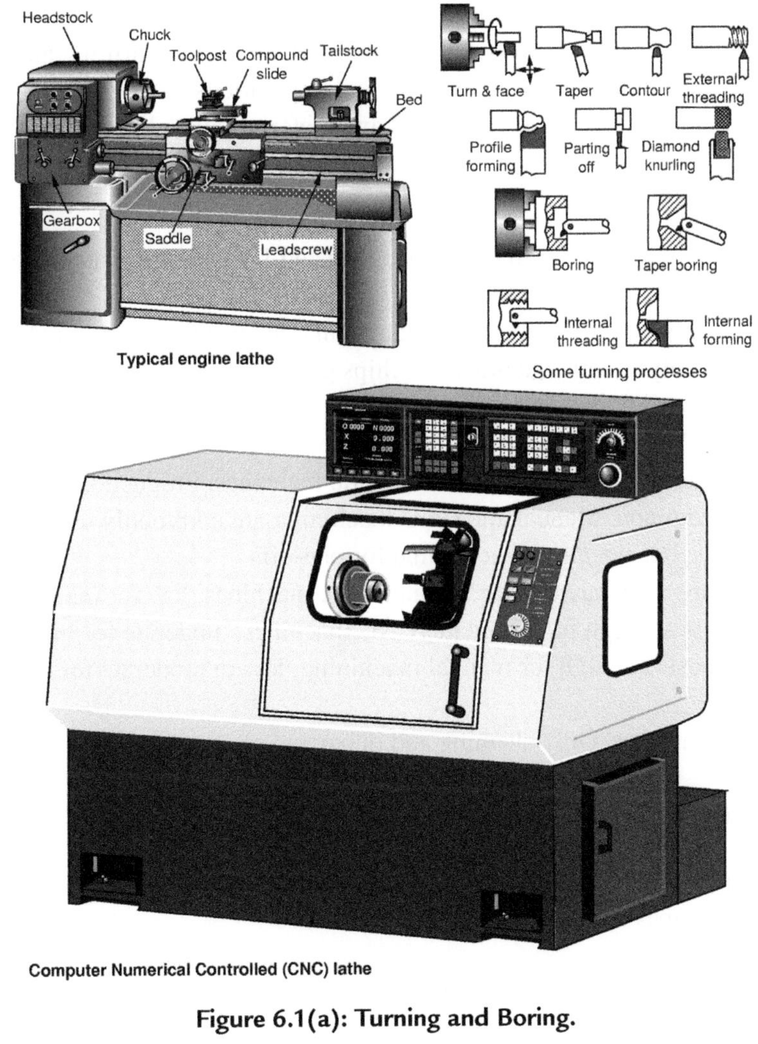

Figure 6.1(a): Turning and Boring.

Manufacturing Process Selection Handbook. http://dx.doi.org/10.1016/B978-0-08-099360-7.00006-9

Materials

All metals (mostly free machining), some plastics, elastomers and ceramics.

Process Variations

- Manually operated machines include: bench lathes (can machine non-standard shape parts) and turret lathes (limited to standard stock material).
- Automatic machines: fully or semi-automated. Follow operations activated by mechanisms on the machine.
- Automatic bar machines: used mainly for the production of screws and similar parts. Single-spindle, multiple-spindle and Swiss types are available.
- CNC machines: movement and control of tool, headstock and saddle are performed by a computer program via stepper motors.
- Machining centres: fully automated, integrated turning, boring, drilling and milling machines capable of performing a wide range of operations.
- Extensive range of cutting tool geometries and materials available.

Economic Considerations

- Production rates are 1–60/h for manual machining and 10–1,000/h for automatic machining.
- Lead times vary from short to moderate.
- Material utilisation is poor to moderate depending on specific operation (10–60% scrap generated, typically). Large quantities of chips generated, which can be recycled.
- Flexibility is low to moderate for automatic machines: change-over and set-up times can be many hours. Manual machines are very flexible.
- Economical quantities are 1,000+ for automatic machines. Production volumes of 100,000+ are common. Manual and CNC machining are commonly used for small production runs, but can also be economic for one-offs.
- Tooling costs are moderate to high for automatic machines, low for manual.
- Equipment costs are high for automatic/CNC machines, moderate for manual machining.
- Direct labour costs are high for manual machining, low to moderate for automatic/CNC machining.
- Finishing costs are low. Only cleaning and deburring required.

Typical Applications

- Any component with rotational symmetrical elements requiring close tolerances.
- Non-standard shapes requiring secondary operations.
- Shafts.
- Screws and fasteners.
- Transmission components.
- Engine parts.

Design Aspects

- Complexity limited to elements with rotational symmetry.
- Little opportunity for part consolidation.
- Can perform many different operations in a logical sequence on the same machine.
- Potential for linking with CAD very high.
- Reduce machining operations to a minimum (for simplicity and lower cycle time).
- Fillet corners and chamfer edges where possible to increase tool life.
- Holes should be drilled with a standard drill point at the bottom for economy.
- Required number of full threads should always be specified.
- Leading threads on both male and female work should be chamfered to assure efficient assembly.
- Special attachments make auxiliary operations possible, for example drilling and milling perpendicular to the length of the work.
- Some special machines allow larger pieces but then operations are restricted.
- Sizes range from 0.5 mm to 2 m+ diameter for manual and CNC machining. Automatic machines usually have a capacity of less than 60 mm diameter.

Quality Issues

- Machinability of the material to be processed is an important issue with regards to: surface roughness, surface integrity, tool life, cutting forces and power requirements. Machinability is expressed in terms of a 'machinability index'[1] for the material.
- Multiple set-ups can be a source of variability.
- Selection of appropriate cutting tool, coolant/lubricant, feed rate, depth of cut and cutting speed with respect to material to be machined is important.
- Coolant also helps flush swarf from cutting area.
- Regular inspection of cutting tool condition and material specification is important for minimum variability.
- Surface detail good to excellent.
- Surface roughness values in the range 0.05–25 μm Ra are obtainable.
- Process capability charts showing the achievable dimensional tolerances for turning/boring (using conventional and diamond-tipped cutting tools) are provided (Figure 6.1(b)). Note that the tolerances on these charts are greatly influenced by the machinability index for the material used and the part geometry.
- Achievable geometrical tolerances are tabulated at the end of the PRIMA.

---

[1] Machinability index for a material is expressed as a percentage based on the relative ease of machining a material with respect to free cutting mild steel, which is 100%, and taken as the standard.

Turned/bored surfaces

Process	Flatness of Surface (μm/m)	Cylindricity (μm/m)	Straightness (Cylinders or Cones) (μm/m)	Parallelism of Flat Surfaces (μm/m)	Parallelism of Cylinders (μm/m)	Roundness (μm/m)
Rough turning/ boring	50	100	100	100	100	40
Fine turning/ boring	30	40	40	50	50	30

Holes (diameter 25 mm)

Process	Taper (μm/40 mm)	Ovality (μm)	Roundness (μm)
Fine turning/boring	22	5	9

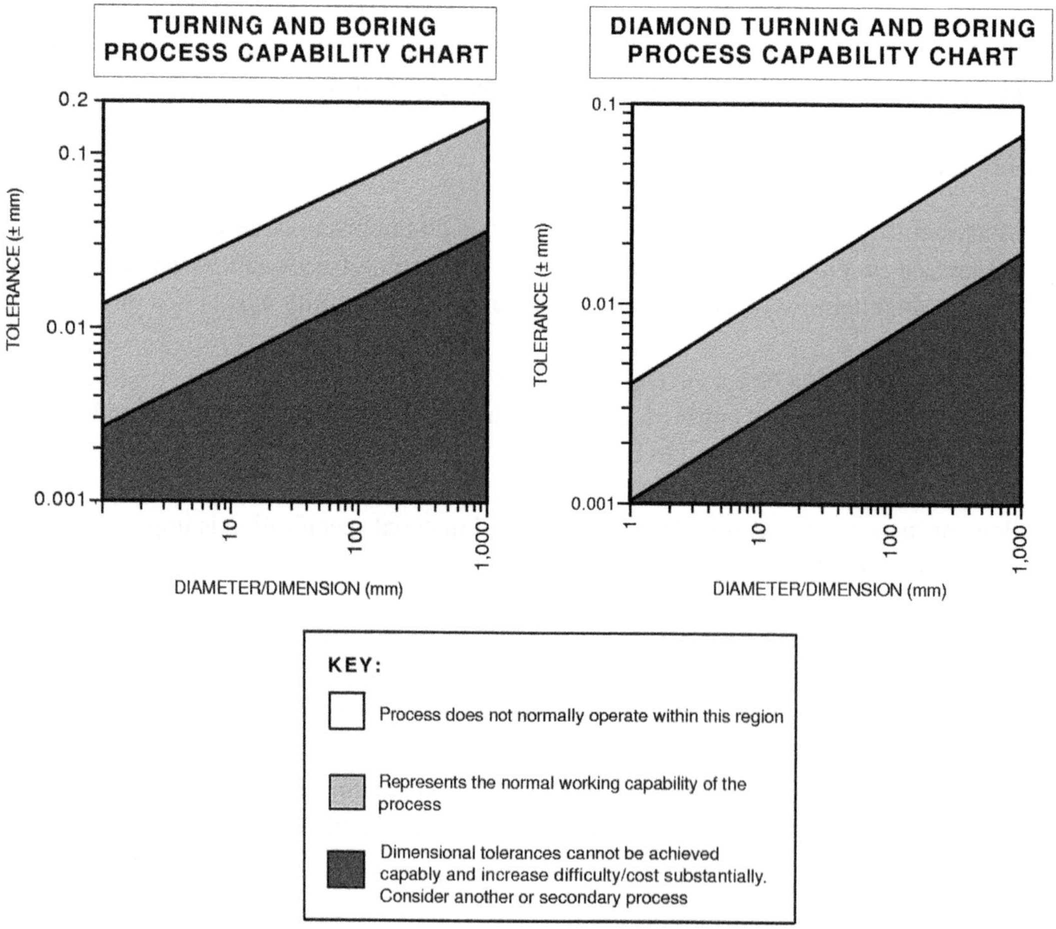

Figure 6.1(b): Turning and Boring Process Capability Charts.

6.2 Milling

Process Description

The removal of material by chip processes using multiple-point cutting tools of various shapes to generate flat surfaces or profiles on a workpiece of regular or irregular section (Figure 6.2(a)).

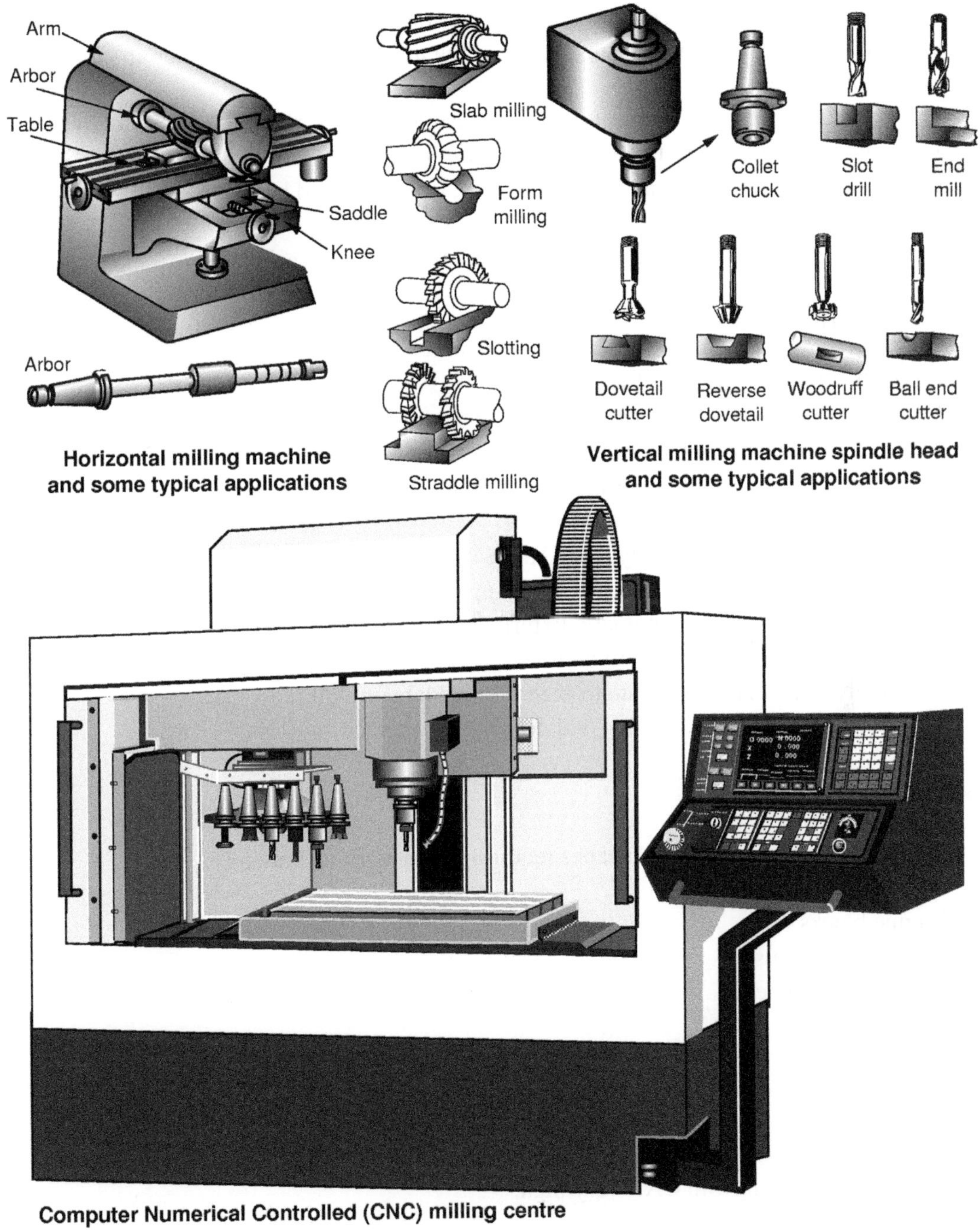

Figure 6.2(a): Milling.

Materials

All metals (mostly free machining) and some plastics and ceramics.

Process Variations

- Horizontal milling: axis of cutter rotation is parallel to surface of workpiece. Includes slab milling, form milling, slotting, gang milling and slitting. Can be either up-cut or down-cut milling.
- Vertical milling: axis of cutter rotation is perpendicular to surface of workpiece. Includes face milling, slotting, dovetail and woodruff milling.
- CNC machines: movement and control of tool, headstock and bed are performed by a computer program via stepper motors.
- Extensive range of cutting tool geometries and tool materials available.

Economic Considerations

- Production rates range from 1 to 100/h.
- Lead times vary from short to moderate. Reduced by CNC.
- Material utilisation is poor. Large quantities of chips generated.
- Recycling of waste material possible but difficult.
- Flexibility is high. Little dedicated tooling.
- Production volumes are usually low. Can be used for one-offs.
- Tooling costs are moderate to high depending on degree of automation (tool carousels, mechanised tool loading, automatic fixturing, etc.).
- Equipment costs are moderate to high.
- Direct labour costs are moderate to high. Skilled labour required.
- Finishing costs are low. Cleaning and deburring required.

Typical Applications

- Any standard or non-standard shapes requiring secondary operations.
- Aircraft wing spars.
- Engine blocks.
- Pump components.
- Machine components.
- Gears.

Design Aspects

- Complexity limited by cutter profiles and workpiece orientation.
- Potential for linking with CAD very high.

- Chamfered edges preferred to radii.
- Use standard sizes and shapes for milling cutter wherever possible.
- Special attachments make auxiliary operations possible, for example gear cutting using an indexing head.
- Minimum section less than 1 mm, but see below.
- Minimum size limited by ability to clamp workpiece to milling machine bed, typically 1.5 m^2, but lengths of 5 m have been milled on special machines.

Quality Issues

- Machinability of the material to be processed is an important issue with regards to: surface roughness, surface integrity, tool life, cutting forces and power requirements. Machinability is expressed in terms of a 'machinability index' for the material.
- Rigidity of milling cutter, workpiece and milling machine important in preventing deflections during machining.
- Selection of appropriate cutting tool, coolant/lubricant, depth of cut, feed rate and cutting speed with respect to material to be machined is important.
- Coolant also helps flush swarf from cutting area.
- Regular inspection of cutting tool condition and material specification is important for minimum variability.
- Surface detail good.
- Surface roughness values in the range 0.2–25 μm Ra are obtainable.
- A process capability chart showing the achievable dimensional tolerances for milling and a chart for positional tolerance capability of CNC milling centres are provided in Figure 6.2(b) (a geometrical tolerance). Note that the tolerances on the milling process capability chart are greatly influenced by the machinability index for the material used.

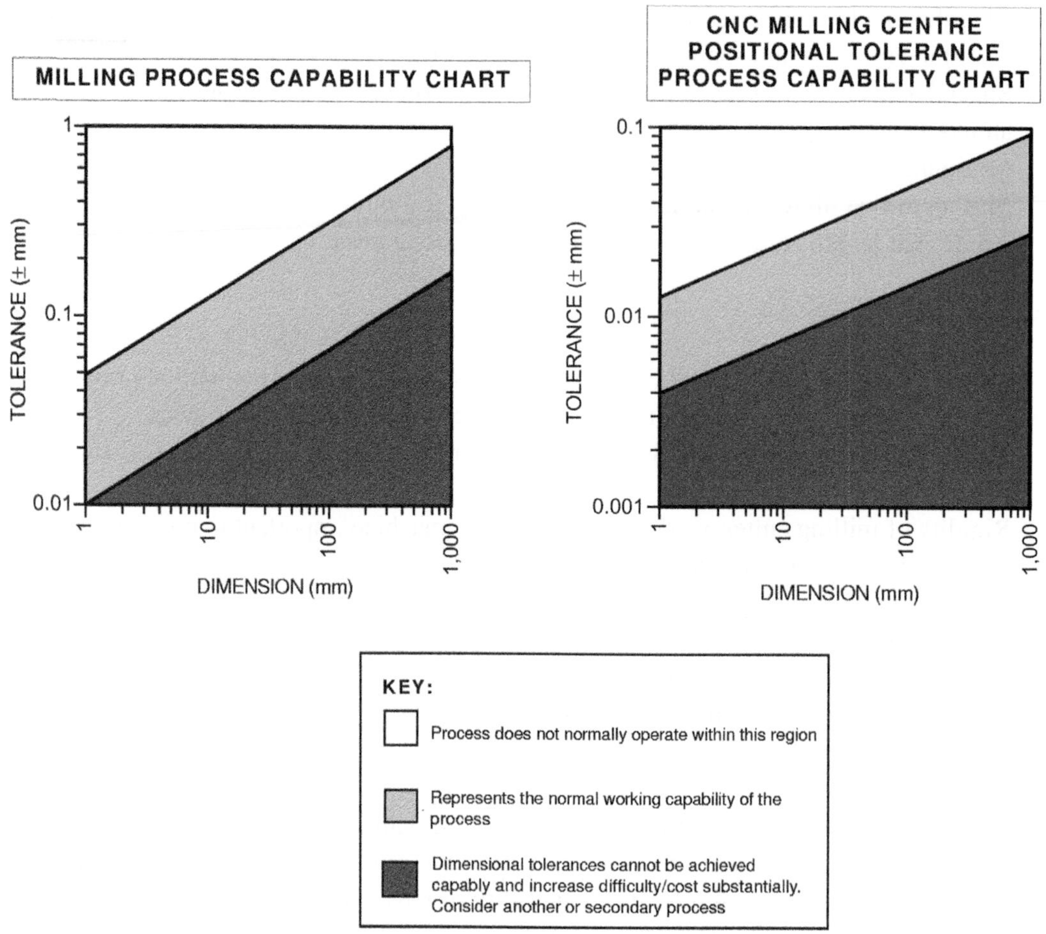

Figure 6.2(b): Milling Process Capability Charts.

6.3 Planing and Shaping

Process Description

The removal of material by chip processes using single-point cutting tools that move in a straight line parallel to the workpiece surface with either the workpiece reciprocating, as in planing, or the tool reciprocating, as in shaping (Figure 6.3(a)). Simplest of all machining processes.

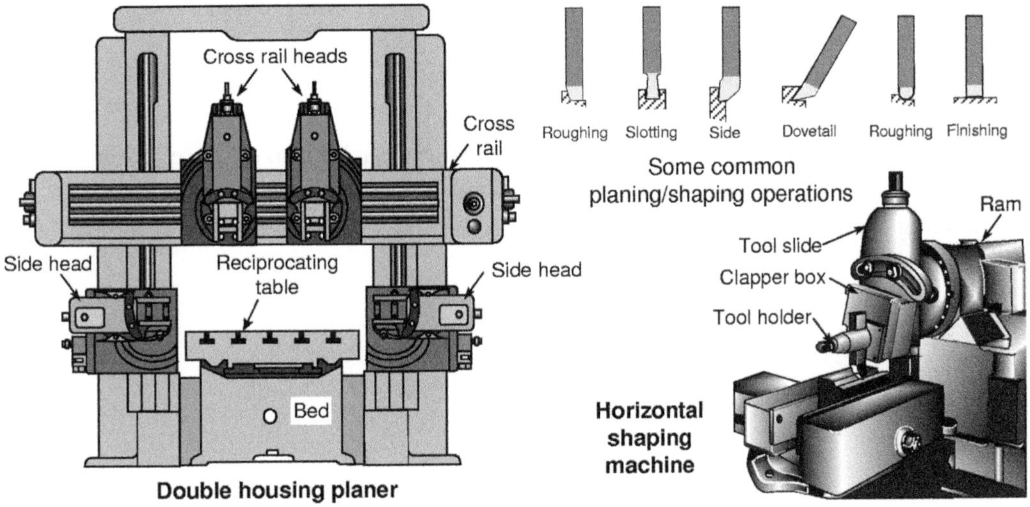

Figure 6.3(a): Planing and Shaping.

Materials

All metals (mostly free machining).

Process Variations

- Double housing planer: closed gantry carrying several tool heads.
- Open side planer: open gantry to accommodate large workpieces carrying usually one tool-head.
- Horizontal shaping: includes push-cut and pull-cut.
- Vertical shaping: includes slotters and key-seaters.
- Wide range of cutting tool geometries and tool materials available.

Economic Considerations

- Production rates range from 1 to 50/h.
- Lead times vary from short to moderate.

- Material utilisation is poor. Large quantities of chips generated, which can be recycled.
- Flexibility is high. Little dedicated tooling, and set-up times are generally short.
- On larger parts, the elapsed time between cutting strokes can be long, making the process inefficient. Can be improved by having the cutting stroke in both directions using several cutting tools and/or machining several parts at once.
- Other processes, for example milling or broaching, may be more economical for larger production runs of smaller parts.
- Planing machines are usually integrated with milling machines to make them more flexible.
- Least economical quantity is one. Production volumes are usually very low.
- Tooling costs are low.
- Equipment costs are moderate to high, depending on machine size and requirements.
- Direct labour costs are high to moderate. Skilled labour may be required.
- Finishing costs are moderate. Normally requires some other machining operations for finishing.

Typical Applications

- Machine tools beds.
- Large castings.
- Die blocks.
- Key seats, slots and notches.
- Large gear teeth.

Design Aspects

- Complexity limited by nature of process, i.e. straight profiles, slots and flat surfaces along length of workpiece.
- As many surfaces as possible should lie in the same plane for machining.
- Rigidity of workpiece design important in preventing vibration.
- Minimum section less than 2 mm, but see below.
- Minimum size limited by ability to clamp workpiece to machine bed.
- Maximum size approximately 25 m long in planing, 2 m long in shaping.

Quality Issues

- Machinability of the material to be processed is an important issue with regards to: surface roughness, surface integrity, tool life, cutting forces and power requirements. Machinability is expressed in terms of a 'machinability index' for the material.
- Adequate clearance should be provided to prevent rubbing and chipping of the cutting tool on return strokes.
- Cutting tools require chip breakers for ductile materials because the strokes can be long during machining and the swarf may tangle and pose a safety hazard.
- Selection of appropriate cutting tool, coolant/lubricant, depth of cut, feed rate and cutting speed with respect to material to be machined is important.
- Coolant also helps flush swarf from cutting area.
- Can produce large, accurate, distortion-free surfaces due to low cutting forces and low local heat generation.
- Surface detail fair.
- Surface roughness values in the range 0.4–25 μm Ra are obtainable.
- A process capability chart showing the achievable dimensional tolerances is provided (Figure 6.3(b)). Note that the tolerances on this chart are greatly influenced by the machinability index for the material used.

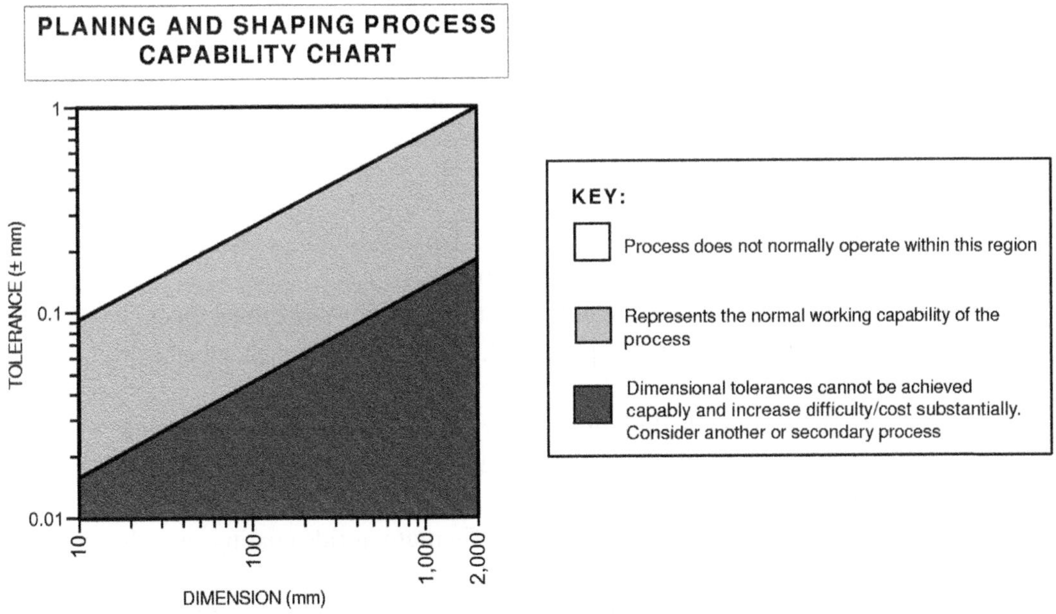

Figure 6.3(b): Planing and Shaping.

6.4 Drilling

Process Description

The removal of material by chip processes using rotating tools of various types with two or more cutting edges to produce cylindrical holes in a workpiece (Figure 6.4(a)).

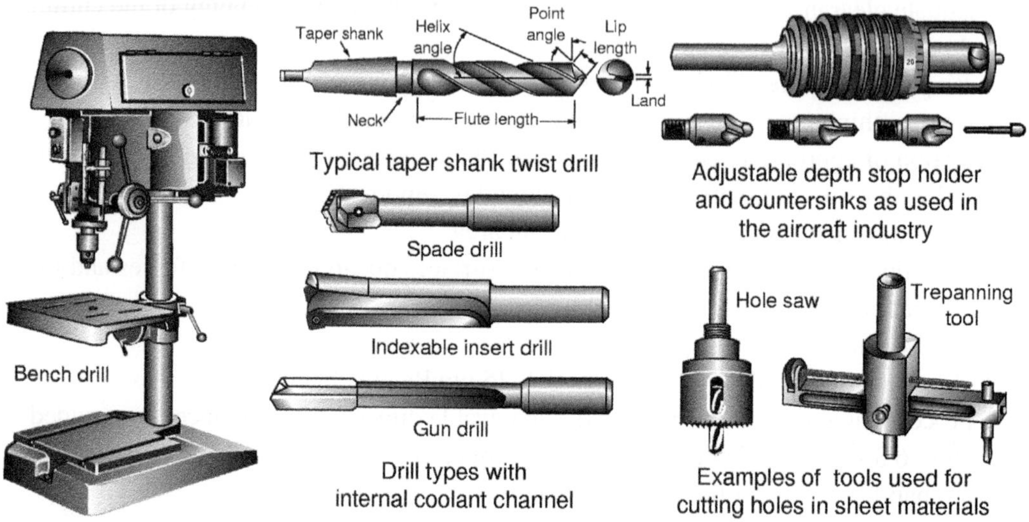

Figure 6.4(a): Drilling.

Materials

All metals (mostly free machining) and some plastics and ceramics.

Process Variations

- Variations on the basic drilling machine include: bench, column, radial arm, gang, multiple spindle, turret and CNC controlled turret.
- Variations on the basic drill types include: twist drill (either three flute, taper shank, bit shank and straight flute), gun drills, spade drill, indexible insert drill, ejector drill, hole saw, trepanning and solid boring drill.
- Variations on conventional drill point geometry are aimed at reducing cutting forces and self-centring capability and include: four facet, helical, Racon, Bickford and split point.
- Wide range of cutting tool materials available. Titanium nitride coatings are also used to increase tool life.
- Drilling can also be performed on lathes, milling machines and machining centres.
- Spot facing, counterboring and countersinking are related drilling processes.

Economic Considerations

- Production rates range from 10 to 500/h.
- Lead times vary from short to moderate. Reduced by automation.
- Material utilisation is very poor. Large quantities of chips generated, which can be recycled.
- Flexibility is high. Little dedicated tooling and generally short set-up times.
- Drill jigs facilitate the reproduction of accurate holes on large production runs.
- Production volumes are usually low to moderate. Can be used for one-offs.
- Production costs significantly reduced with multiple spindle machines when used on large production runs.
- Tooling costs are low.
- Equipment costs are low to moderate, depending on degree of automation and simultaneous drilling heads.
- Direct labour costs are low to moderate. Low operator skill required.
- Finishing costs are low. Cleaning and de-burring required.

Typical Applications

- Any component requiring cylindrical holes, either blind or through.
- Engine blocks.
- Pump components.
- Machine components.

Design Aspects

- Complexity limited to cylindrical blind or through-hole.
- Use standard sizes wherever possible.
- Faces to be drilled are usually required to be perpendicular to the drilling direction unless spot faced and adequate clearance should be provided for.
- Exit surfaces should be perpendicular to hole.
- Through-holes preferred to blind holes.
- Allowances should be made for drill point depths in blind holes.
- Flat-bottomed holes should be avoided.
- Centre drilling usually required before drilling unless special drill point geometry used.
- Holes with a length-to-diameter ratio of greater than 70 have been produced but problems with hole straightness, coolant supply and chip removal may cause drill breakage.
- Sizes range from 0.1 mm diameter for twist drills to 250 mm diameter for trepanning.

Quality Issues

- Machinability of the material to be processed is an important issue with regards to: surface roughness, surface integrity, tool life, cutting forces and power requirements. Machinability is expressed in terms of a 'machinability index' for the material.
- Hard spots, oxide layers and poor surfaces can cause drill point to blunt or break.
- Accurate re-grinding of the drill point geometry is required to maintain correct hole size and balance cutting forces to avoid drill breakage.
- Rigidity of drilling machine, workpiece and drill holder, and concentricity of drill spindle important in preventing oversize holes, chatter and poor surface finish.
- Selection of appropriate drill geometry (including relief and rake angles), coolant/lubricant, size of cut/hole, feed rate and cutting speed with respect to material to be machined, is important.
- Drills may require chip breakers for ductile materials to efficiently remove swarf from cutting area.
- Coolant also helps flush swarf from cutting area in long through-holes and blind holes.
- Surface detail is fair.
- Surface roughness values in the range 0.4–12.5 μm Ra are obtainable.
- A process capability chart showing the achievable dimensional tolerances is provided (Figure 6.4(b)). Note that the tolerances on this chart are greatly influenced by the machinability index for the material used.
- Achievable geometrical tolerances are tabulated at the end of the PRIMA.

Geometric Property	Taper (μm/40mm)	Ovality (μm)	Roundness (μm)
25-mm-diameter holes	36	13	23

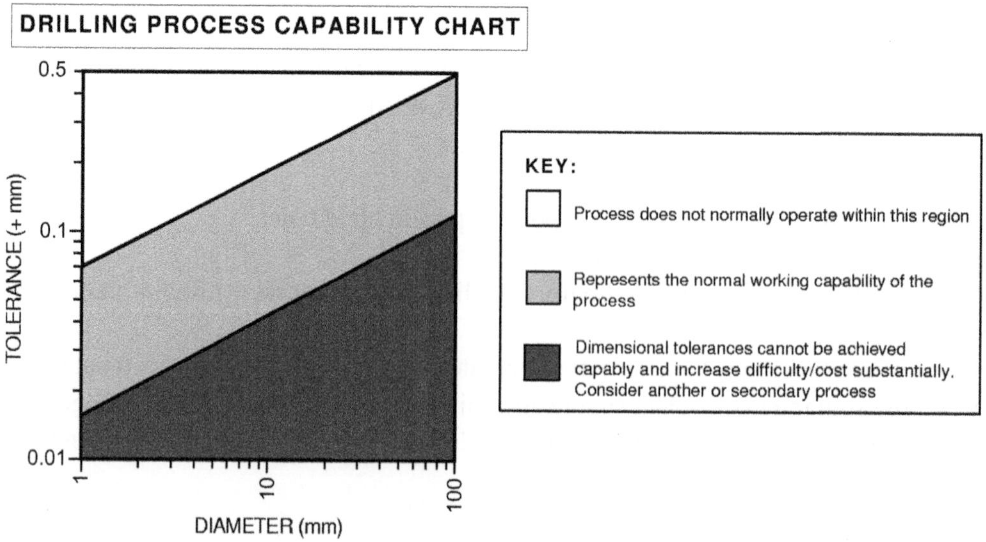

Figure 6.4(b): Drilling Process Capability Charts.

6.5 Broaching

Process Description

The removal of material by chip processes using a multiple-point cutting tool that is pushed or pulled across the workpiece surface. With successively deeper cuts, the desired profile is gradually generated in a single pass (Figure 6.5(a)).

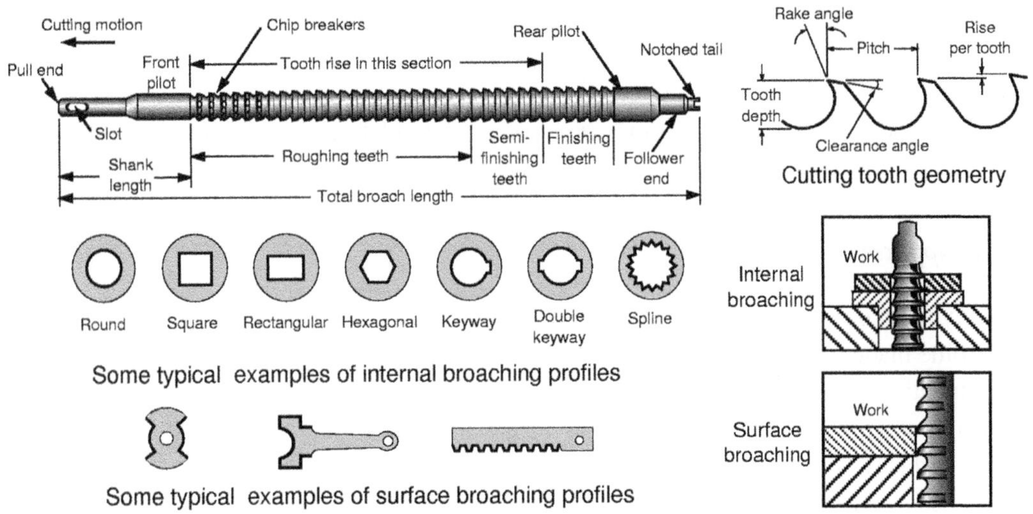

Figure 6.5(a): Broaching.

Materials

All metals (mostly free machining).

Process Variations

- Horizontal, vertical or rotary broaching machines with push and/or pull capability.
- Broaching tools can be single or combination types, internal or external, performing either roughing or finishing operations.
- Some indexible insert broaches are available for surface broaching and titanium nitride coatings are also used to increase tool life.

Economic Considerations

- Production rates up to 400/h.
- To improve production rates, many parts can be machined at once, called stacking. Stacking is best-suited to internal features.
- Automation possible to improve production rates.

- Lead times are moderate.
- Material utilisation is poor. Large quantities of chips generated, which can be recycled.
- Flexibility is high. Little dedicated tooling and set-up times are generally short.
- Accurate re-grinding of the broaching tool is required on large production runs, which uses expensive fixtures and grinding machines.
- Production volumes are usually very high, 10,000–100,000.
- Tooling costs are high. Broaching tools are very expensive due to their complexity and the economics of this process must be carefully studied on this basis.
- Equipment costs are low to moderate.
- Direct labour costs are low to moderate. Some skilled labour may be required.
- Finishing costs are low. Some deburring may be required.

Typical Applications

- Many regular or irregular, internal or external profiles.
- Turbine blade root forms.
- Connecting rod ends.
- Rifling on gun barrels.
- Flat surfaces.
- Key seats and slots.
- Splines, both straight and helical.
- Gear teeth.

Design Aspects

- Complexity limited by nature of process, i.e. straight, curved and complex profiles; slots and flat surfaces along length of workpiece.
- Part design should allow for sufficient clamping area and clearance for broaching tool.
- A hole is initially required for internal broaching for broaching tool access. This can be achieved by either punching, boring or drilling the blank.
- Ideally, between 0.5 and 6 mm should be removed by the broaching tool on any one surface.
- More than one surface can be cut simultaneously.
- Workpiece must be strong enough to withstand the pressure of continuous cutting action of broach.
- Large surfaces, blind holes and sharp corners should be avoided.
- Chamfers are preferred to radiused corners.
- Minimum stroke = 25 mm.
- Maximum stroke = 3 m.

Quality Issues

- Machinability of the material to be processed is an important issue with regards to: surface roughness, surface integrity, tool life, cutting forces and power requirements. Machinability is expressed in terms of a 'machinability index' for the material.
- For materials with high surface hardness, the first tooth on the broach should cut beneath this layer to improve tool life.
- Soft or non-uniform materials may tear during machining.
- Adequate clearance should be provided to prevent rubbing and chipping of the broaching tool on return strokes.
- Broaching tools may require chip breakers for very ductile materials to efficiently remove swarf from cutting area.
- Selection of appropriate cutting tool material, coolant/lubricant, depth of cut per tooth and cutting speed with respect to material to be machined is important.
- Coolant also helps flush swarf from cutting area.
- Surface detail excellent.
- Surface roughness values in the range 0.4–6.3 μm Ra are obtainable.
- A process capability chart showing the achievable dimensional tolerances is provided (Figure 6.5(b)). Note that the tolerances on this chart are greatly influenced by the machinability index for the material used and geometry complexity.
- Achievable geometrical tolerances are tabulated at the end of the PRIMA.

Geometric Property	Taper (μm/40 mm)	Ovality (μm)	Roundness (μm)
25-mm-diameter holes	1	1	2

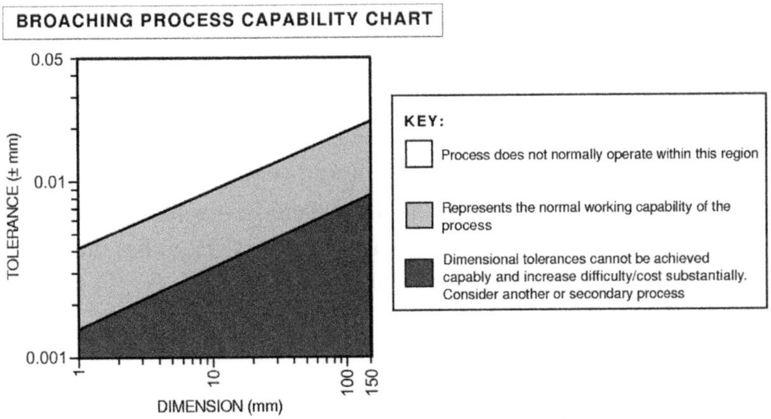

Figure 6.5(b): Broaching Process Capability Chart.

6.6 Reaming

Process Description

The removal of small amounts of material by chip processes using tools of various types with several cutting edges to improve the accuracy, roundness and surface finish of existing cylindrical holes in a workpiece. The tool or the work can rotate relative to each other (Figure 6.6(a)).

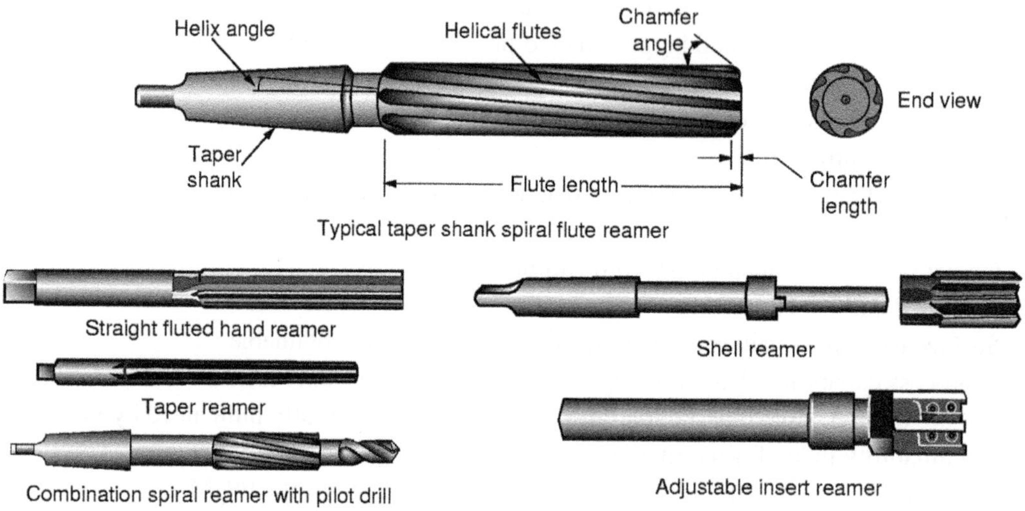

Figure 6.6(a): Reaming.

Materials

All metals (mostly free machining).

Process Variations

- No special machines are used for reaming. Reaming can be performed on drilling machines, lathes, milling machines and machining centres, or by hand.
- Basic reamer types include: hand (straight and tapered), machine (rose and fluted), shell, expansion, adjustable and indexible insert reamers. Titanium nitride coatings are sometimes used to increase tool life. Combination drills and reamers are also available.

Economic Considerations

- Production rates range from 10 to 500/h.
- Lead times vary from short to moderate. Reduced by automation.

- Minimum amount of material removed.
- Flexibility is high. Little dedicated tooling and generally short set-up times.
- Production volumes are usually low to moderate.
- Can be used for one-offs.
- Production costs significantly reduced with multiple-spindle machines.
- Tooling costs are low.
- Equipment costs are low.
- Direct labour costs are low to moderate. Low operator skill required.
- Finishing costs are low. Cleaning and deburring required.

Typical Applications

Any component requiring accurate, cylindrical or tapered holes with good surface finish, either blind or through, after a primary hole-making operation, typically drilling.

Design Aspects

- Complexity limited to straight or tapered cylindrical blind or through-holes.
- Ideally, reaming allowances should be 0.1 mm per 5 mm of diameter, i.e. for a finished reamed hole of 20 mm diameter, the pilot hole should be approximately 19.6 mm in diameter. However, drilled holes prior to reaming should be a standard size wherever possible.
- Allowances should be made for reamer end chamfers and the slight taper on some reamers when machining blind holes, although more suited to through-holes.
- Use standard sizes wherever possible.
- Through-holes preferred to blind holes.
- Sizes range from 3 to 100 mm diameter.

Quality Issues

- Machinability of the material to be processed is an important issue with regards to: surface roughness, surface integrity, tool life, cutting forces and power requirements. Machinability is expressed in terms of a 'machinability index' for the material.
- Any misalignment between workpiece and reamer will cause chatter, oversize holes and bell-mouthing of hole entrance. Piloted reamers ensure alignment of the workpiece and reamer.
- Most accurate holes are centre drilled, drilled, bored and reamed to finished size.
- Proper maintenance and reconditioning of reamers is required to maintain correct hole size and surface finish requirements. To work efficiently, a reamer must have all its teeth cutting.
- Pick-up or galling are caused by too much material being removed by the reamer.

- Selection of appropriate reamer geometry (including relief and rake angles), coolant/lubricant (if required), size of hole, feed rate and cutting speed with respect to material to be machined, is important.
- Reaming is performed at one-third the speed and two-thirds the feed rate of drilling, for optimum conditions.
- Coolant also helps flush swarf from cutting area in long through-holes and blind holes.
- Surface detail good.
- Decreasing feed rate improves surface finish.
- Surface roughness values in the range 0.4–6.3 μm Ra are obtainable.
- A process capability chart showing the achievable dimensional tolerances is provided (Figure 6.6(b)). Note that the tolerances on this chart are greatly influenced by the machinability index for the material used.
- Achievable geometrical tolerances are tabulated at the end of the PRIMA.

Geometric Property	Taper (μm/40 mm)	Ovality (μm)	Roundness (μm)
25-mm-diameter holes	10	1	3

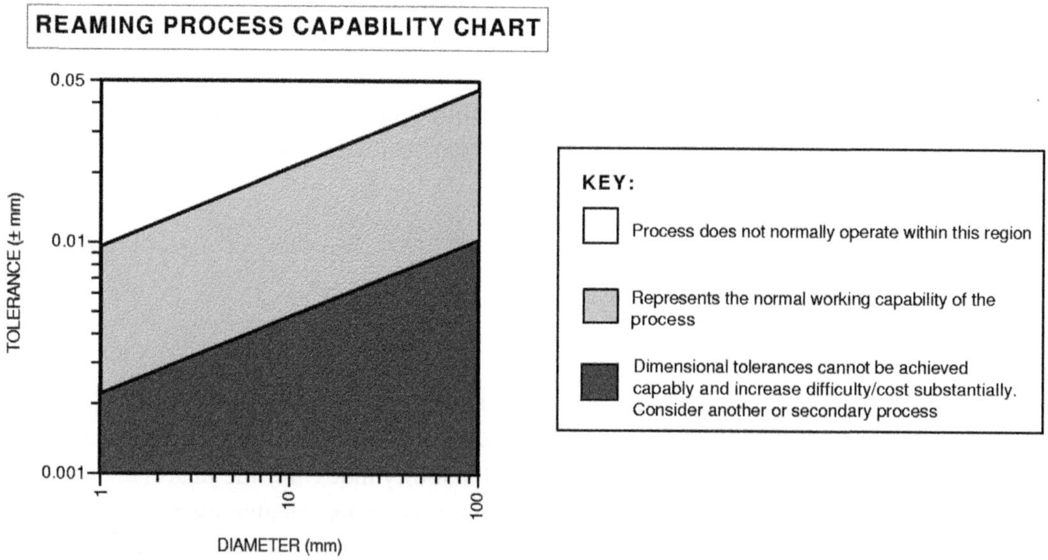

Figure 6.6(b): Reaming Process Capability Chart.

6.7 Grinding

Process Description

The removal of small layer material by the action of an abrasive spinning wheel on a rotating or reciprocating workpiece (Figure 6.7(a)).

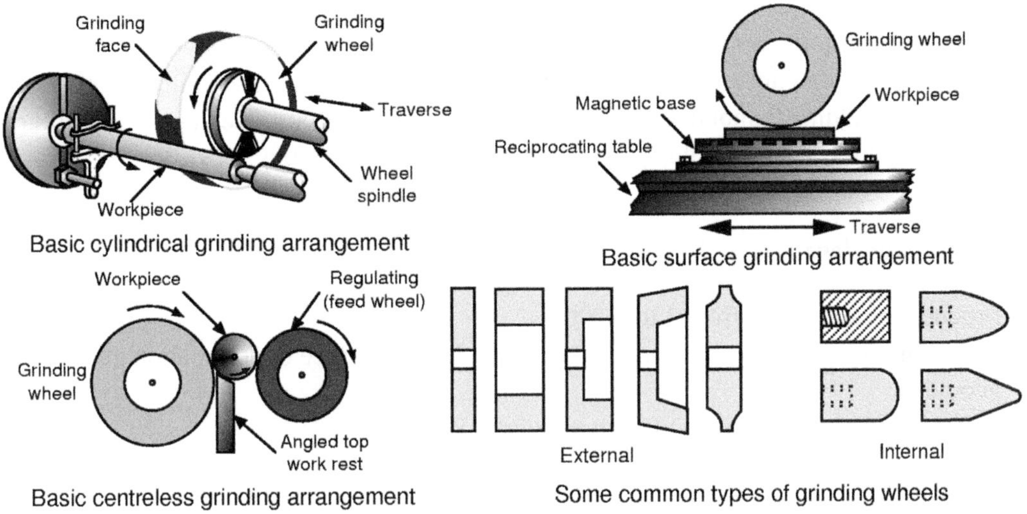

Figure 6.7(a): Grinding.

Materials

All hard materials. Not suitable for soft or flexible materials.

Process Variations

- Surface grinding: workpiece is mounted on a reciprocating or rotating bed and a rotating abrasive wheel (either horizontal or vertical axis of rotation) is fed across the surface.
- Cylindrical grinding: rotating abrasive wheel fed along the periphery of a slower rotating cylindrical workpiece. Also includes: thread, form and plunge grinding.
- Internal grinding: small rotating abrasive wheel fed into the bore of a cylindrical rotating workpiece.
- Centreless grinding: workpiece is supported on a work rest blade and ground between two wheels, one of which is a regulating wheel operating at 5% of the speed of the other.
- Creep-feed grinding: grinding operation performed in a single pass with a large depth of cut.

- Low stress grinding: precision grinding machinery with the application of low feeds and speeds to introduce beneficial compressive residual stresses into the surface.
- Tool grinder: precision bench grinding unit for tool dressing.
- Off-hand grinding: a fixed grinding machine (either bench or pedestal) where the work is manually presented to the grinding wheel.
- Portable grinding: a hand-held unit used for fettling and cutting.
- CNC machines: movement and control of abrasive wheel and workpiece are performed by a computer program via stepper motors.
- Extensive range of abrasive wheel geometries, abrasive materials (aluminium oxide, emery, corundum, diamond, Cubic Boron Nitride (CBN)), grain size, hardness grading and bond types (resin, vitrified glass, rubber, metal) are available.

Economic Considerations

- Production rates range from 1 to 1,000/h.
- Lead times vary from short to moderate, depending on degree of automation and geometry.
- Material utilisation is poor. Difficult to recycle waste material.
- Flexibility of grinding is high.
- Turning can compete with grinding in some situations.
- Suitable for all quantities.
- Tooling costs are low to moderate.
- Equipment costs are moderate to high depending on degree of automation.
- Direct labour costs range from high to low depending on degree of automation and part complexity.
- Finishing costs are very low. Cleaning required.

Typical Applications

- Grinding is used for the generation of basic geometric surfaces and finishing of a wide range of components.
- Parts requiring fine surface roughness and/or close tolerances.
- Bearing surfaces.
- Valve seats.
- Gears.
- Cams.
- Keys and key seats.
- Mould and die cavities.
- Cutting teeth.

Design Aspects

- Complexity limited to nature of workpiece surface, i.e. cylindrical or flat, unless profiled wheels and/or special machines are used.
- Grinding should be used to remove the minimum amount of material.
- Surface features should be kept simple to avoid frequent dressing of the wheel.
- Fillets and corner radii should be as liberal as possible.
- Deep holes and recesses should be avoided.
- Parts should be mounted securely to avoid deflections as high forces can be generated during the grinding process.
- May not be suitable for delicate workpieces.
- For best results use the largest wheel possible for the relevant workpiece.
- Minimum section = 0.5 mm.
- Sizes range from 0.5 mm to 2 m+ diameter for cylindrical grinding. Maximum size for surface grinding approximately 6 m in length. Less than 1 m diameter for centreless grinding.

Quality Issues

- Interruptions on the workpiece surface, for example key seats and recesses, may cause vibration and chatter.
- Unit pressures vary with area of contact. High pressures use hard grade, fine grit abrasive wheels.
- Surface tensile residual stresses remain in the workpiece due to localised high-temperature gradients. This may be critical in heat-sensitive applications or when fatigue strength is important. Low stress grinding can impart beneficial compressive stresses.
- The final size of the workpiece is determined by the speed of response of the gauging system and the forces built up in machine as a result of cutting loads.
- Gauging may be contact or non-contact; this will probably be dictated by the part.
- The properties of the wheel may change in the course of the process. Grinding wheels require occasional dressing to ensure uniform cutting properties.
- Use of grinding fluid is important for chip removal and cooling of the workpiece.
- Grinding wheels need careful storage and to be visually inspected for cracks before use.
- Grinding wheels require balancing before use to minimise vibration because of the high rotational speeds.
- Surface roughness is controlled by the wheel grading, wheel condition, feed rate at finish size and cleanliness of the cutting fluid.
- Surface detail is excellent.
- Surface roughness values in the range 0.025–6.3 μm Ra are obtainable.

- Process capability charts showing the achievable dimensional tolerances for surface and cylindrical grinding are provided (Figure 6.7(b)).
- Achievable geometrical tolerances are tabulated at the end of the PRIMA.

Ground surfaces

Process	Flatness of Surface (μm/m)	Cylindricity (μm/m)	Straightness (Cylinders or Cones) (μm/m)	Parallelism of Flat Surfaces (μm/m)	Parallelism of Cylinders (μm/m)	Roundness (μm/m)
Rough grinding	30	50	50	50	50	20
Fine grinding	20	20	20	30	20	10

Holes (diameter 25 mm)

Process	Taper (μm/40 mm)	Ovality (μm)	Roundness (μm)
Rough grinding	25	3	5
Fine grinding	4	1	6

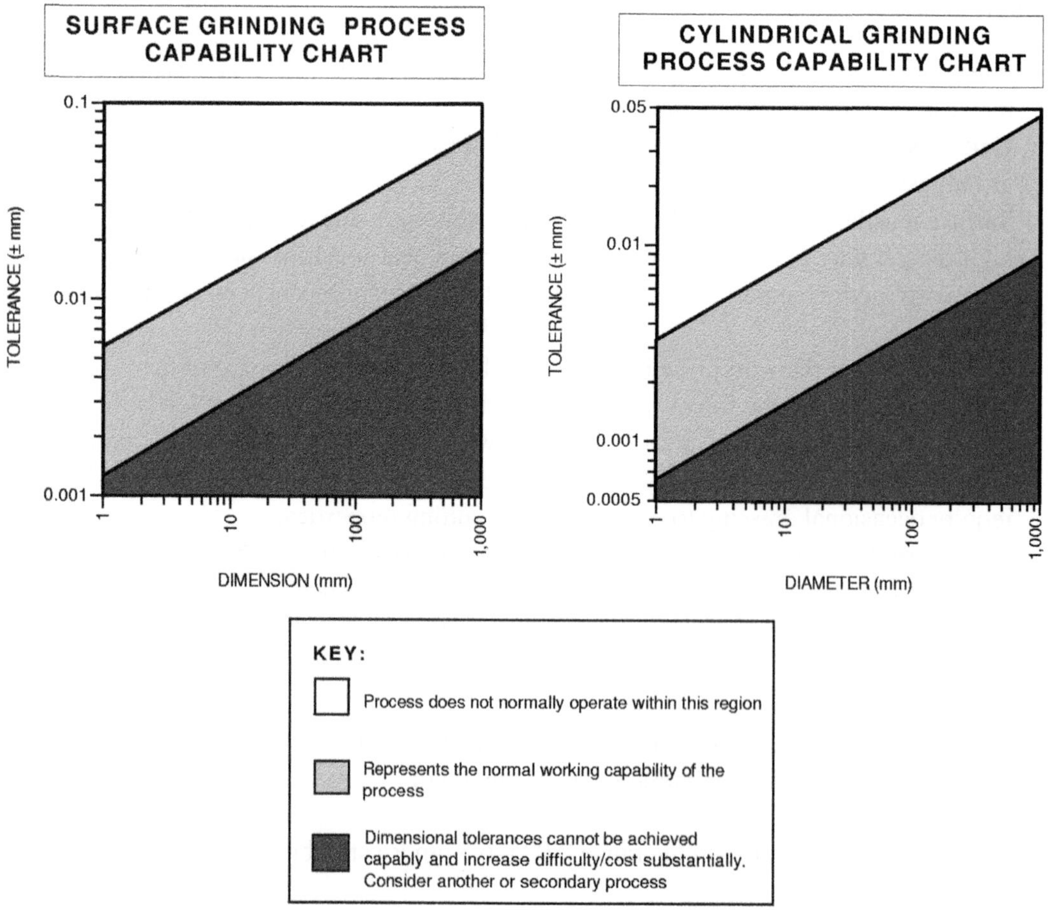

Figure 6.7(b): Grinding Process Capability Charts.

6.8 Honing

Process Description

The removal of small amounts of material by floating segmented abrasive stones mounted on an expanding mandrel, which rotates with low rotary speed and reciprocates along the surface of the workpiece (Figure 6.8(a)).

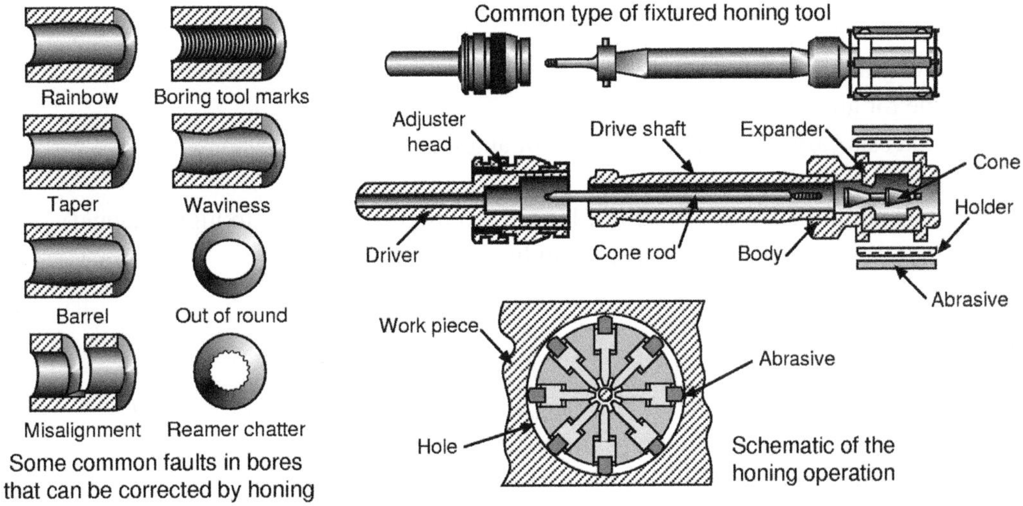

Figure 6.8(a): Honing.

Materials

All materials, including some ceramics and plastics.

Process Variations

- Horizontal and vertical honing machines with single or multiple spindles, with either short or long stroke capability.
- Honing can also be performed on lathes and drilling machines.
- Internal and external cylindrical surfaces are honed commonly. Also, spherical, toroidal and flat surfaces can be honed, but are less common applications.
- Single-stroke bore finishing and superfinishing using 'superabrasives' such as diamond and Cubic Boron Nitride (CBN) are related processes.
- Superfinishing is similar to honing, but performed on outside surfaces previously ground or lapped to improve finish.
- Laser honing: for precise surface topographies.
- Large range of stone geometries, abrasive materials, grain size, hardness grading, bond types and methods (co-axial and match honing) are available.

- Automation aspects include in-process gauging and adaptive control to optimise cutting conditions and control accuracy.
- Workpieces can be manually presented to honing mandrel.

Economic Considerations

- Production rates range from 10 to 1,000/h depending on number of spindles. Typically 60/h for single spindle machines.
- Lead times are short.
- Very little material removed.
- Suitable for all quantities.
- Tooling costs vary depending on degree of automation and size.
- Equipment costs are moderate.
- Direct labour costs are moderate. Skill level required is moderate to high (manual).
- Finishing costs are very low. Cleaning only required.

Typical Applications

- Any component where superior accuracy, surface finish and/or improvement of geometric features required on cylindrical features.
- Bearing surfaces.
- Pin and dowel holes.
- Engine cylinder bores.
- Rifle bores.

Design Aspects

- Honing is performed to remove the minimum amount of material, usually between 0.02 and 0.2 mm.
- Complexity limited to nature of workpiece surface, i.e. cylindrical (internal and external), spherical, flat or toroidal.
- Honing logically follows the grinding process to produce precision surfaces.
- Surface features should be kept simple.
- Chamfers are required on entrance to bores to facilitate easy access of honing tool.
- Blind holes should have undercuts.
- For small holes less than 15 mm in diameter, the maximum length that can be honed is 20 times the diameter of the hole. For best results, a length to diameter ratio of 1 is recommended.
- Maximum length for large holes is 12 m.
- Sizes range from 6 to 750 mm+ diameter for cylindrical honing.

Quality Issues

- Interruptions on the workpiece surface, for example key seats and holes, reduce the quality of finish. Can be offset by increasing rotary speed of honing stone.
- The process has the ability to correct geometrical inaccuracies, for example bell-mouthing, barrelling, tapers and waviness in holes, as well as removing machining marks.
- Surface finish and accuracy is controlled by the stone grain size, feed pressure, area of contact, coolant access, stroke length, rotary speed and stone reciprocation speed, which when optimised ensure breakdown of the stone and good self-dressing characteristics.
- Coolant also helps flush swarf from cutting area in long through-holes and blind holes.
- Little heat generated at surface, therefore original surface characteristics of the component not altered.
- Surface detail excellent.
- Surface roughness values in the range 0.025–1.6 μm Ra are obtainable.
- Relatively soft and flexible materials tend to give inferior surface finish to hard materials.
- A process capability chart showing the achievable dimensional tolerances is provided (Figure 6.8(b)).
- Achievable geometrical tolerances are tabulated at the end of the PRIMA.

Geometric Property	Taper (μm/40 mm)	Ovality (μm)	Roundness (μm)
25-mm-diameter holes	4	1	0.5

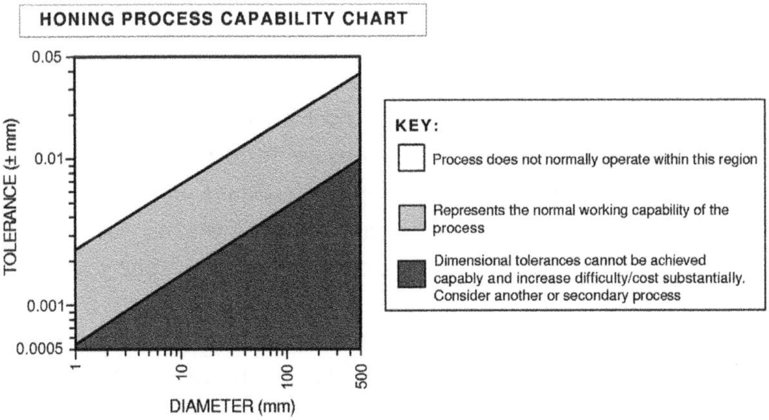

Figure 6.8(b): Honing Process Capability Chart.

6.9 Lapping

Process Description

The removal of very small amounts of material by the relative motion of fine abrasive particles embedded in a soft material (the lap), with the aid of a lubricating and carrier fluid (Figure 6.9(a)).

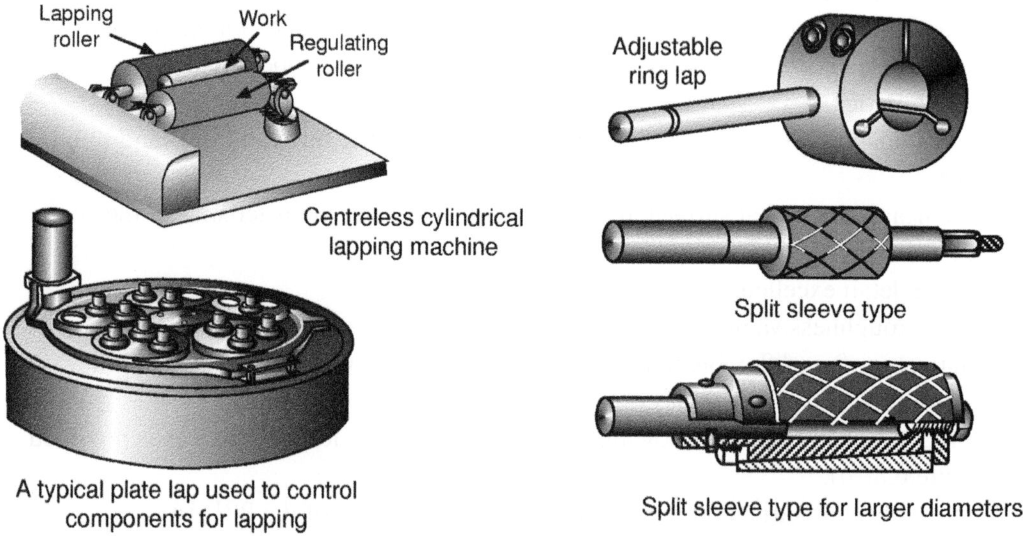

Figure 6.9(a): Lapping.

Materials

All materials, but materials of low hardness or high flexiblity present problems.

Process Variations

- Hand lapping: operator moves the workpiece over a grooved surface plate in an irregular rotary motion, turning the part frequently to ensure uniformity.
- Machine lapping: horizontal and vertical lapping machines with a variety of floating work holding devices that can carry many parts at once over the rotating plate lap.
- Centreless lapping: used for internal and external cylindrical, spherical and contoured surfaces.
- Pressure jet lapping: uses a low-viscosity mix of abrasive grit and water applied at high speed to the surface using compressed air. Similar to Abrasive Jet Machining (AJM) (see PRIMA 7.7).
- Range of lap materials, abrasive materials, grain size and carrier fluids are available for different materials.

Economic Considerations

- Production rates range from 10 to 3000/h, depending on level of automation.
- Lead times are short.
- Very little material removed.
- Suitable for all quantities.
- Tooling costs vary depending on degree of automation and size.
- Equipment costs are moderate.
- Direct labour costs are low to moderate. Operator skill required for hand lapping.
- Finishing costs are very low. Cleaning only required.

Typical Applications

- Any component where superior surface finish is required on flat, cylindrical or contoured surfaces.
- Bearing surfaces.
- Gauge blocks.
- Piston rings.
- Balls for ball bearings.
- Piston pins.
- Valve seats.
- Glass lenses.
- Pump gears.

Design Aspects

- Complexity limited to nature of workpiece surface, i.e. flat, cylindrical (internal and external) or spherical.
- Lapping is performed to remove the minimum amount of material, usually between 0.005 and 0.01 mm.
- Lapping should not be specified if the surface finish on the component is not critical and can be produced by other processes.
- Lapping logically follows the grinding or honing process to produce precision surfaces.
- Parts that are required to provide lapping pressure under their own weight should have a low centre of gravity and be stable.
- Surface features should be kept simple.
- Sizes range from 1 to 500 mm for flat lapping.
- Centreless lapping sizes range from 0.75 to 300 mm diameter.
- Maximum lengths are 4 m+ for up to 75 mm diameter.

Quality Issues

- Soft materials are difficult to lap due to abrasive particles becoming embedded in workpiece material.
- Low lapping speeds can introduce beneficial compressive residual stresses into the surface of workpiece to improve fatigue resistance.
- Choice of abrasive, lap and carrier important for specific material types.
- Surface detail excellent.
- Surface roughness values in the range 0.012–0.8 μm Ra are obtainable.
- A process capability chart showing the achievable dimensional tolerances is provided (Figure 6.9(b)).

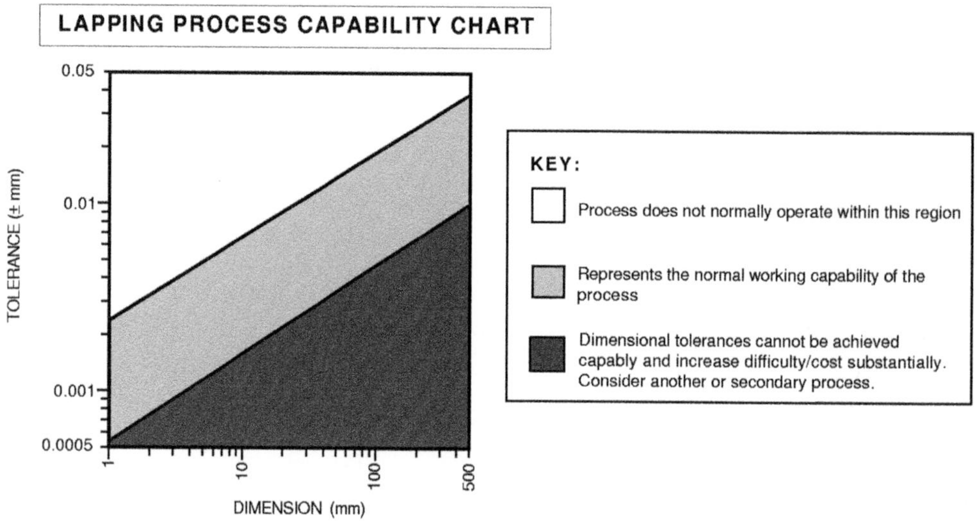

Figure 6.9(b): Lapping Process Capability Chart.

CHAPTER 7

Non-traditional Machining Processes

7.1 Electrical Discharge Machining (EDM)

Process Description

The tool, usually graphite, and the workpiece are essentially electrodes, the tool being the negative of the cavity to be produced. The workpiece is vaporised by spark discharges created by a power supply. The gap between the workpiece and tool is kept constant and a dielectric fluid is used to cool the vaporised 'chips' and then flush them away from the workpiece surface (Figure 7.1). Usually CNC control.

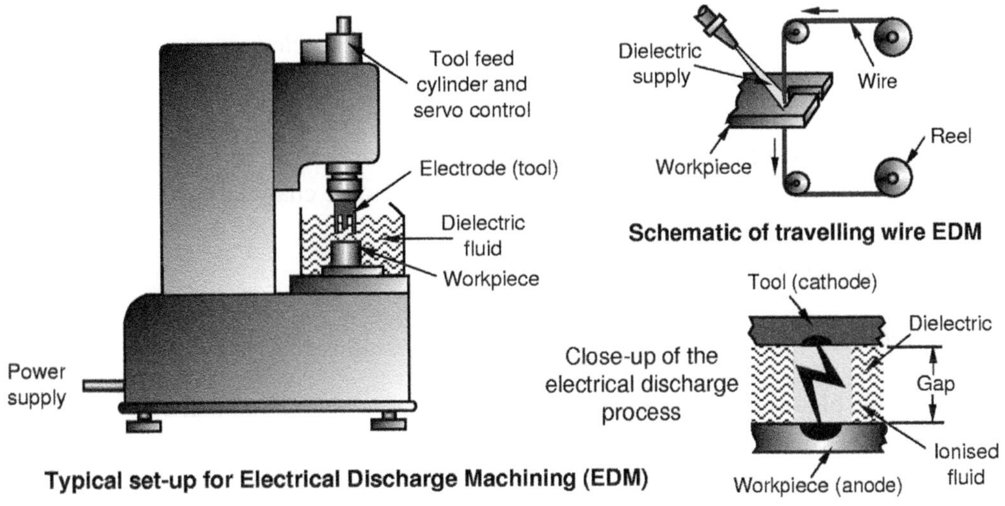

Figure 7.1: Electrical Discharge Machining (EDM).

Materials

- Any electrically conductive material irrespective of material hardness, commonly tool steels, carbides, polycrystalline diamond (PCD) and ceramics, but not cast iron.
- Melting point and latent heat of melting are important properties partially determining the material removal rate.

Manufacturing Process Selection Handbook. http://dx.doi.org/10.1016/B978-0-08-099360-7.00007-0

Process Variations

- Travelling wire EDM: wire moves slowly along the prescribed path on the workpiece and cuts the metal with sparks, creating a slot of 'kerf'.
- No-wear EDM: minimising tool wear of steels by reversing the polarity and using copper tools.
- Electrical Discharge Grinding (EDG): graphite or brass grinding wheel rotates relative to the rotating workpiece and removes material by spark erosion (no abrasives involved).
- Ultrasonic EDM: increases production rate and gives less surface damage.

Economic Considerations

- Production rates very low.
- Material removal rates up to 1.6 mm^3/min.
- Cutting rate for travelling wire EDM is approximately 0.635 mm/s.
- Material removal/cutting rates are a function of the current rate and material properties.
- Lead time days to several weeks depending on complexity of electrode tool.
- Tools can be of segmented construction for high complexity work.
- Material utilisation very poor. Scrap material cannot be recycled.
- Disposal of sludge and chemicals used can be costly.
- High degree of automation possible.
- Economical for low production runs. Can be used for one-offs.
- Tooling costs are high. High tool wear rates mean periodic changing.
- Equipment costs generally high.
- Direct labour costs are low to moderate.

Typical Applications

- Tool and die blocks for forging, extrusion, casting, punching, blanking, etc.
- Honeycomb structures and irregular shapes.
- Prototype parts.
- Burr-free parts.

Design Aspects

- High degree of shape complexity possible, limited only by ability to produce tool shape.
- Travelling wire EDM can create 2D and 3D profiles.
- Suitable for small-diameter, deep holes with length-to-diameter ratios up to 20:1. Can be up to 100:1 for special applications.
- Undercuts are possible with specialised tooling.
- No mechanical forces are used for cutting, therefore simple fixtures can be used.

- Possible to machine thin and delicate sections due to minimal machining forces.
- Minimum radius = 0.025 mm.
- Minimum hole/slot size = 0.05 mm.
- Travelling wire EDM can cut sections up to 150 mm.

Quality Issues

- Burr-free part production.
- Produces slightly tapered holes, especially if blind, and some overcut.
- Optimum tool-to-workpiece gap ranges from 0.012 to 0.51 mm.
- Surface layer is altered metallurgically and chemically due to high thermal energies.
- A hard skin, or recast layer, produced, may offer longer life, lower friction and lubricant retention for dies, but can be removed if undesirable.
- Beneath the recast layer is a heat-affected zone, which may be softer than the parent material.
- Finishing cuts are made at lower removal rates.
- Tool wear is related to the melting points of the materials involved and this affects accuracy. May require changing periodically.
- Being a thermal process, residual stresses and fine cracks may form.
- Removal rate can be increased with the expense of a poorer surface finish.
- Surface detail good.
- Surface roughness values are in the range 0.4–25 μm Ra. Dependent on current density, material being machined and rate of removal.
- Achievable tolerances are in the range ±0.01 to ±0.125 mm. (Process capability charts have not been included. Capability is not primarily driven by characteristic dimension but by the material being processed.)

7.2 Electrochemical Machining (ECM)

Process Description

Workpiece material is removed by electrolysis. A tool, usually copper (negative electrode), of the desired shape is kept a fixed distance away from the electrically conductive workpiece (positive electrode), which is immersed in a bath containing a fast-flowing electrolyte and connected to a power supply. The workpiece is then dissolved by an electrochemical reaction to the shape of the tool. The electrolyte also removes the 'sludge' produced at the workpiece surface (Figure 7.2).

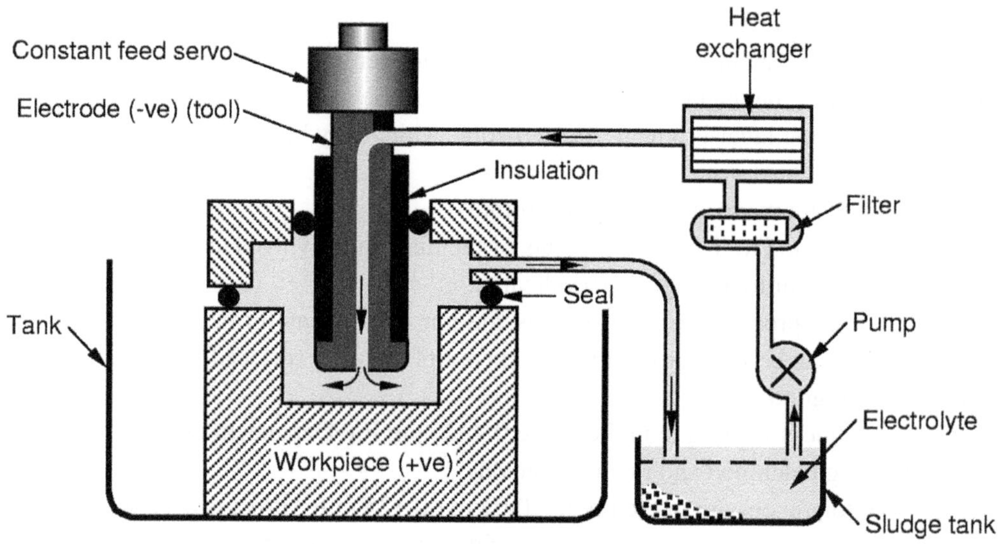

Figure 7.2: Electrochemical Machining.

Materials

Any electrically conductive material irrespective of material hardness, commonly tool steels, nickel alloys and titanium alloys. Ceramics and copper alloys are also processed, occasionally.

Process Variations

- Electrochemical Grinding (ECG): combination of electrochemical reaction and abrasive machining of workpiece.
- Shaped Tube Electrolyte Machining (electrochemical drilling): uses an insulated titanium tube as a cathode, through which the electrolyte is pumped. Can create very-small-diameter holes with large length-to-diameter ratios. Slow process cycle.
- Electrochemical polishing: for deburring and honing.

Economic Considerations

- Production rates are moderate.
- Material removal rates are typically 50–250 mm^3/s.
- Linear penetration rates up to 0.15 mm/s.
- Dependent on current density, electrolyte, and gap between tool and workpiece.
- High power consumption.
- Lead time can be several weeks. Tools are very complex.
- Set-up times can be short.
- Material utilisation very poor. Scrap material cannot be recycled.
- Disposal of sludge and chemicals used can be costly and hazardous.
- High degree of automation possible.
- Economical for moderate to high production runs.
- Tooling costs are very high. Dedicated tooling.
- Equipment costs generally high.
- Direct labour costs are low to moderate.

Typical Applications

- Hole (circular and non-circular) production, profiling and contouring of components.
- Engine casting features.
- Turbine blade shaping.
- Dies for forging.
- Gun barrel rifling.
- Honeycomb structures and irregular shapes.
- Burr-free parts.
- Deep holes.

Design Aspects

- High degree of shape complexity possible, limited only by ability to produce tool shape.
- Can be used for material susceptible to heat damage.
- Suitable for small-diameter, deep holes with length-to-diameter ratios up to 50:1. Electrochemical drilling can increase this to 300:1.
- Suitable for parts affected by thermal processes.
- Undercuts are possible with specialised tooling.
- Possible to machine thin and delicate sections due to no processing forces.
- Cannot produce perfectly sharp corners.
- Minimum radius = 0.05 mm.
- Minimum hole size = 0.1 mm diameter.

Quality Issues

- Burr-free part production.
- Produces slightly tapered holes, especially if deep, and some overcut possible.
- Finishing cuts are made at low material removal rates.
- Deep holes will have tapered walls.
- No stresses introduced, either thermal or mechanical.
- Virtually no tool wear.
- Arcing may cause tool damage.
- Some electrolyte solutions can be corrosive to tool, workpiece and equipment.
- Surface detail good.
- Surface roughness values are in the range 0.2–12.5 μm Ra. Dependent on current density and material being machined.
- Achievable tolerances are in the range ±0.013 to ±0.5 mm. (Process capability charts have not been included. Capability is not primarily driven by characteristic dimension but by the material being processed.)

7.3 Electron Beam Machining (EBM)

Process Description

An electron gun bombards the workpiece with electrons up to 80% of the speed of light, generating localised heat and evaporating the workpiece surface. Magnetic lenses focus the electron beam and electromagnetic coils control its position. The workpiece is typically contained within a vacuum chamber (Figure 7.3).

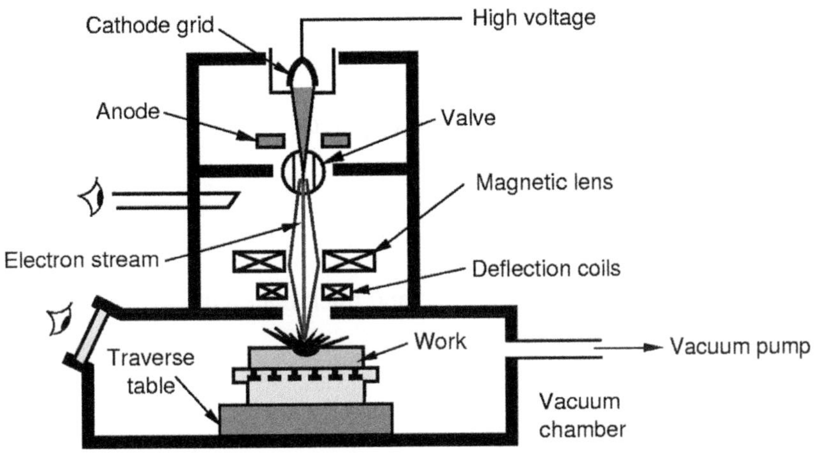

Figure 7.3: Electron Beam Machining.

Materials

Any material regardless of its type, electrical conductivity and hardness.

Process Variations

- Electron Beam Welding (EBW; see PRIMA 11.5): used to weld a range of material of varying thicknesses giving a small weld area and heat-affected zone, with no flux or filler.
- The electron beam process can also be used for cutting, profiling, slotting and surface hardening, using the same equipment by varying process parameters.

Economic Considerations

- Production rates are dependent on size of vacuum chamber and by the ability to process a number of parts in batches at each loading cycle (less than 1 s per hole cycle time on thin workpieces).
- Parts should closely match size of chamber.

- Material removal rates are low, typically 10 mm^3/min. Penetration speeds up to 10 mm/s possible.
- Lead times can be several weeks.
- Set-up times can be short, but the time to create a vacuum in the chamber at each loading cycle is an important consideration.
- Material utilisation is good.
- High degree of automation possible.
- High energy consumption process.
- Economical with low to moderate production runs for thin parts requiring small cuts.
- Tooling costs are very high.
- Equipment costs are very high.
- Direct labour costs are high. Skilled labour required.
- Finishing costs are very low.

Typical Applications

- Multiple small-diameter holes in very thin and thick materials.
- Injector nozzle holes.
- Small extrusion die holes.
- Irregular-shaped holes and slots.
- Engraving.
- Features in silicon wafers for the electronics industry.

Design Aspects

- Electron beam path can be programmed to produce a desired pattern.
- Suitable for small-diameter, deep holes with length to diameter ratios up to 100:1.
- Possible to machine thin and delicate sections due to no mechanical processing forces.
- Sharp corners are difficult to produce.
- Better to have more small holes requiring less heat than a few large holes requiring considerable heat.
- Maximum thickness = 150 mm.
- Minimum hole size = 0.01 mm diameter.

Quality Issues

- Localised thermal stresses giving very small heat-affected zones, small recast layers and low distortion of thin parts possible.
- Integrity of vacuum important. Beam dispersion occurs due to electron collision with air molecules.

- The reflectivity of the workpiece surface is important. Dull and unpolished surfaces are preferred.
- Hazardous X-rays are produced during processing and require lead shielding.
- Produces slightly tapered holes, especially if deep holes are required.
- Critical parameters to control during process are: voltage, beam current, beam diameter and work speed.
- The melting temperature of the material may also have a bearing on quality of surface finish.
- Surface roughness values are in the range 0.4–6.3 μm Ra.
- Achievable tolerances are in the range ±0.013 to ±0.125 mm. (Process capability charts have not been included. Capability is not primarily driven by characteristic dimension.)

7.4 Laser Beam Machining (LBM)

Process Description

A controlled laser beam is focused on to the workpiece surface, causing it vaporise locally. The material then leaves the surface in the vaporised or liquid state at high velocity (Figure 7.4).

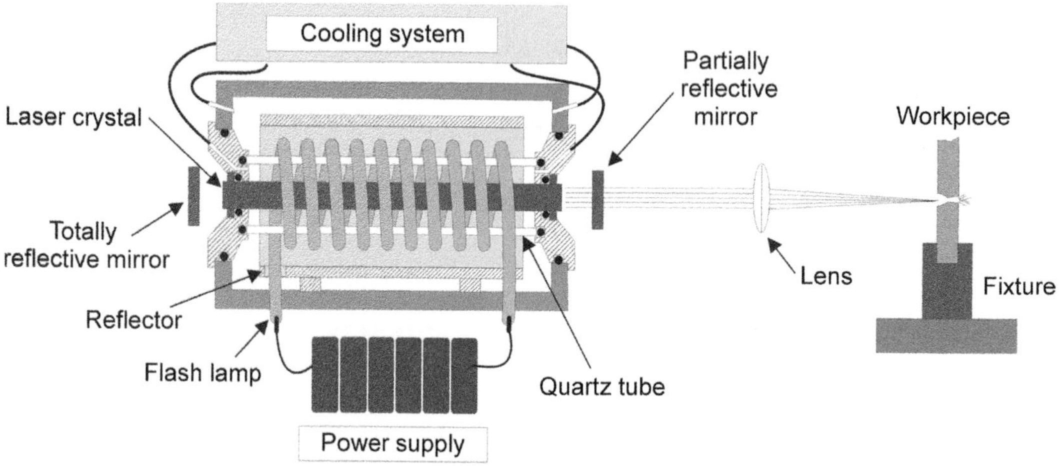

Figure 7.4: Laser Beam Machining.

Materials

Most materials, but dependent on thermal diffusivity and to a lesser extent the optical characteristics of material, rather than chemical composition, electrical conductivity or hardness.

Process Variations

- Many types of laser are available, used for different applications. Common laser types available are: CO_2, Nd:YAG, Nd:glass, ruby and excimer. Depending on economics of process, pulsed and continuous wave modes are used.
- High-pressure gas streams are used to enhance the process by aiding the exothermic reaction process, keeping the surrounding material cool and blowing the vaporised or molten material and slag away from the workpiece surface.
- Laser beam machines can also be used for cutting, surface hardening, welding (LBW) (see PRIMA 11.6), drilling, blanking, honing, engraving and trimming, by varying the power density.

Economic Considerations

- Production rates are moderate to high: 100 holes/s possible for drilling.
- Higher material removal rate than conventional machining.
- Material removal rates are typically 5 mm^3/s and cutting speeds 70 mm/s.
- High power consumption.
- Lead times can be short, typically weeks.
- Set-up times are short.
- Material utilisation is good.
- High degree of automation possible.
- High flexibility. Integration with CNC punching machines is popular, giving greater design freedom.
- Possible to perform many operations on same machine by varying process parameters.
- Economical for low to moderate production runs.
- Tooling costs are very high.
- Equipment costs are very high.
- Direct labour costs are medium to high. Some skilled labour required.

Typical Applications

- For holes, profiling, scribing, engraving and trimming.
- Non-standard shaped holes, slots and profiling.
- Prototype parts.
- Small-diameter lubrication holes.
- Features in silicon wafers in the electronics industry.

Design Aspects

- Laser can be directed, shaped and focused by reflective optics permitting high spatial freedom in two and three dimensions with special equipment.
- Suitable for small-diameter, deep holes with length to diameter ratios up to 50:1.
- Special techniques are required to drill blind and stepped holes, but not accurate.
- Minimal work holding fixtures are required.
- Sharp corners are possible, but radii should be provided for in the design.
- Maximum thicknesses: mild steel = 25 mm, stainless steel = 13 mm, aluminium = 10 mm.
- Maximum hole size (not profiled) = 1.3 mm.
- Minimum hole size = 0.005 mm diameter.

Quality Issues

- Difficulty of material processing is dictated by how close the material's boiling and vaporisation points are.
- Localised thermal stresses, heat-affected zones, recast layers and distortion of very thin parts may be produced. Recast layers can be removed if undesirable.
- No cutting forces, so simple fixtures can be used.
- Possible to machine thin and delicate sections due to no mechanical contact.
- The cutting of flammable materials is usually inert-gas assisted. Metals are usually oxygen-assisted.
- Control of the pulse duration is important to minimise the heat-affected zone, depth and size of molten metal pool surrounding the cut.
- The reflectivity of the workpiece surface is important. Dull and unpolished surfaces are preferred.
- Hole wall geometry can be irregular. Deep holes can cause beam divergence.
- Surface detail fair.
- Surface roughness values are in the range 0.4–6.3 μm Ra.
- Achievable tolerances are in the range ±0.015 to ±0.125 mm. (Process capability charts have not been included. Capability is not primarily driven by characteristic dimension.)

7.5 Chemical Machining (CM)

Process Description

Selective chemical dissolution of the workpiece material by immersion in a bath containing an etchant (usually acid or alkali solution). The areas that are not required to be etched are masked with 'cut and peel' tapes, paints or polymeric materials (Figure 7.5(a)).

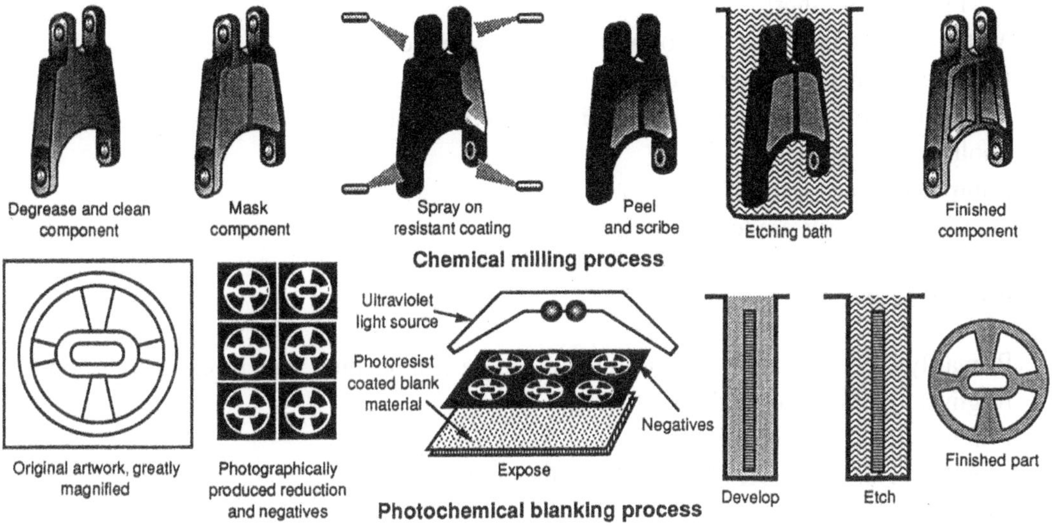

Figure 7.5(a): Chemical Machining.

Materials

Most materials can be chemically machined with the correct chemical etchant selection, commonly: ferrous; nickel; titanium; magnesium and copper alloys; and silicon.

Process Variations

- Chemical milling: chemical removal of material to a specified depth on large areas.
- Chemical blanking: used for thin parts requiring penetration through thickness.
- Photochemical blanking: uses photographic techniques to blank very thin sheets of metal, primarily for the production of printed circuit-boards.
- Thermochemical machining: uses a hot corrosive gas.
- Chemical jet machining: uses a single jet of etchant for deburring.

Economic Considerations

- Production rates are low to moderate. Can be improved by machining a large sheet before cutting out the individual parts. Parts can also be etched on both sides simultaneously.
- Linear penetration rate is very slow, typically 0.0025–0.1 mm/min, but dependent on material.
- Lead times are short.
- Set-up times are short.
- Material utilisation poor. Scrap material cannot be recycled.
- Disposal of chemicals used can be costly.
- Economical for low production runs. Least economical quantity is one.
- Tooling costs are low.
- Equipment costs generally low.
- Direct labour costs are low.

Typical Applications

- Primarily used for weight reduction in aerospace components, panels, extrusions and forgings by producing shallow cavities.
- Printed circuit-board tracks.
- Features in silicon wafers for the electronics industry.
- Decorative panels.
- Printing plates.
- Honeycomb structures.
- Irregular contours and stepped cavities.
- Burr-free parts.

Design Aspects

- High degree of shape complexity possible in two dimensions.
- Suitable for parts affected by thermal processes.
- Undercuts are always present. The etch factor for a material is the ratio of the etched depth to the size of undercut.
- Controlling the size of small holes in thin sheet is difficult.
- Compensation for the undercut should be taken into account when designing the masking template.
- Inside edges always have radii. Outside edges have sharp corners.
- Possible to machine thin and delicate sections due to no processing forces.
- Minimum thickness = 0.013 mm.
- Maximum depth of cut = 13 mm.
- Maximum size = 3.7 m × 15 m, but dependent on bath size.

Quality Issues

- Residual stresses in the part should be removed before processing to prevent distortion.
- Surfaces need to be clean and free from grease and scale to allow good masking adhesion and uniform material removal.
- Masking material should not react with the chemical etchant.
- Parts should be washed thoroughly after processing to prevent further chemical reactions.
- Porosity in castings/welds and intergranular defects are preferentially attacked by the etchant. This causes surface irregularities and non-uniformities.
- Room temperature and humidity, bath temperature and stirring need to be controlled to obtain uniform material removal.
- Surface detail is good.
- Surface roughness values are in the range 0.4–6.3 μm Ra and are dependent on the material being processed.
- Process capability charts showing the achievable dimensional tolerances for various materials are provided (Figure 7.5(b)).

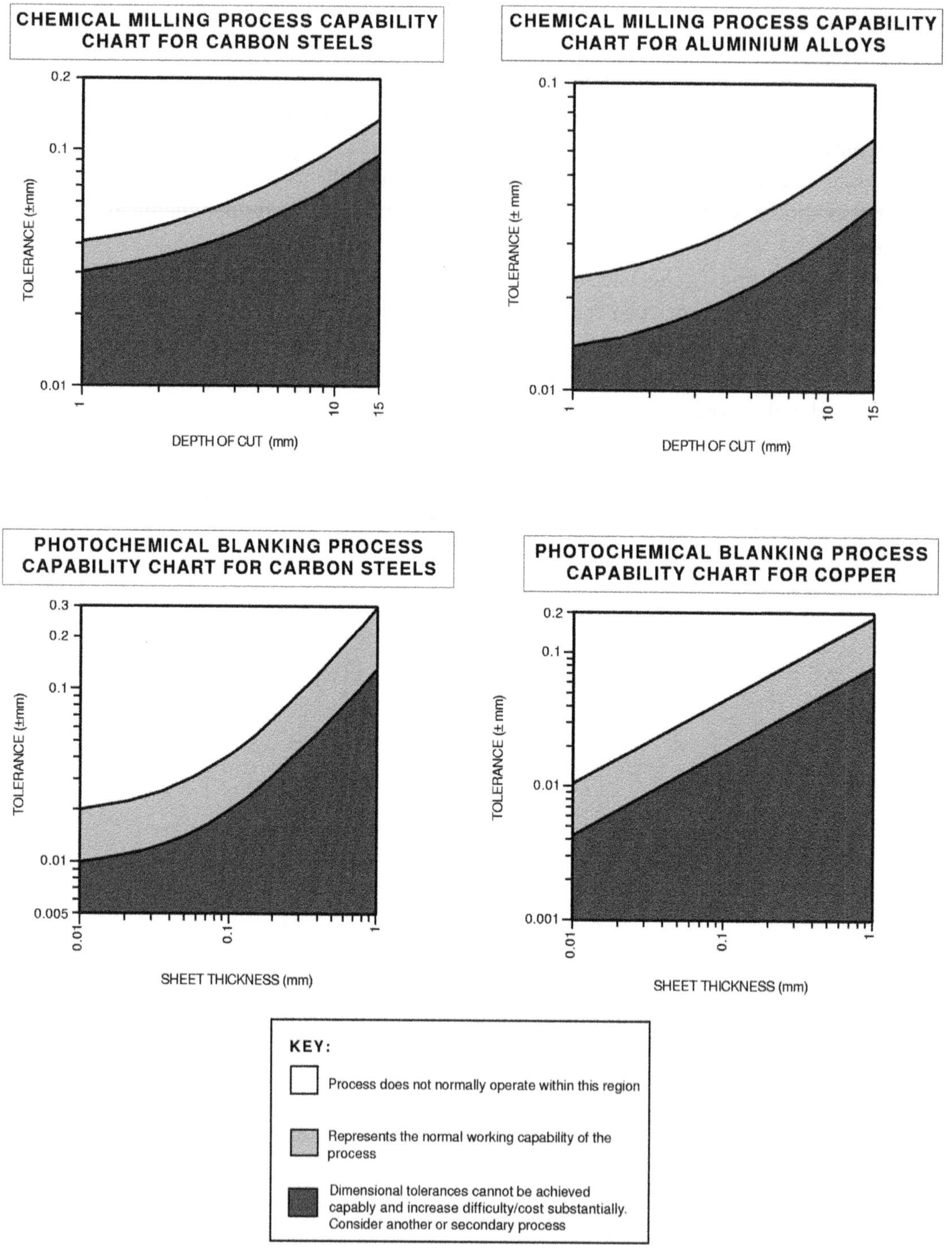

Figure 7.5(b): Chemical Machining Process Capability Charts.

7.6 Ultrasonic Machining (USM)

Process Description

The tool, which is negative of the workpiece, is vibrated at around 20 kHz with an amplitude between 0.013 and 0.1 mm in an abrasive grit slurry at the workpiece surface. The workpiece material is removed by essentially three mechanisms: hammering of the grit against the surface by the tool, impact of free abrasive grit particles (erosion), and microcavitation. The slurry also removes debris away from the surface. The tool is gradually moved down, maintaining a constant gap between the tool and workpiece surface (Figure 7.6).

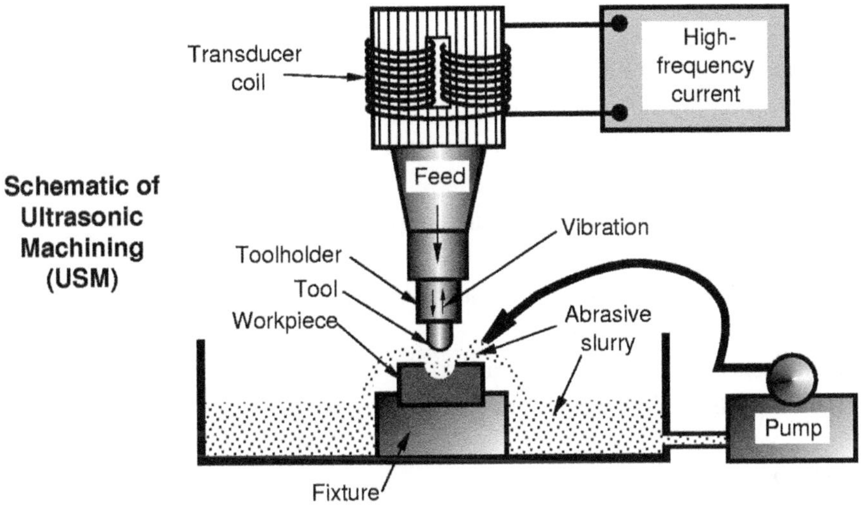

Figure 7.6: Ultrasonic Machining.

Materials

Any material; however, brittle hard materials are preferred to ductile, for example ceramics, precious stones, tool steels, titanium and glass.

Process Variations

- Vibrations are either piezo-electric or magnetostrictive transducer generated.
- Tool materials vary with application and allowable tool wear during machining. Common tool materials are: mild steel, stainless steel, tool steel, aluminium, brass and carbides (higher wear rates are experienced with aluminium and brass).

- Abrasive grit available in many grades and material types. Materials commonly used are: boron carbide, aluminium oxide, diamond and silicon carbide.
- Liquid medium can be water, benzine or oil. Higher viscosity media decrease material removal rates.
- Rotary USM: a rotating diamond-coated tool is used for drilling and threading, but with no abrasive involved.
- Ultrasonic cleaning: uses high-frequency sound waves in a liquid causing cavitation that cleans the surface of the component, similar to a scrubbing action. Used to remove scale, rust, etc.

Economic Considerations

- Production rates very low.
- Material removal rates are low, typically 13 mm^3/s.
- Linear penetration rates up to 0.4 mm/s.
- Lead time typically days, depending on complexity of tool. Special tooling required for each job.
- Material utilisation poor. Scrap material cannot be recycled.
- High degree of automation possible.
- Economical for low production runs. Can be used for one-offs.
- Tooling costs are high.
- Equipment costs are generally moderate.
- Direct labour costs are low to moderate.

Typical Applications

- Burr-free holes and slots in hard, brittle materials.
- Complex cavities.
- Coining operations.

Design Aspects

- Limited to shape of tool and control in two dimensions.
- Tool and tool holder designed with mass, shape and mechanical property considerations.
- Avoid sharp profiles, corners and radii as abrasive slurry erodes them away.
- Overcut will be produced that is approximately twice the grit size.
- Suitable for small-diameter holes with length to diameter ratios typically 3:1. Up to 4:1 using special equipment.

- Waste removal limits hole depths.
- Maximum hole size = 90 mm.
- Minimum hole size = 0.08 mm.

Quality Issues

- Tapering of slots and holes occurs.
- Through-holes in brittle materials should have a backing plate.
- Amplitude and frequency of vibration, tool material, impact force, abrasive grit grade and slurry viscosity and concentration all impact on accuracy, surface roughness and material removal rate.
- Finishing cuts are made at lower material removal rates.
- Tool wear problematic. Tool changes can be frequent.
- Part is burr free with no residual stresses, distortion or thermal effects.
- Difference in wear rate between the tool and workpiece materials should be as high as possible.
- Surface detail is good.
- Surface roughness values are in the range 0.2–1.6 μm Ra.
- Finer surface roughness values are obtained with finer grit grades.
- Achievable tolerances are in the range ±0.005 to ±0.05 mm. (Process capability charts have not been included. Capability is not primarily driven by characteristic dimension.)

7.7 Abrasive Jet Machining (AJM)

Process Description

Erosive action of an abrasive in a fluid is focused to a high-velocity (150–300 m/s) jet through a sapphire nozzle. The abrasive and fractured particles are carried away from the cutting area by the jet (Figure 7.7).

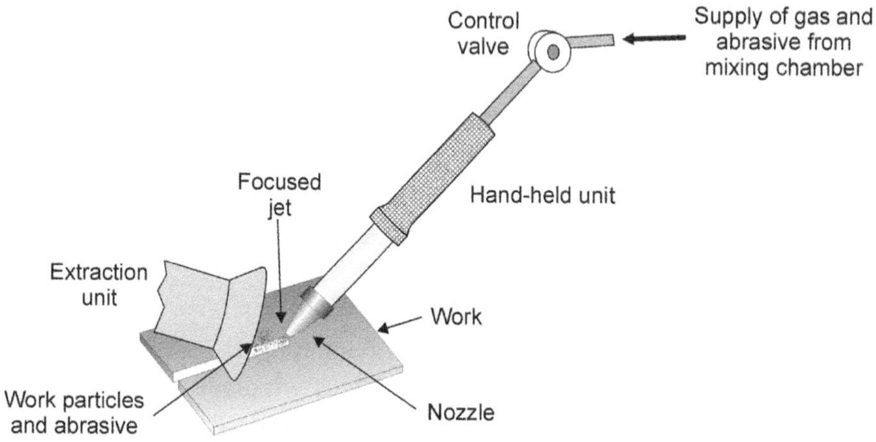

Figure 7.7: Abrasive Jet Machining.

Materials

- Suitable for brittle and/or fragile materials.
- Refractory metals, titanium alloys, ceramics, metallic honeycomb structural materials, acrylic, composites, glass, silicon and graphite.

Process Variations

- Two systems are used for introducing the abrasive to the jet stream:
  - Entrainment system: pressurised water jet pulls abrasive particles into the stream, they are mixed in a tube and exit the nozzle.
  - Abrasive slurry system: mixing of fluid medium and abrasive particles takes place prior to pressurisation in a separate chamber to create the slurry. Higher wear rates throughout the equipment experienced using this system, but less expensive.
- Fluid medium: either water or a gas (air or CO_2).
- Abrasive types: aluminium oxide and silicon carbide use.
- Tungsten can also be used for the nozzle, but has a higher wear rate than sapphire.

- Nozzle orifice can be round or square.
- Water jet machining: very-high-pressure focused jet of water used for cutting food, leather, paper and foamed plastics.
- Chemical jet machining: uses a single jet of etchant for deburring.

Economic Considerations

- Production rates are moderate.
- Material removal rates are low, typically 15 mm^3/min.
- Penetration rate ranges from 10 to 1200 mm/min.
- Removal rate depends on the hardness of material and process parameters.
- Material utilisation is poor. Scrap material cannot be recycled.
- Can be fully automated using robots. Added flexibility.
- Small power requirements needed.
- Economical for low production runs.
- Tooling costs are high.
- Equipment costs are generally high.
- Direct labour costs are low to moderate, depending on degree of automation.

Typical Applications

- Through-holes, slots and profiles in hard, brittle materials.
- For cutting, slitting, drilling, contouring, etching, cleaning, deburring and polishing.
- Electronic component etching.
- Etching and cutting glass.
- Cutting metal foils.

Design Aspects

- Features limited to profiles, holes and slots.
- Depth of cut can be increased with jet pressure.
- Blind holes not possible.
- Long holes have tapered walls.
- Slot widths range from 0.12 to 0.25 mm.

Quality Issues

- No heat and therefore no heat-affected zone. Part free from metallurgical effects and residual stresses.
- Minimal dust, toxicity and fire hazard, but high noise levels.

- Less than 1 mm focus length from work should be maintained so no loss of definition and stray abrasion occurs.
- Minimal tool dulling.
- Inclination of jet angle to work can be less than 90°, but at increased jet divergence on work and therefore less control over material being cut.
- Abrasive size, slurry composition and flow rate are important control variables of the process for consistency.
- Abrasive slurry cannot be recycled due to abrasive grit blunting reducing effectiveness.
- Abrasive can become embedded in work surface.
- Surface detail is good to excellent.
- Surface roughness values are in the range 0.1–1.6 μm Ra.
- Surface roughness depends on abrasive particle size.
- Achievable tolerances are in the range ±0.001 to ±0.013 mm. (Process capability charts have not been included. Capability is not primarily driven by characteristic dimension.)

CHAPTER 8

Rapid Prototyping Processes

8.1 Stereolithography (SLA)

Process Description

A low-power laser beam is directed by a mirror to trace a thin 2D cross-section in a liquid photopolymer resin that is stored in a build chamber. Through the process of photopolymerisation, a layer of hardened and cured 3D pixels of polymer are created, a process which only takes place near the surface due to absorption and scattering of the laser beam. The build platform is lowered down an amount equal to the thickness of the previously cured layer, and a sweeper blade moves across replenishing a layer of resin. The laser traces out the next 2D cross-section and the process is repeated. The self-adhesive property of the material used causes the layers to bond and, after many layers have been cured, a 3D part is built up. The completed part is drained, washed in solvent to remove excess resin and subjected to UV light to cure the component completely, after which support constructions are removed by cutting (Figure 8.1).

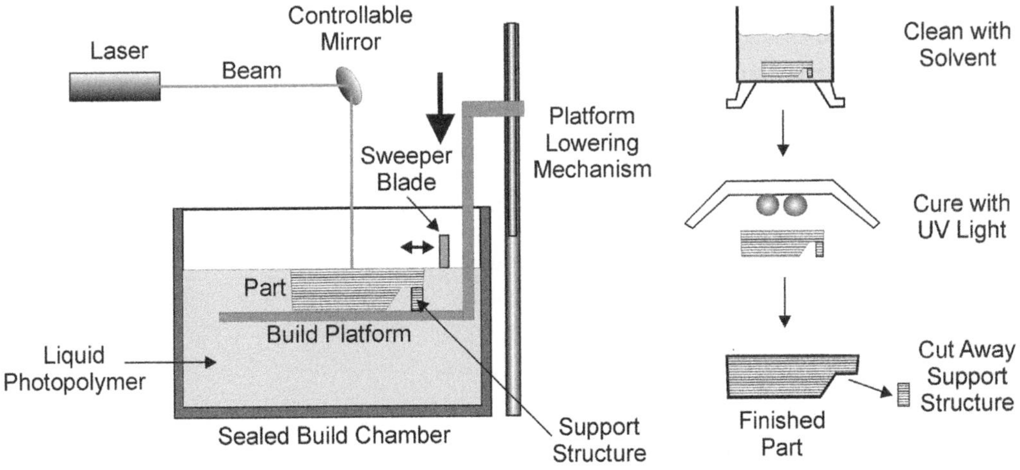

Figure 8.1: Stereolithography.

Manufacturing Process Selection Handbook. http://dx.doi.org/10.1016/B978-0-08-099360-7.00008-2

Materials

- Liquid form of photopolymers such as elastomer, epoxy, urethane, acrylate and vinyl ether.
- Liquid photopolymer can be clear or white.

Process Variations

- Laser types can be UV He:Cd or argon ion, up to 1 W in power.
- Solid Ground Curing (SGC): uses UV light to cure an entire layer of light-sensitive resin, the area being pre-defined by a photomask.
- Solid Object Ultraviolet Laser Plotter (SOUP): the laser is mounted and moved by a controlled mechanism in the horizontal plane above the build chamber of photopolymer, rather than being static and directed by mirrors.

Economic Considerations

- Moderately fast build speed, though large parts can take a day to fully complete.
- Automated process, but skilled labour required.
- Lead times are 1–2 weeks.
- Material utilisation is high. Support structures represent waste.
- Waste material must be disposed of properly. Cannot be recycled.
- Reworking costs are high.
- Photopolymers are expensive and the process requires that the whole build chamber is filled.
- Economical for low production runs 1–20 parts.
- Equipment costs are moderate.
- Direct labour costs are low.
- Some finishing may be required, but final curing also adds time.

Typical Applications

- Non-functional prototypes.
- Form and fit parts in assemblies.
- Product concept models.
- Medical models.
- Casting patterns.
- Snap fits and hinges.

Design Aspects

- Complex and intricate 3D parts created in CAD.
- Undercut features, cantilevered walls and overhangs require support structures to be designed as part of the CAD model or manually on the machine.

- Part should be orientated in the build chamber in order to reduce the amount of support structures.
- Supports can take the form of gussets, ceilings, webs or points.
- Anisotropy in material properties exist due to the additive layer method. Strength weakest in vertical build direction.
- Tensile strengths up to 75 MPa are possible, depending on material used.
- Increased strength can be achieved at certain locations by increasing layer thickness.
- Conventional machining processes can be used for non-standard details, e.g. threads.
- The finished part is translucent.
- Layer thickness = 0.025–0.15 mm.
- Typical maximum dimensions of part = 600 mm × 500 mm × 500 mm.
- Minimum section = 0.1 mm.

Quality Issues

- Calibration of the laser is required periodically for some systems.
- Green creep distortion tests can help assess the dimensional stability of the laser-cured resins.
- Laser exposure and scan speeds need to balance full curing and process speed.
- Total thickness layer is related to the scanning speed and depth penetration factor of the photopolymer resin.
- A short delay called a 'dip delay' is required after laser curing of a layer, so that liquid photopolymer can settle flat and evenly, inhibiting bubble formation.
- The liquid photopolymer is difficult to work, toxic and pungent, and requires safe handling procedures.
- The selection of the correct photopolymer is associated with strength, shrinkage and distortion of the part that is tolerable post-curing.
- Material properties may change over time, accelerated by light, moisture, heat and chemical environments.
- The machine is sealed to prevent toxic fumes from escaping, generated by the laser solidifying the resin.
- Avoid the location of support structures on planes where good surface finish is required.
- Support structures must be removed from part manually, but the process is simple and quick.
- Additional post-processing may be required for part finishing.
- The durability of the part is limited when exposed to sunlight.
- Finished part can be painted.
- Tolerances achievable = ±0.1 to ±0.2 mm.
- Typical surface roughness = 2 μm Ra.
- Surface finish is excellent in comparison to other rapid prototyping technologies.

8.2 3D Printing (3DP)

Process Description

A printing head (similar to those found in inkjet printers) deposits a liquid binder on to a powder in a build chamber. The powder particles become bonded together and the build platform is lowered down an amount equal to the thickness of the layer created. The powder is replenished in the build chamber from a similar powder supply chamber adjacent to it, compacted and levelled on top of the last bonded layer using a roller. The process is repeated, building up a 3D part. The completed part is cleaned of excess powder and typically impregnated with a sealant (Figure 8.2).

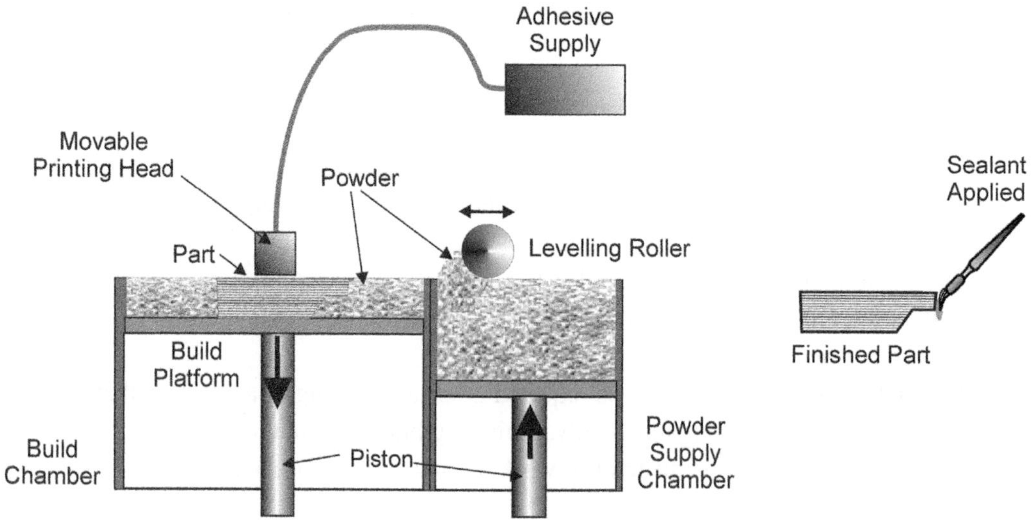

Figure 8.2: 3D Printing.

Materials

- Powder form of stainless steel, bronze, ceramics, moulding sand, plaster and starch.
- Liquid form of binder such as wax, epoxy resin, elastomer and polyurethane.

Process Variations

- Single-head (takes several passes) or multi-head (just one pass) printing devices used.
- A sintering stage can be added to the process to further bond the powders.
- Jetted Photopolymer System (JPS): wide array of printing heads are used to deposit a photopolymer resin, as used in Stereolithography (see PRIMA 8.1), which is cured through exposure to UV light.
- Thermal Phase Change Inkjet Printing: uses two separate printing heads, one dispensing a thermoplastic melt and the other hot wax support material, to create a 2D layer that hardens on contact. A milling tool machines the surface level and the wax is dissolved or melted out.

Economic Considerations

- Fast build speed.
- Automated and very flexible process, but skilled labour required.
- Lead times are typically less than 1 week.
- Materials are low cost.
- Material utilisation is high.
- Excess powder material is reusable.
- Equipment costs are low.
- Direct labour costs are low.
- Finishing processes are generally required, which adds cost.

Typical Applications

- Product concept models.
- Models for ergonomic testing.
- Surgical planning models.
- Coloured models, e.g. stress patterns from Finite Element Analysis (FEA).
- Architectural models.
- Non-functional prototypes.
- Consumer goods.
- Packaging.
- Models for casting and moulding.
- Patterns and cores for casting processes.

Design Aspects

- Complex and intricate 3D parts created in CAD.
- Any powder material not bonded and unused in the build chamber acts as a support medium for any overhanging or undercut features on the part. No additional support structures are therefore needed.
- Parts should be orientated in the build chamber with the height being the smallest dimension in order to reduce the number of layers created.
- Anisotropy in material properties exist due to additive layer method. Strength weakest in vertical build direction.
- Tensile strength depends on the material being used, e.g. plaster powder models have <5 MPa tensile strength.
- Binder and powder each contribute approximately 50% of material by volume to a component.
- Layer thickness = 0.05–0.15 mm.
- Typical maximum dimensions of part = 600 mm × 500 mm × 400 mm.
- Minimum section = 0.2 mm.

Quality Issues

- Parts are fragile direct from the process.
- Epoxy resin and cyanoacrylate adhesive can be used to impregnate the finished part to improve finish, durability and part strength. Wax also used but not as strong as resins or adhesives.
- Can introduce multiple colours to the model.
- Accuracy is dependent on binder droplet size, powder size, the diffusion of the binder into the powder, and printer head positional resolution.
- 'Stair-stepping' features on sloping surfaces in the vertical build plane can be created.
- Not particularly suitable where an accurate fit is required for the part due to relatively poor tolerances and strength.
- Tolerances achievable = ±0.1 to ±0.5 mm.
- Typical surface roughness = 60 μm Ra.
- Finish is grainy and typically requires additional finishing processes.

8.3 Selective Laser Sintering (SLS)

Process Description

A high-power laser beam directed by a mirror is used to sinter powdered material in thin 2D cross-sections. The powder is in a sealed chamber with a nitrogen atmosphere to prevent oxidation. It is pre-heated using infrared heaters to a temperature just below the melting temperature of the powdered material. The build platform is lowered down an amount equal to the thickness of the sintered layer. A roller replenishes the layer of powder from adjacent powder supply chambers, the laser traces out the next 2D cross-section, and the process is repeated until a 3D structure is built up. Excess powder not sintered on each layer acts as a support for the part being built. The part is removed and excess powder is removed by brushing or vacuuming (Figure 8.3).

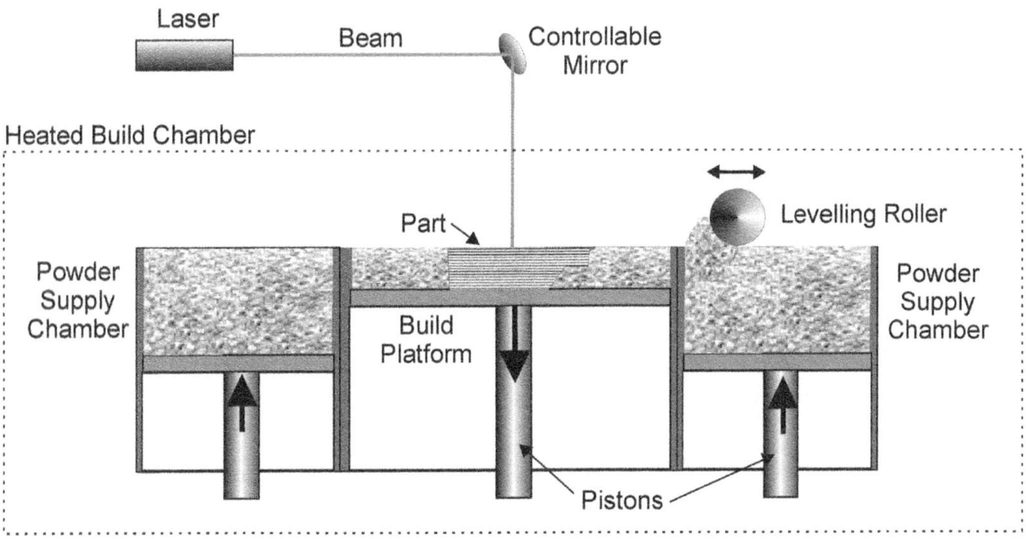

Figure 8.3: Selective Laser Sintering (SLS).

Materials

- Powder form of heat-fusible thermoplastics (including glass filled), elastomers and wax.
- Powdered metals with binder, e.g. stainless steel, tools and alloy steels, titanium, tungsten, copper alloy, aluminium and nickel super alloys.
- Powdered ceramics and moulding sand (with binder).

Process Variations

- Direct Metal Laser Sintering (DMLS): used for processing metal and ceramic composite parts using very fine powders with no binders, and high-powered lasers to sinter the particles directly in an inert atmosphere. Achieves very high part densities, but support structures are required for overhanging features. Typically parts will go through a number of post-processing stages, including support removal, shot peening (see PRIMA 9.6) and polishing, to improve fatigue resistance and finish.
- Laser types: Yb-fibre or CO_2, up to 100 W in power. Up to 200 W for DMLS.

Economic Considerations

- Medium build speed.
- Automated process, but skilled labour required.
- Lead times are typically 1 week.
- Material utilisation is high.
- Leftover powder material is directly reusable.
- Powdered material cost is low to moderate, depending on type.
- Economical for production runs of 1–50+, depending on size and complexity. DMLS can be viable for larger volumes, 1,000+.
- Equipment costs are high.
- High energy requirement.
- Direct labour costs are low.
- Some finishing may be required.

Typical Applications

- Final functional components in low to moderate volumes, e.g. impellers, fuel nozzles for aerospace sector.
- Functional prototypes, e.g. for experimental testing, wind tunnels, etc.
- Form and fit parts in assemblies.
- Product concept models.
- Patterns, moulds and cores for casting and moulding.
- Rapid tooling.
- Medical and dental implants.
- Models for ergonomic testing.
- Snap fits and hinges.

Design Aspects

- Complex and intricate 3D parts created with CAD.
- Any powder material not sintered in the build chamber acts as a support medium for any overhanging or undercut features on the part. No additional support structures are therefore needed.

- Parts should be orientated in the build chamber with the height being the smallest dimension in order to reduce the number of layers created.
- Anisotropy in material properties exist due to additive layer method. Strength weakest in vertical build direction.
- Approximately 50 MPa tensile strength for some thermoplastics, 300 MPa for bronze-impregnated steel, 1 GPa+ for titanium using DMLS.
- Parts can be machined using conventional processes.
- Layer thickness = 0.075–0.15 mm, depending on material used.
- Thinner layers can be achieved with DMLS >0.02 mm.
- Typical maximum dimensions of part = 450 mm × 375 mm × 325 mm.
- Minimum section = 0.4 mm.

Quality Issues

- Control of build chamber temperature just below the material's melting point is important to facilitate fusion between layers as the heat from the laser only needs to raise the temperature a small amount for sintering. Also important in order to minimise thermal distortions.
- Large parts can take many hours to cool before handling is possible.
- Machine needs thorough cleaning when changing material powder types to avoid contamination.
- Stress relieving and annealing may be required for metal parts to reduce internal residual stresses.
- Additional curing may be required when using ceramic materials.
- The nitrogen atmosphere maintained in the build chamber also prevents possibility of explosion of the powder.
- 'Stair-stepping' features on sloping surfaces in the vertical build plane can be created.
- Greater than 99% densities can be achieved after sintering; 100% density is achievable when impregnated with another material.
- Impregnation improves mechanical properties and surface finish. Typically a lower melting temperature metal alloy such as an alloy of copper is used for impregnation.
- Finished part can be painted and surface coated.
- Tolerances achievable = ±0.05 to ±0.25 mm.
- Typical surface roughness = 7–10 μm Ra.
- Finish is related to coarse powders used, and may require additional finishing processes. Powder diameter is usually between 20 and 100 μm.

8.4 Laminated Object Manufacturing (LOM)

Process Description

Sheet material coated with an adhesive is moved into the build area using a feed roll and pressure is applied using a heated roller to bond to the layer below. The sheet is cut using a CO_2 laser beam directed by a mirror and optic heads to create the required 2D profile. The build platform is lowered down an amount equal to the thickness of the layer created and the process is repeated, building up a 3D part. Excess sheet surrounding the 2D profile is cross-hatched with the laser for easier removal (chopped away in sections later) and the remaining sheet is moved away on a waste take-up roll. The finished part is removed and is typically sanded down to improve the surface finish and then sealed (Figure 8.4).

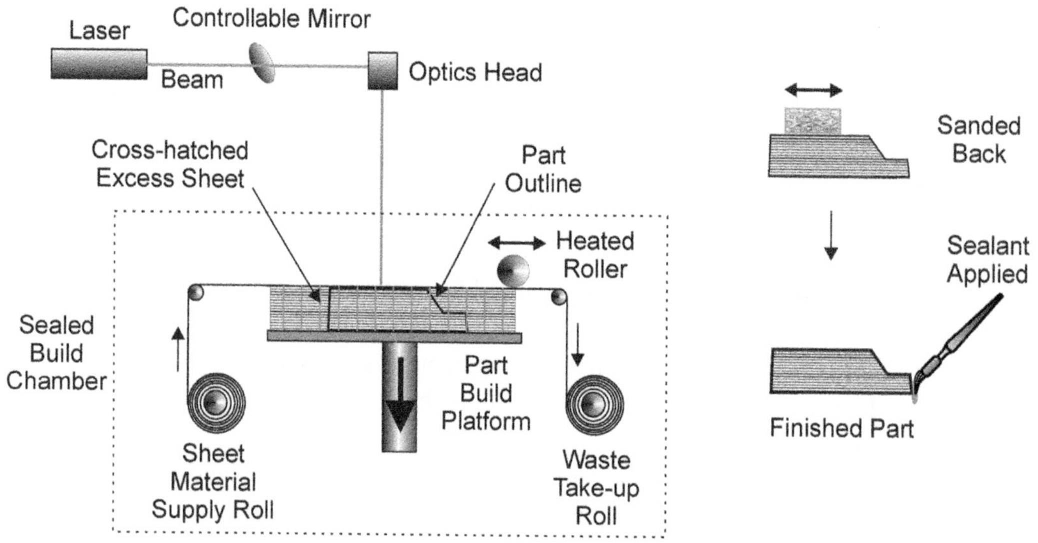

Figure 8.4: Laminated Object Manufacturing (LOM).

Materials

Thin sheet form of paper, some thermoplastics, metal foils and ceramics.

Process Variations

- Several commercial variants use a knife instead of a laser to cut the sheets.
- Solvents rather than adhesives can be used to bond sheets of thermoplastics.
- CO_2 laser power varies between 25 and 50 W depending on material being cut.

Economic Considerations

- Medium speeds, but varies widely depending on material and complexity.
- Lead times are less than 1 week.
- Material costs are very low.
- Material utilisation is low to moderate.
- Waste material is not reusable, although possible to recycle.
- Economical for one-offs.
- Equipment costs generally moderate.
- Direct labour costs are moderate. Skilled labour required at all times.
- Finishing time and costs can be moderate to high.

Typical Applications

- Large product concept models.
- Non-functional prototypes.
- Casting patterns and cores.
- Tooling models.

Design Aspects

- Complex and large 3D parts created with CAD.
- Cannot create hollow parts.
- No additional support structures are required.
- Undercuts and re-entrant features are very difficult to create without manual intervention.
- Parts should be orientated in the build chamber with the height being the smallest dimension, in order to reduce the number of layers created.
- Tensile strength of finished part is highly anisotropic, e.g. for paper <5 MPa tensile strength at right angles to the lay-up direction, >60 MPa in line with the lay-up direction.
- Not suitable for conventional machining processes due to possibility of delamination.
- Layer thicknesses range = 0.05–0.2 mm.
- Typical maximum dimensions of part = 800 mm × 550 mm × 500 mm.
- Minimum section = 0.2 mm.

Quality Issues

- Little shrinkage and distortion during processing.
- Can be complex and time consuming to create certain geometries.
- Not suitable for parts subjected to any shear forces in the axis of the build plane due to relatively weak bonded nature of sheets.
- 'Stair-stepping' features on sloping surfaces in the vertical build plane can be created.
- Laser is carefully modulated to penetrate to a depth of exactly one layer thickness.

- Combustible materials such as paper represent a fire hazard.
- Build chamber must be sealed to, as the process produces smoke that needs extraction and fire extinguishing equipment located in the build chamber.
- Post-processing is needed to protect the component from ingress of moisture through the sheet layers, which can cause swelling.
- Variety of sealants used on finished part: wax, paint, varnish, urethane, silicon or epoxy resin depending on protection needed.
- Finished part can also be painted to improve appearance and to add colour.
- Tolerances achievable = ±0.1 to ±0.25 mm.
- Typical surface roughness = 30–40 μm Ra.
- Surface finish is poor in comparison to other rapid prototyping processes, and requires additional finishing processes.

8.5 Fused Deposition Modelling (FDM)

Process Description

Solid material, usually in filament form, is melted and extruded through a heated nozzle to create a molten bead of build material. The build chamber is maintained at a temperature just below the melting point of the build material. The controllable nozzle is moved in the horizontal plane, depositing the molten bead to create a thin layer of the required 2D profile. The molten bead solidifies and effectively cold welds on contact with the previous layer. The build platform is lowered down an amount equal to the thickness of the solidified layer, and the process is repeated, building up a 3D part. Additional support material for overhangs and undercuts is simultaneously deposited during the build process using a second nozzle. The support material can be dissolved away after the part is removed from the build chamber (Figure 8.5).

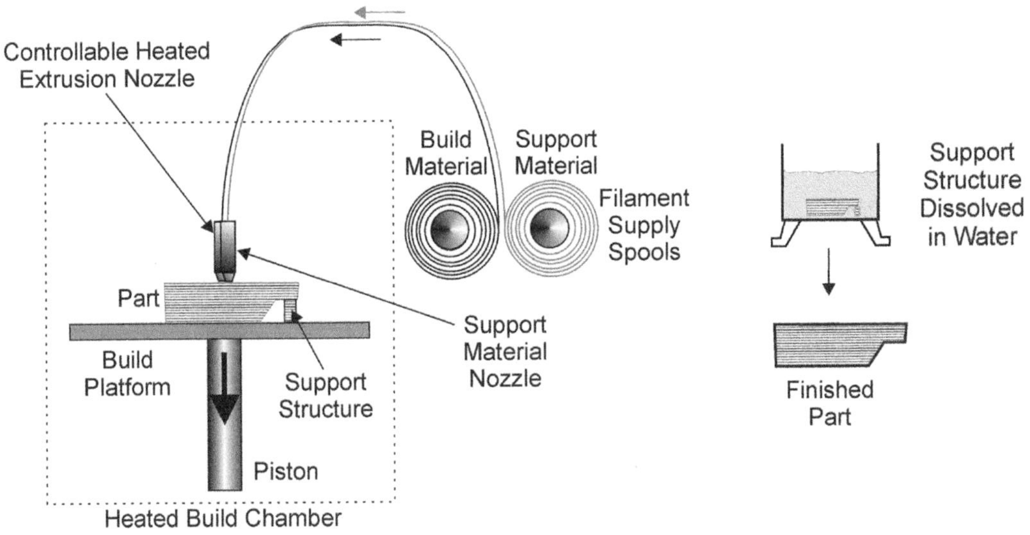

Figure 8.5: Fused Deposition Modelling (FDM).

Materials

- Build material supplied in filament coils (diameter 1.5 mm).
- Commonly, wax, elastomers and a number of thermoplastics are used as the build material.
- Ceramics (with binder material), eutectic metals and glass-fibre-reinforced materials have been used to produce components on a limited basis.
- Support materials include wax and nylon-like material.

Process Variations

- Some process variants use pellet form of the build material rather than filament.
- Support materials are either water-soluble or broken away.

Economic Considerations

- Build speeds are slow (to medium), but dependent on cross-section size.
- Lead times are 1–2 weeks.
- Material utilisation is high.
- Both support and build material are moderately expensive.
- Economical for low production runs of small components and one-offs.
- Equipment costs generally low to moderate.
- Direct labour costs are low.
- Finishing costs are low to moderate.
- Support material removal and surface finishing are typical post-processing operations.

Typical Applications

- Functional prototypes, e.g. for experimental testing, wind tunnels, etc.
- Product concept models.
- Patterns and cores for casting processes.
- Rapid tooling.
- Medical models.

Design Aspects

- Complex and intricate 3D parts created with CAD.
- Support structures must be designed and fabricated for undercuts and overhanging features.
- Supports can take the form of boxes, ceilings or webs.
- Part should be orientated in the build chamber in order to reduce the amount of support structures.
- More suited to small-volume, medium section components.
- Difficult to build thin wall sections, acute angles or sharp edges in vertical plane due to contact pressure with extruded bead from nozzle, which could cause deformation.
- Can produce internal structures in the part to save weight/volume, e.g. a lattice structure. Soluble support material is required.
- Conventional machining processes can be used for non-standard details, e.g. threads.
- Anisotropy in material properties exist due additive layer method. Strength weakest in vertical build direction.
- Typical tensile strength is approximately two-thirds of the strength of the same thermoplastic that has been injection-moulded.

- Tensile strength for thermoplastics: ABS = 35 MPa, PC = 60 MPa.
- Layer thickness = 0.05–0.75 mm.
- Typical maximum dimensions of part = 600 mm × 600 mm × 500 mm.
- Minimum section = 0.3 mm.

Quality Issues

- Control of build chamber temperature is important to minimise energy needed to melt the filament at the extrusion nozzle.
- Layer thickness and build accuracy is related to the nozzle diameter.
- Nozzle speed and material extrusion rate require control to provide consistent deposition rate and layer thickness.
- 'Stair-stepping' features on sloping surfaces in the vertical build plane can be created.
- Components exhibit virtually no porosity.
- Process can be installed anywhere being non-toxic and environmentally safe.
- Different colour build material can be supplied.
- Uses materials with high structural stability, temperature, chemical and water resistance properties.
- Tolerances achievable = ±0.1 to ±0.25 mm.
- Typical surface roughness = 6–12 μm Ra.
- Some finishing may be required to improve surface finish, depending on application.

CHAPTER 9

Surface Engineering Processes

9.1 Carburising

Process Description

A variety of processes used to diffuse extra carbon (up to 0.8%) into the surface of steels that initially have low carbon content by heating to a temperature at which austenite in the steel has a high solubility for carbon introduced into the process atmosphere. Subsequent quenching creates a hard martensitic structure in the surface layer, maintaining a ductile core in the part (Figure 9.1).

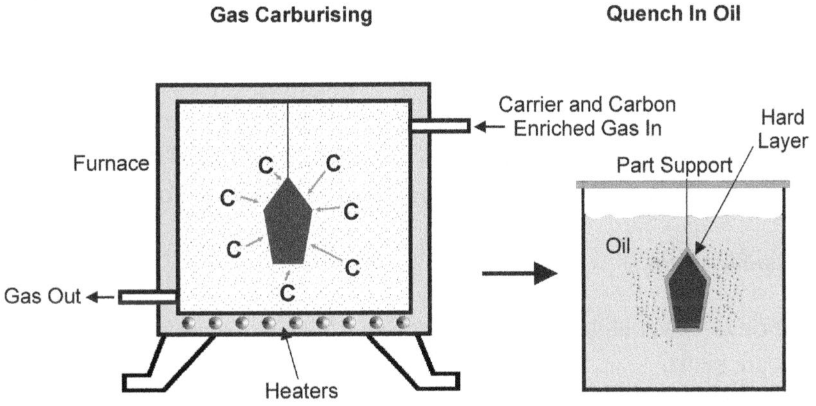

Figure 9.1: Carburising.

Materials

Wide range of steel compositions, but generally low carbon steel and low alloy steels where carbon content is less than 0.2%.

Process Variations

- Gas Carburising: the carbon availability of the carburising atmosphere is increased using a carrier gas, e.g. nitrogen, carbon monoxide and hydrogen, with a carbon-enriched gas, e.g. methane, propane or natural gas. Process temperature up to 980°C. Good control of case depth and can be a continuous operation with integrated quenching facilities in the furnace.

Manufacturing Process Selection Handbook. http://dx.doi.org/10.1016/B978-0-08-099360-7.00009-4

- Pack Carburising: parts are packed in a box with charcoal and barium carbonate (accelerates the diffusion process), which produces a carbon-rich atmosphere mainly composed of carbon monoxide. Process temperature up to 1090°C. Poor control of carbon depth, but relatively inexpensive though less utilised process. Quenching not possible directly.
- Salt Bath Carburising: parts are immersed in a molten salt bath containing mixtures of sodium cyanide and barium chloride. Process temperature is up to 950°C. Relatively fast process.
- Fluidised Bed Carburising: high heat transfer rates provide a rapid process and fuel efficiency, but difficulty in controlling carbon diffusion rates.
- Vacuum Carburising: a low vacuum pressure but higher temperature carburising process up to 1050°C giving good process control, but relatively expensive and low production rates.
- Plasma Carburising: plasma is created in an inert gas environment at about 1050°C increasing the carburising rate and depositing carbon on part surface. Effective process control and efficiency.
- Quenching in water, brine or oil is used to harden the surface after carburising. Typically immersed in a bath of the quenching medium, the choice of which is related to the cooling rates obtainable to minimise part distortion.
- Carbonitriding: lower temperature process generally, (up to 900°C) in which small amounts of nitrogen (0.5%) from ammonia gas, in addition to carbon, are diffused into the steel, producing lower distortion than carburising and a hard surface layer. There are many process variants, similar to carburising, e.g. salt bath, gas or plasma technologies.

Economic Considerations

- Production rates vary, but generally long processing time (several hours).
- Lead times are short.
- Equipment costs are high for vacuum carburising, low for pack carburising.
- Tooling costs are generally low.
- Applicable to low to medium production volumes, generally.
- Labour costs are moderate. Requires some skilled labour.
- Some variant processes can be automated, e.g. gas carburising can be a continuous process.
- Finishing costs are low unless parts are highly distorted.
- Post-processing costs include quenching and cleaning.

Typical Applications

- Transmission shafts.
- Gears and sprockets.
- Bearing races.
- Bushes.

- Clutch plates.
- Cams and camshafts.
- Mandrels.
- Cold forgings.

Design Aspects

- Primarily for the improvement of wear resistance (through an increase in hardness) and fatigue resistance.
- All geometric shapes are possible, with some limitations.
- Non-uniform sections and abrupt changes in section should be avoided.
- Sharp corners should be avoided.
- Generous radii on shafts should be provided.
- Difficult to create uniform case layer for internal part features.
- Treatment thickness from 50 μm for salt bath carburising to 5 mm+ for gas carburising.
- Hardness range = 600–1200 HV.
- Part size limited by equipment size only.

Quality Issues

- Quality of the carburising atmosphere is difficult to continuously maintain.
- Difficult to control case depth using some carburising process variants.
- There are operational safety requirements associated with salt bath and gas carburising processes.
- Distortion can be considerable due to high process temperature, particularly around unsymmetrical features.
- Quench cracking can occur in the part with high cooling rates.
- Slow cooling then reheating to the austenising temperature removes high residual stresses whilst maintaining hard surface properties.
- Dimensional changes, that the process of carburising and quenching imparts on the part, are related to process used, and size, shape and material for the part, typically less than ±0.05 mm.
- Surface finish is poor to average without post-processing such as grinding.

9.2 Nitriding

Process Description

A variety of processes used to diffuse nitrogen into the surface of the alloy steel being treated, creating a hard layer of nitrides with the material's original stable alloying elements such as aluminium, vanadium, titanium, molybdenum and chromium. No quenching is necessary (Figure 9.2).

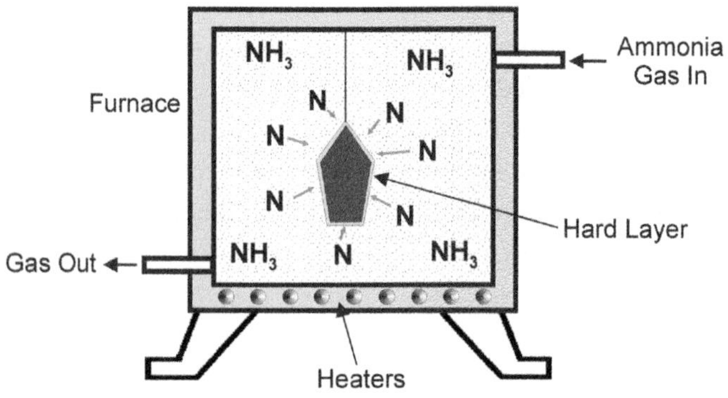

Figure 9.2: Nitriding.

Materials

Low alloy steels, stainless steels and high-speed tool steels.

Process Variations

- Gas Nitriding: the part is heated in dry ammonia gas (NH_3) at temperatures up to 530°C. The nitrogen dissociates, resulting in atomic nitrogen diffusing into the part surface. Slow process taking up to 80 hours, which is more suited to low volumes.
- Salt Bath Nitriding: preheated part is immersed in a bath of molten salts of sodium cyanide at a temperature of up to 570°C. For low volumes due to long treatment times.
- Plasma Nitriding: part is made the cathode and the reactor wall is the anode. Positively charged nitrogen ions from a plasma bombard the part, causing nitrogen to be absorbed into the surface. Lower temperature process down to 340°C caused by ion bombardment on part. Good process control of thickness treatment and faster process more suitable for highly complex parts.
- Nitrocarburising: nitrogen and carbon diffused into the surface using gas, salt bath or plasma technologies. Higher temperature process variant at 570°C and achieves only small depths of treatment in low carbon steels and cast irons.

Economic Considerations

- Processing times can be long.
- Lead times are short.
- Equipment costs are moderate to high.
- Tooling costs are low.
- Labour costs are low to moderate.
- Applicable to a wide range of production volumes.
- Finishing costs are low. No quenching necessary for hardening, but removal of white layer necessary.

Typical Applications

- Transmission shafts.
- Gears.
- Sprockets.
- Crankshafts.
- Valve stems.
- Diesel injectors.
- Pumps.
- Cutting tools.
- Extruder screws.
- Dies and moulds.
- Bolts and nuts.
- Machine guides.

Design Aspects

- Primarily for the improvement of fatigue resistance through the creation of high compressive surface residual stresses.
- Hardness also improved, which improves wear properties.
- Corrosion resistance improved (except for stainless steels).
- All geometric shapes are possible.
- Treatment thickness ranges from 2.5 μm to 0.75 mm.
- Typical hardness 500–1100 HV depending on process variant used.
- Part size limited by equipment size only.

Quality Issues

- Parts for nitriding should ideally be finish machined before the treatment and cleaned thoroughly.
- Careful process control required to avoid the creation of brittle white layer (iron nitrides), which can be up to 20 μm thick from gas nitriding and 8 μm for plasma nitriding. Can be ground away.
- Hardness developed is temper resistant up to approximately 560°C service temperature.
- Little or no distortion due to low temperatures used (compared to carburising, for example).
- Original dimensions are not altered by the treatment.
- Surface finish is good.

9.3 Ion Implantation

Process Description

Impurity ions are accelerated by an electromagnetic field and guided by electrostatic plates towards the surface of the part, which is contained in a vacuum chamber. The high-energy ions penetrate the surface of the part and interact with the atoms, modifying the chemical, physical and electrical properties of a thin layer whilst the bulk of the part is not altered (Figure 9.3).

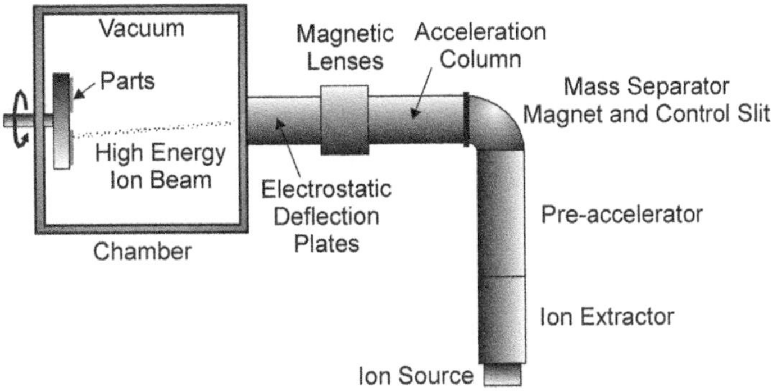

Figure 9.3: Ion Implantation.

Materials

- Most metals, e.g. aluminium alloys, stainless steel, tool steel and die steel. Also ceramics and polymers.
- Large choice of impurity ions, e.g. nitrogen is used for ion implantation of steels to create fine hard nitrides in steels; boron is implanted into silicon wafers for dosing.

Process Variations

- Process operates through a range of low temperatures (200–400°C).
- Plasma Surface Ion Implantation: direct implantation of high-energy ions from a plasma contained in a vacuum chamber with the part to be coated.

Economic Considerations

- Production rates are low.
- Lead times are moderate.
- Equipment costs are very high due to ion and vacuum equipment.
- Tooling costs are moderate. Special fixtures are required to uniformly treat part geometries.

- Very versatile process across a range of materials.
- Applicable to a wide range of production volumes.
- Labour costs are high. Requires skilled labour.
- No finishing required.

Typical Applications

- Silicon wafers.
- Localised hardening of injection mouldings.
- Medical implants.
- Cutting tools.
- Dies and moulds.
- Machine parts.

Design Aspects

- Primarily improves resistance to wear through an increase in hardness.
- Improves resistance to corrosion.
- Reduces the coefficient of friction of the surface.
- Line of sight process, therefore complex geometries are more difficult to treat uniformly.
- Treatment thickness ranges from 0.01 to 1 μm.
- Hardness ranges from 500 to 650 HV.
- Part size limited by vacuum chamber size only.

Quality Issues

- Very controllable and repeatable process.
- Selection of ion beam parameters is crucial, e.g. impurity ion type, energy levels and number of ions that impact the surface in a given time (dose).
- A high vacuum must be maintained for process to be effective.
- High compressive residual stresses are created due to ion implantation damage.
- No distortion of the part is experienced due to low processing temperatures, therefore can be applied to already finished machined parts.
- Does not suffer from delamination as in surface coating processes.
- The process does not adversely affect component dimensions or bulk material properties.
- Positional tolerance of ion implantation = ±0.001 mm.
- Surface finish is not altered from the original.

9.4 Anodising

Process Description

A process by which the naturally existing, but readily damaged, oxide layer on the surface of aluminium parts is increased. The aluminium part (anode), previously prepared with a number of pre-treatments, and lead (cathode) are immersed in an electrolyte (dilute sulphuric acid). A d.c. current passed between them breaks down the oxygen in the water, which combines with the aluminium to form aluminium oxide at the surface of the parts. The acid in the electrolyte keeps the oxide layer porous until the part is rinsed and then immersed in boiling water, which seals the surface pores (Figure 9.4).

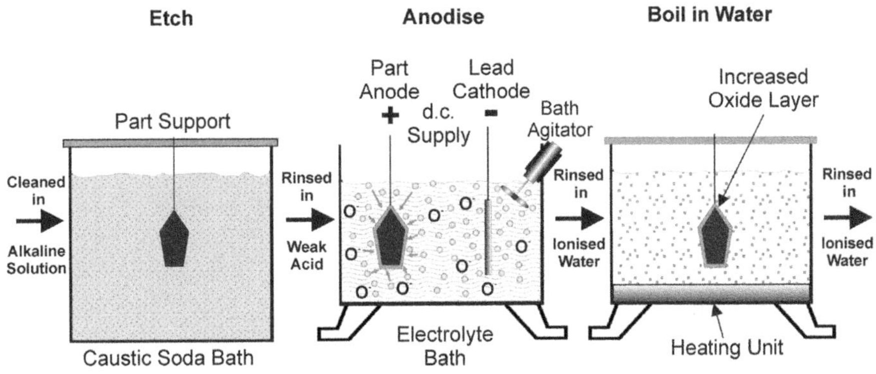

Figure 9.4: Anodising.

Materials

- Aluminum and its alloys typically.
- Other elemental metals can also be treated in a similar way, e.g. zinc, magnesium and titanium.

Process Variations

- Range of dilute acids used as electrolyte, e.g. 15% sulphuric acid gives a soft, easily dyed coating. Organic acids give a hard coating.
- The temperature of the electrolyte gives different properties of the anodised surface, e.g. sulphuric acid at 20°C gives a soft, transparent, easily dyed coating; at 5°C it gives a hard, dense, dull grey coating (hard anodising).

- Colour can be introduced to a part's surface in a number of ways: immersion in a bath that contains dyes or inorganic pigments that are absorbed into the porous surface and later sealed in boiling water; immersion into a special electrolyte with time of immersion, temperature and electrical current controlled; colouring of the intermetallic particles that are spread throughout the depth of the anodic film; electrolytically deposit metal and metal oxide particles into the anodised layer and seal in.

Economic Considerations

- Processing rates are moderately slow.
- Lead times are short.
- Equipment costs are moderate.
- Tooling costs are low.
- Labour costs are low to moderate.
- Process can be automated for small parts.
- Applicable to a wide range of production volumes.
- Associated pre- and post-process and cleaning costs, with many stages involved.

Typical Applications

- Architectural parts.
- Food processing equipment.
- Fluid power cylinders and parts.
- Pistons.
- Nozzles.
- Pumps and valves.
- Gears.
- Brake mechanisms.
- Hinges and swivel joints.
- Keys.
- Aircraft wheels.
- Automotive trim.
- Picture frames.
- Moulds.

Design Aspects

- Primarily improves corrosion resistance.
- Provides a decorative finish on the surface.
- Electrically insulating treatment created.
- Improves emissivity of surface.

- Line of sight process, therefore complex geometries are more difficult to treat uniformly.
- Hardness ranges from 50 to 500 HV.
- Treatment thickness ranges from 0.25 to 75 μm (natural oxide layer on aluminium is 0.5 μm thick).
- Part size limited by bath size only, typically <2 m^3.

Quality Issues

- Surface preparation of the part is very important before anodising can take place. Oil, grease, oxides and other impurities must be removed using alkali and acid solutions.
- A bath agitator can help with consistency of ions in solution.
- Contamination of the baths and their solutions must be avoided. Disposal must be properly undertaken.
- Anodised coatings should be washed down regularly (at least once every 6 months in less severe applications and more often in marine and industrial environments).
- Organic dyes may fade with exposure to UV light.
- Can use a preparation for improved surface adhesion for paints, adhesives and lubricants.
- Does not create part distortion due to low processing temperatures.
- Bending of the part is not recommended due to brittle nature of, and possibility of cracking, of the anodised layer.
- Surface finish is primarily a function of the part's original surface finish.
- Not suitable for castings as they initially have a poor surface finish and porosity.

9.5 Thermal Hardening

Process Description

A variety of processes are used to locally heat the surface of the part (usually steel) until it reaches the austenising temperature before being rapidly quenched to form a hard martensite structure (Figure 9.5).

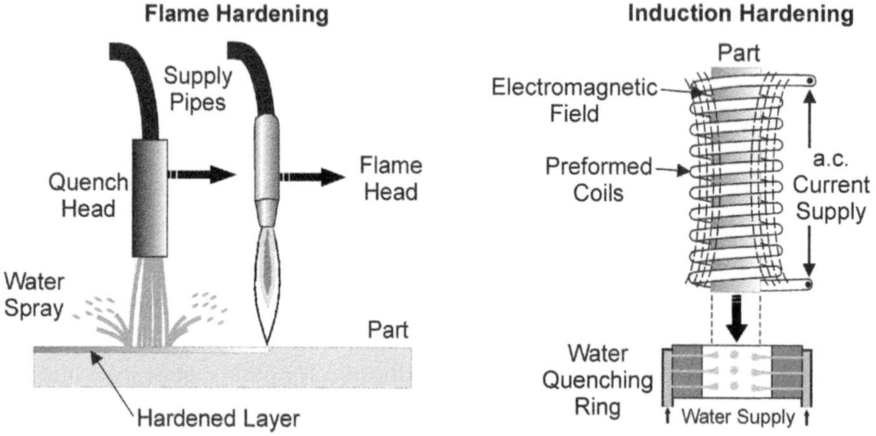

Figure 9.5: Thermal Hardening.

Materials

Limited to ferrous alloys with sufficient carbon content to be heat treated, e.g. medium carbon steels, low alloy steels and cast irons.

Process Variations

- Flame Hardening: a gas flame composed of a single (for localised heating) or multiple flame heads (for large surfaces) moves across the surface of the part, raising the temperature up to 850°C, followed by a water quenching head that sprays water at the surface previously heated. Choice of fuels, e.g. acetylene, propane or natural gas. The flame head can be static and the part moves at a predetermined speed in another process variation, or as in spin gardening, where the part spins in front of the flame head and is then quenched. The quenching system may be incorporated into the torch itself. The preferred process for larger parts and lower volumes.

- Induction Hardening: water-cooled, preformed copper coils in close proximity to the part surface have an a.c. electric current passed through them, generating an electromagnetic field that subsequently heats up the part surface using eddy currents. Temperatures up to 850°C can be achieved. High frequencies give a shallower depth of heat (500 kHz for 0.5 mm depth), whilst low frequencies give a deeper depth of heat (1 kHz for 5 mm depth). Generally, lower depth of hardness layer created than, for example, flame hardening, with a short cycle time (seconds to several minutes), making it suitable for smaller parts. Quenching with water, oil or air should quickly follow using a separate quench ring integrated with the coil assembly. Several types of inductor shape, e.g. pancake, coil (both internal and external) and scanning inductor, limits part complexity to match these inductor shapes.
- Electron Beam Hardening: a beam of electrons is used to heat up the surface of the part (see PRIMA 7.3). Highly repeatable and localised process that produces no distortion. Expensive equipment costs and is more suitable for smaller parts due to vacuum chamber size limitations. No quenching is necessary.
- Laser Beam Hardening: a high-power laser heats up the surface of the part (see PRIMA 7.4). Rapid and highly controllable localised process that produces no distortion, although expensive equipment costs and more suitable for smaller parts.
- Tungsten Inert-gas (TIG) Hardening: TIG welding torch used to locally harden cast iron by surface melting (see PRIMA 11.1).

Economic Considerations

- Fast process but usually applied to individual parts.
- Lead times are short generally.
- Equipment costs are low for flame hardening/high-frequency hardening, to very high for Electron Beam hardening.
- Tooling costs are moderate. Induction hardening requires preformed coils matched to the part geometry.
- Labour costs can be high with flame hardening, significantly reducing with other processes.
- Process allows the hardening of selected areas of a part that are more economical.
- Applicable to a wide range of production volumes.
- High potential for automation and integration of quenching and tempering.
- Finishing costs are low.

Typical Applications

- Gears and sprockets.
- Splines.

- Crankshafts.
- Con rods.
- Cam shafts.
- Bearing races.
- Cutting tools.
- Machine beds.

Design Aspects

- Primarily for the improvement of wear resistance and abrasion resistance on selected areas of a part's surface, through an increase in hardness, following quenching.
- Better suited to part shapes that are flat, cylindrical (internal and external features) or have conical surfaces, although more complex shapes can be hardened with improved control.
- Thickness of treatment ranges from 0.3 to 10 mm+.
- Hardness ranges from 420 to 720 HV.
- No practical part size restriction for flame hardening, but part size limited for electron beam hardening due to vacuum chamber size, and induction hardening due to coil size.

Quality Issues

- The depth of hardening depends on control of the process parameters and travel.
- Parts can be rotated in the electromagnetic field in induction hardening to improve uniformity of heating.
- Distortion is minimal due to localised heating of surface.
- Dimensional changes of the initial part are negligible.
- Surface finish is not altered from that of the original except through oxidation of the surface, which can be removed by abrasive cleaning.

9.6 Shot Peening

Process Description

A stream of small high-velocity pellets (shot) are accelerated and directed at the surface of the part contained in a chamber at ambient temperatures. Compressive residual stresses are induced at the surface by the impingement of the shot, effectively cold working it through plastic deformation, visually evident by a series of overlapping dimples. The compressed layer is deep enough to be able to stop cracks and improve fatigue resistance (Figure 9.6).

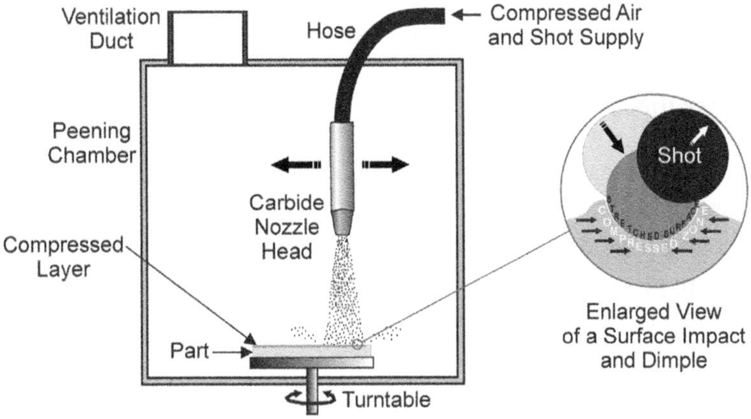

Figure 9.6: Shot Peening.

Materials

Most ductile metals, e.g. aluminium alloys, carbon and low alloy steels, ductile irons, nickel alloys and titanium alloys.

Process Variations

- Shot Peening uses a variety of material types, sizes and hardness of shot, e.g. iron, steel, ceramic (zirconium oxide) and glass beads ranging from approximately 0.2 to 5 mm diameter. Also wire (cut or conditioned) in carbon and stainless steel.
- A variety of machines are used to provide the high shot velocity necessary, e.g. direct wheel throwing machine and compressed air with a carbide nozzle.
- Laser Shock Peening: a laser beam is used to create short pulses of energy directed through a layer of flowing water and on to an ablative layer of paint or tape placed on the surface of the part. The laser explodes the ablative layer, creating a shockwave that is cushioned by the water (called inertial tamping), imparting compressive stresses in the direction of the surface of the part. Expensive and slow process, but depth of treatment greater than shot peening.

- Water Jet Peening: high-pressure water accelerated through a small nozzle (see PRIMA 7.7) and on to the part surface, imparting compressive residual stresses.
- Ultrasonic Peening: a small amount of shot is contained with the part and vibrated at high frequency using special tooling (see PRIMA 7.6), imparting compressive residual stresses.
- Surface (Deep) Rolling: hard balls or rollers under high forces and repetitive motion are used to impart compressive residual stresses on the surface of the part, e.g. large fillet radii on shafts, valve stems and roller bearings.
- Other related mechanical processes that create compressive residual stresses beneficial to fatigue resistance are expansion of rivet and bolt holes using a tapered drift tool, known as caulking (see PRIMA 11.16), and peening of welds using a hammer.

Economic Considerations

- Localised treatment means the process can be rapid and cost effective.
- Lead times are short.
- Equipment costs are moderate.
- Tooling costs are generally low, except for ultrasonic peening, which has dedicated tooling in relation to the part geometry.
- Labour costs can be high.
- Some automation possible in part transit though chamber.
- Applicable for low to moderate production volumes.
- Finishing costs are low.

Typical Applications

- Axles and shafts.
- Gears.
- Con rods.
- Crankshafts.
- Torsion bars.
- Chain saw blades.
- Joints and pins.
- Coil and leaf springs.
- Oil well drilling parts.
- Turbine and compressor blades.

Design Aspects

- Primarily used to improve fatigue resistance (at high cycles rather than low cycles).
- Effective in reducing stress concentration effects of notches, fillets, forging pits and other surface defects.
- Improves resistance to wear, galvanic corrosion, stress corrosion cracking and cavitation erosion.
- Occasionally used for a decorative finish on parts.
- Part complexity can be moderately high, with a number of limitations.
- Fillet radii should be at least two times the shot radius used.
- Features such as grooves may prevent effective peening by disruption of shot stream.
- Holes with large length-to-diameter ratios will make it more difficult to control the uniformity of peening in these features.
- Sharp corners and feathered edges should be avoided or removed before peening (0.25 mm minimum feature size) due to possibility of fracture.
- Treatment thickness ranges from 0.025 to 1 mm.
- Part size limited by chamber size only when used, otherwise no practical limitations for portable peening equipment.

Quality Issues

- Part material must be mechanically stable in order to resist bombardment of shots.
- Peening improves the distribution of residual stresses in the part surface created by machining processes, particularly grinding.
- Typically the last process to be carried on a part, although low stress grinding can be used to finish the surface post-shot peening (see PRIMA 6.7).
- Areas not required to be shot peened can be masked off using moulded rubber.
- Parts require cleaning before peening takes place, e.g. removal of dust, grease and oxides.
- Peening intensity (calibrated to the Almen test) is governed by the shot velocity, hardness, size, weight, angle of the stream against the surface of the part and time of exposure.
- Guaranteeing a uniform coverage of the shot and therefore uniform compressive residual stresses requires each point on the surface of the part to be impacted at least two to four times. At 98% coverage, the surface is said to be fully peened.
- The magnitude of the compressive residual stress depends mainly on the material being peened, its ability to be cold worked and peening intensity.
- For equilibrium, compressive residual stresses at the surface will be balanced by higher tensile stresses at the core of the part.
- High peening intensities cause excessive tensile stresses at the part's core.
- The beneficial compressive residual stresses will be relaxed if the component is subjected to high service temperatures, greater than 300°C for steels and 175°C for aluminium alloys.

- Flat parts such as sheet or plate need to be peened on both sides equally to avoid distortion.
- Shot peening damage in the form of laps and folds creates stress concentrations that reduce fatigue performance of the part.
- A brittle layer can be created on nickel alloy and titanium alloy parts due to over cold working of the surface.
- The peening shot will be damaged through use, (and therefore broken and worn shot), and dust needs to be removed periodically as part of the shot recycling system.
- Parts produced using EDM methods (see PRIMA 7.1) require the recast layer to be removed before peening.
- Adequate exhaust and ventilation is required in enclosed peening chambers.
- Allowance should be made for increase of external diameters and decrease of internal diameters within a tolerance of ±0.05 mm.
- Surface finish is average. Creates a series of overlapping dimples as the surface feature.
- High peening intensities may produce undesirable surface roughness values.

9.7 Chromating

Process Description

The part to be coated is first cleaned in a solvent then dipped in a heated bath containing chromic acid solution with ions of chromium (Cr^{6+}). The acid reacts and partially dissolves the surface depositing a very thin layer of complex chromium compounds. The part is then rinsed in water (Figure 9.7).

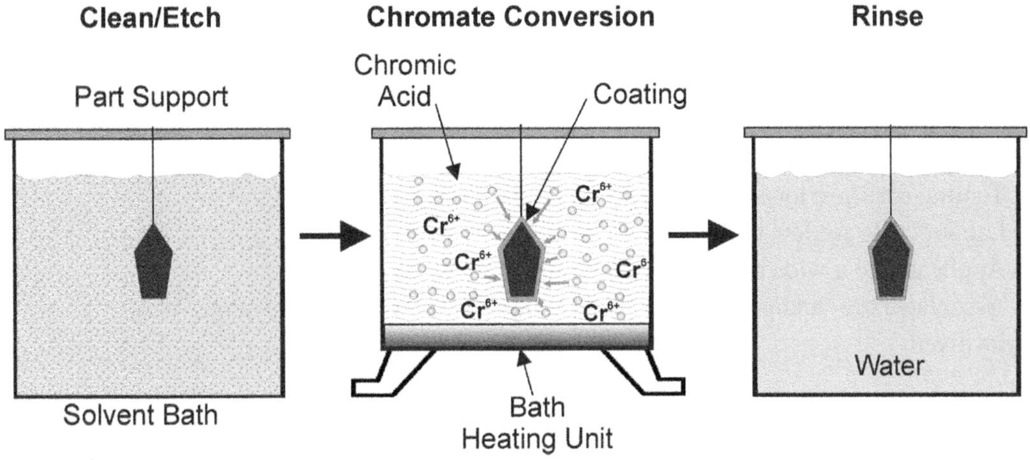

Figure 9.7: Chromating.

Materials

- Low carbon steels, stainless steel and electrodeposited, zinc-plated steel.
- Non-ferrous metals, e.g. aluminium alloys, zinc alloys, magnesium alloys, tin alloys, copper, silver, cadmium and manganese.

Process Variations

- Also known as Chromate Conversion Coating.
- The composition of acid solutions varies widely depending on the material to be coated, typically aqueous solutions of chromic acid, chromium salts (sodium or potassium chromate or dichromate), hydrofluoric acid or phosphoric acid.
- Different pH levels are required for different applications: pH 1.5 for medium corrosion resistance; pH 3.5 for medium/strong corrosion resistance (and to impart changes in coating colour); pH 6 for strong corrosion resistance.
- Can be applied using dipping, spraying, brushing and rolling.
- Processing time ranges from a few seconds to several minutes.

- Process operating temperature ranges from 20 to 175°C. Higher temperatures enhance the process through acceleration of the acidic reaction.
- Variety of coating colours available, e.g. gold, blue, yellow, black and transparent.
- Phosphating: the part is immersed in a bath with a solution of zinc phosphate and dilute phosphoric acid. A crystalline layer of zinc phosphate is grown on the part surface. Cost-effective process for coating steel. Can also use magnesium phosphate salts for magnesium phosphate coating.

Economic Considerations

- Production rates are moderately fast.
- Lead times are short.
- Equipment costs are low.
- Tooling costs are low.
- Labour costs are low to moderate.
- Applicable to a wide range of production volumes.
- Associated pre- and post-process and cleaning costs, with several stages involved.

Typical Applications

- Transmission parts.
- Mandrels.
- Gears.
- Extrusions.
- Tools.
- Fasteners.
- Hinges.
- Electronic components.
- Waveguides.
- Pre-treatment for paints.

Design Aspects

- Primarily used to improve the corrosion resistance of parts.
- Provides a decorative finish.
- Can provide some moderate improvements to wear resistance.
- Can provide electrical resistance coatings.
- All geometric shapes are possible.
- Sharp corners should be avoided.

- Can be used for threaded parts.
- Typical coating thickness ranges from 0.01 to 5 μm.
- Part size limited by bath size only, typically 1 m^3.

Quality Issues

- Parts require extensive cleaning in solvents and etchants prior to processing.
- Process is controlled by acid type/strength, material being coated, shape complexity, immersion time and temperature, and effectiveness of part cleaning and rinsing.
- Some processes use a degassing treatment at 200°C to prevent hydrogen embrittlement.
- Post-process rinsing cools the part and removes acid from the surface to effectively neutralise the acid attack.
- Can provide good paint, powder, lacquer and plastic coat adhesion properties on the surface of parts.
- Final coating thickness is difficult to measure, therefore correct process parameters and their control are required for repeatability.
- After a certain thickness the coating becomes porous and does not adhere satisfactorily, compromising corrosion protection.
- Coatings do not delaminate under bending stresses.
- The coating will slowly dissolve if in direct continuous contact with water, and so more suitable for corrosion protection in high-humidity or marine service environments.
- Damaged coatings can be easily renewed, although small imperfections, rubs or scratches have the ability to self-heal.
- There are several safety issues: chromate increases the risk of skin cancer; inhalation of acid vapours must be avoided by operators; personal protective equipment must be used for eyes, nose, face and hands.
- Disposal of acids must be properly conducted.
- Tolerance on thickness ranges from ±1 to ±2 μm.
- Surface finish is excellent, typically in the range 0.2–1 μm Ra.

9.8 Chemical Vapour Deposition (CVD)

Process Description

The part to be coated is placed in a reaction chamber and is heated to a high temperature. A chemical carrier gas composed of halides or carbonyls of the coating material passes around the part and deposits the coating metal or compound on the part surface through a chemical reaction from the vapour phase (Figure 9.8).

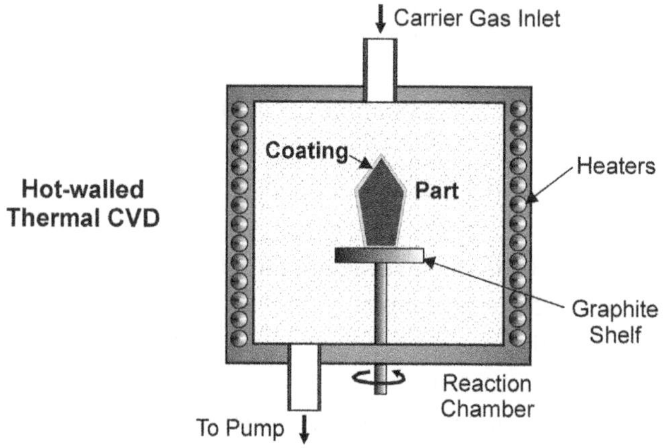

Figure 9.8: Chemical Vapour Deposition (CVD).

Materials

- Materials coated: most metals, ceramics, glass and thermosetting plastics.
- Coating materials: elemental metals, e.g. aluminium, cobalt, copper, iridium, titanium, tungsten; carbides, e.g. titanium carbide, chromium carbide, silicon carbide, zirconium carbide; nitrides, e.g. titanium nitride, silicon nitride, zirconium nitride; oxides, e.g. aluminium oxide, zirconium oxide; carbon fibres, carbon nanotubes and diamond.

Process Variations

- Thermal CVD: typically operates between 800 and 2000°C, and different pressures, e.g. atmospheric, low pressure and ultra-high vacuum. Can be hot-walled or cold-walled reaction chamber. Part is usually rotated in the horizontal plane to improve coating uniformity or transferred using a conveyor through the reactor.
- Plasma-enhanced CVD: a plasma in a vacuum is used to ionise and dissociate the chemical carrier gas, enhancing the chemical reaction and providing heat. Operates at lower temperature than thermal CVD and improved deposition rates.

- Flame-assisted CVD: uses a flame as the heat source in an open atmosphere chamber for depositing oxides. Low-cost process.
- Laser-assisted CVD: a focused laser beam provides the localised heat source on the part. The carrier gas undergoes a thermally induced chemical reaction, depositing the coating on the part. Operates at lower temperature than thermal CVD, but with higher deposition rates.

Economic Considerations

- Deposition rates of up to 1 kg/h.
- Lead times are moderate.
- Equipment costs are high due to complexity of reaction chamber.
- Tooling costs are low.
- Suitable for low to moderate volumes.
- A number of parts can be coated in one operation.
- Labour costs are moderate. Some skilled labour required.
- Typically no finishing required.

Typical Applications

- High-purity and high-density monolithic parts such as semiconductor materials, ultrafine powders and high-strength fibres.
- Superconducting films.
- Dielectric films.
- Photovoltaic films.
- Decorative parts.
- Optical parts.
- Heat exchanger parts.
- Rocket and nozzle parts.
- Gas turbines.
- Cutting tools.
- Bearing races.

Design Aspects

- Primarily improves wear and corrosion/oxidation resistance.
- Produces a decorative finish.
- All shape complexities possible, but with several limitations.
- Non-uniform sections and abrupt changes in section should be avoided due to creation of residual stresses and distortion at high temperatures.
- High-aspect-ratio holes and underside of features can be completely coated.

- Sharp corners should be avoided.
- Maximum section thickness = 10 mm.
- Typical coating thickness ranges from 1 μm to 2 mm+.
- Hardness ranges from 200 to 2500 HV depending on coating material used.
- Part size is only limited by chamber size.

Quality Issues

- Controlling factors are the mass transport, temperature and pressure parameters, as well as the chemistry of the vapour phase reaction.
- Density of the coating is 99.99%+ of the source material.
- Variation in density and purity of coatings can be controlled in the process.
- Good coating adhesion.
- Part surfaces need to be cleaned thoroughly before coating using grit blasting and solvents in extreme cases.
- Some chemical carrier gases are toxic, corrosive or explosive, which necessitates a closed system and safety procedures.
- Has the potential to reduce fatigue life due to CVD creating columnar grains that act as nucleation sites for fatigue cracks in bending.
- Distortion can occur due to relatively high temperatures experienced by the part.
- Tolerances are good to fair.
- Surface finish is excellent.

9.9 Physical Vapour Deposition (PVD)

Process Description

The formation of a coating on the part by the condensation of the vaporised form of the coating material in a vacuum chamber (Figure 9.9).

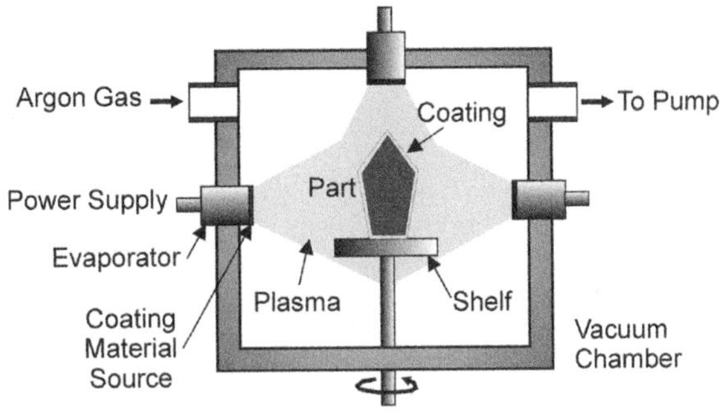

Figure 9.9: Physical Vapour Deposition (PVD).

Materials

- Materials coated: most metals, plastics, ceramics and glass.
- Coating materials: aluminum, copper, nickel and zirconium.

Process Variations

- Process operating temperature ranges from 50 to 700°C.
- Sputtering: plasma discharge of argon ions bombard the coating material source dislodging it away, some of it as vapour, and it is carried to the surface of the part where it condenses to form a coating. Can be done in a reactive gas to deposit carbides and nitrides. Low deposition rates, but improved coating bond strength and density.
- Resistive Heating PVD: material to be deposited is heated to a high vapour pressure by electrically resistive heating in a low vacuum. Suitable for lower melting temperature metals.
- Electron Beam PVD: electron bombardment of the coating material source in a high vacuum creates the vapour.
- Cathodic Arc Deposition: high-power arc creates the vapour.
- Ion Plating: evaporated atoms become ionised in a plasma and attracted to the part. Can be performed in a reactive gas to deposit carbides and nitrides.
- Pulsed Laser Deposition: high-power laser ablates the coating material source, creating a vapour.

Economic Considerations

- Deposition rates of up to 0.5 kg/h.
- Lead times are moderate.
- Equipment costs are high.
- Tooling costs are low.
- Suitable for high production volumes.
- A number of components can be coated in one operation.
- Labour costs are moderate. Some skilled labour required.
- Typically no finishing needed.

Typical Applications

- Drills and milling cutters.
- Gears and pistons.
- Bearings races.
- Moulds and dies.
- Gas turbine parts.
- Architectural, decorative and ornamental parts.
- Jewellery and watches.
- Toys.
- Window and picture frames.
- Kitchenware and utensils.
- Bathroom fixtures.
- Printed circuit-boards.
- Semi- and superconductor wafers and devices.
- Metalised plastic parts.
- Optical films, e.g. photoelectric and anti-reflective coatings on solar panels.

Design Aspects

- Primarily for the improvement of corrosion and wear resistance.
- Decorative finish.
- Line of sight process, therefore complex geometries are more difficult to treat uniformly.
- Recesses and slots difficult to coat.
- Typical thickness ranges from 0.05 to 100 μm.
- Hardness ranges from 100 to 2000 HV.
- Component size is limited by chamber size.

Quality Issues

- Can produce high-performance materials and coatings that cannot be produced by other processes.
- Excellent process control.
- Part surfaces need to be cleaned thoroughly before coating, using grit blasting and solvents in extreme cases.
- Can suffer from weak coating adhesion.
- Density of the coating is almost 100% of the parent material.
- More environmentally friendly than other surface coating techniques.
- Distortion is negligible due to low processing temperatures.
- Tolerances are good.
- Surface finish is excellent.

9.10 Electroless Nickel

Process Description

The part is immersed in a metallic salt solution of nickel, e.g. nickel sulphate and a chemical reducing agent, which effectively substitutes for an electric current in reducing the metallic salt to a nickel coating material. The nickel is then deposited on the catalysed surface of the part (Figure 9.10).

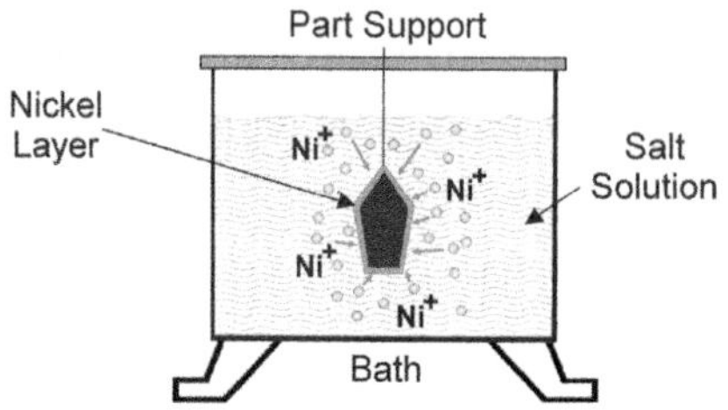

Figure 9.10: Electroless Nickel.

Materials

- Materials coated: ferrous (steels) and non-ferrous metals (nickel, cobalt and aluminum), plastics and ceramics.
- Coating materials: nickel chloride or nickel sulphate salt solutions used to create nickel–boron or nickel–phosphorus coatings.
- Copper coatings are also feasible.

Process Variations

- Chemical reducing agents either sodium hypophosphate or sodium borohydride.
- Process operating temperature ranges from 30 to 100°C (using bath heaters).

Economic Considerations

- Low deposition rates at <5 μm/h for copper, <20 μm/h for nickel.
- Lead times are low.
- Equipment costs are moderate.
- Tooling costs are low.
- No direct electrical energy source needed, though cost of chemicals is high.
- Low labour costs.
- Overall higher coating costs than electroplating.
- Applicable to higher volume production.
- High pre-treatment costs.
- Low finishing costs.

Typical Applications

- Metalisation of plastic parts.
- Printed circuit-boards.
- Heat sinks.
- Wave guides.
- Bearing journals.
- Gears.
- Hydraulic pistons.
- Pumps.
- Impeller blades.
- Optical parts.
- Fasteners.

Design Aspects

- Primarily improves resistance to corrosion and wear.
- All geometric shapes are possible, with few limitations.
- Sharp edges should be avoided.
- Minimum radius for features = 0.4 mm.
- Typical coating thickness ranges from 10 μm to 0.5 mm.
- Hardness ranges from 600 to 1100 HV.
- Part size limited by bath size only, typically 1 m^3.

Quality Issues

- Excellent uniformity of coating.
- Controlling process parameters are temperature, pH level and composition (impurities) of metallic salt solution.
- Bath solution is continuously filtered to remove by-products of earlier treatments.
- Cleaning of parts is required for good coating adhesion, e.g. etching.
- Coating produced is brittle and of high density.
- The coating contains about 15% phosphorus.
- Post-coating heat treatments improve hardness due to nickel phosphate precipitation.
- Parts can be brazed and soldered readily post-treatment, but possess poor welding characteristics.
- Heat treatment may be required to develop optimum material properties.
- Low friction polymers can be incorporated in solution.
- No distortion of part due to very low processing temperature.
- Process better suited to parts where dimensional accuracy needs maintaining compared to the use of electroplating.
- Surface finish is excellent.

9.11 Electroplating

Process Description

The part (the cathode) is immersed in an ionised electrolytic solution together with the coating material (the sacrificial anode). A d.c. current is passed through the solution causing the metallic ions to transfer from anode to cathode, depositing a thin coating of the metallic ions on the part (Figure 9.11).

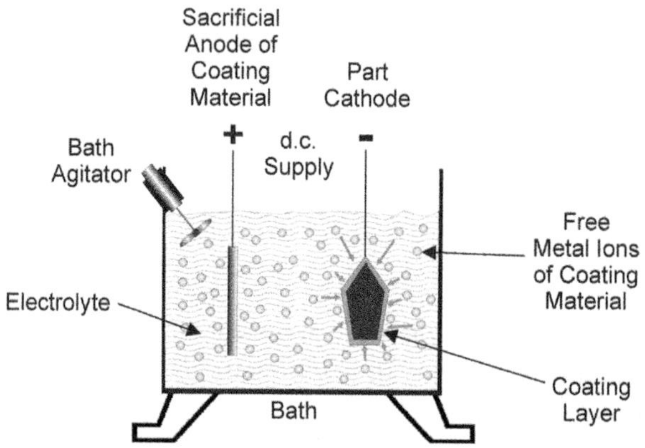

Figure 9.11: Electroplating.

Materials

- Materials coated: carbon steels, low alloy steels, stainless steels, aluminium and copper alloys. Some thermoplastics and glass, pre-coated with an electrically conductive coating.
- Coating materials: aluminum, brass, bronze, cadmium, chromium, copper, gold, indium, iron, lead, nickel, silver, tin, tin–lead alloys and zinc. A variety of forms are available for the sacrificial anode, e.g. rods, balls and foils.

Process Variations

- Process operating temperature ranges from 20 to 60°C.
- Processing time ranges from several minutes to several hours, depending on coating thickness required.
- Brush Electroplating: the cathode is the part and the anode is a brush dipped in plating solution. High portability, low masking requirements, but cannot produce a thick coating.

Economic Considerations

- Deposition rates of up to 0.5 kg/h.
- Lead times are short.

- Equipment costs are high.
- Tooling costs are low.
- Labour costs are moderate.
- Applicable to a wide range of production volumes.
- Associated pre- and post-process and cleaning costs, with several stages involved.

Typical Applications

- Cooking utensils.
- Kitchenware.
- Lamps and ornaments.
- Jewellery.
- Trophies and medals.
- Hand tools.
- Surgical and dental instruments.
- Food processing equipment.
- Automotive trim parts.
- Fasteners.

Design Aspects

- Primarily for the improvement of corrosion and wear resistance.
- Provides a decorative finish.
- Improved electrical and optical properties.
- Relatively complex parts possible with limitations, but more suited to simple shapes that are mostly flat, e.g. plates and sheets.
- Sharp edges and grooves should be rounded.
- Blind cavities cannot be coated uniformly.
- Typical coating thickness ranges from 10 μm to 0.25 mm.
- Hardness ranges from 25 to 1,000 HV.
- Maximum part size is limited by the tank size, typically $<3\ m^3$.

Quality Issues

- Cleaning of parts is required for good coating adhesion, e.g. etching.
- Coating quality is dependent on electrolyte solution composition, its strength and temperature, part material and its shape complexity, current/voltage levels, time of immersion and effectiveness of part cleaning.
- Adverse part geometry can cause non-uniform coating.
- A bath agitator can help with consistency of ions in solution.

- Electroplating with chromium or nickel creates high tensile residual stresses and microcracking in the coating, reducing fatigue resistance.
- Density of the coating is 100% of the parent material. No porosity.
- There are safety and environmental concerns associated with electroplating bath processes.
- Hydrogen embrittlement in electroplated steel can occur which accelerates failure.
- No distortion of part due to low temperature processing.
- Tolerance on thickness ranges from ±1 to ±15 μm.
- Surface finish is very good to excellent, typically in the range 0.03–2 μm Ra.

9.12 Hot Dip Coating

Process Description

A coating is applied to a part by immersing it in a bath of the molten coating metal. A brittle intermetallic compound layer is formed at the surface, which provides the necessary adherence to fix the coating metal to the part metal (Figure 9.12).

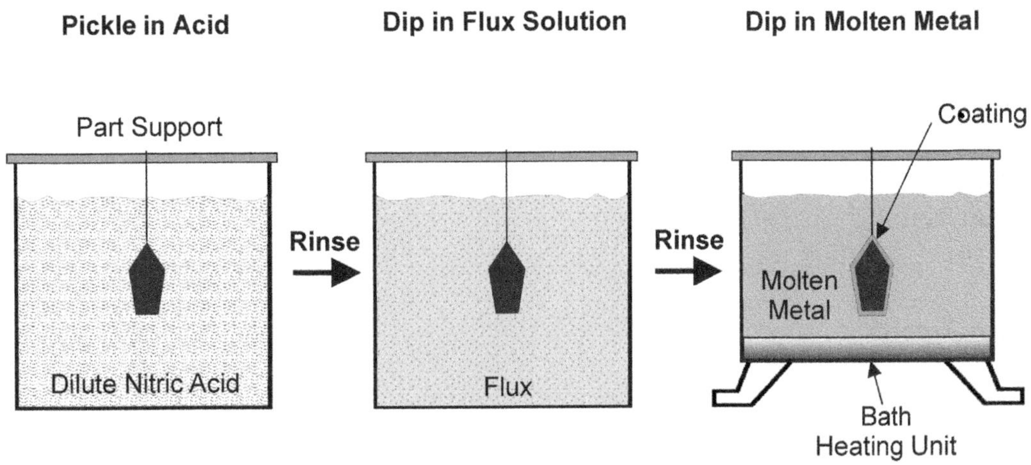

Figure 9.12: Hot Dip Coating.

Materials

- Materials coated: ferrous metals.
- Coating materials: zinc, tin, lead and aluminium.
- Zinc coating of steels is popularly known as galvanising.

Process Variations

Process operating temperature varies depending on the coating material's melting temperature, typically 325–700°C.

Economic Considerations

- Deposition rates are high.
- Lead times are short.
- Equipment costs are moderate.
- Tooling costs are low.
- Labour costs are moderate.

- Applicable to a wide range of production volumes.
- Automation and continuous operation possible for wire, sheet, etc.
- Associated pre-treatment costs with many stages involved.
- Finishing costs are very low.

Typical Applications

- Heater tubes.
- Water tanks.
- Boiler cases.
- Pipe fittings.
- Fencing wire.
- Sheet metal.
- Fasteners.
- Nails.

Design Aspects

- Primarily for the improvement of corrosion resistance.
- All geometric shapes are possible, with several limitations.
- Sharp edges should be rounded.
- Blind cavities cannot be coated uniformly.
- Treatment thickness ranges from 10 to 130 μm.
- Hardness depends on coating material used but generally less than 15 HV.
- Part size limited only by bath size typically <30 m^3.

Quality Issues

- Melting temperature of the metal part has to be greater than that of the coating metal.
- Parts require extensive cleaning prior to coating such as pickling, rinsing and fluxing.
- Immersion time is determined by the weight, mass and configuration of the part being galvanised.
- Coating thickness is controlled by the composition of the steel substrate and the immersion time.
- Distortion of the part can occur due to the relatively high temperatures required to melt some coating materials.
- Embrittlement and delamination of coating can occur.
- Draining is required to remove excess coating material from the part. Dimensional tolerances are therefore difficult to establish.
- Surface finish can be poor, characterised by large grain patterns.

9.13 Thermal Spraying

Process Description

Droplets of molten coating material are accelerated through a spray torch or gun on to the surface of the part using a high-speed gas stream. The molten droplets, which are just above melting temperature, overlap and interlock, rapidly adhering to the part's surface, creating a coating (Figure 9.13).

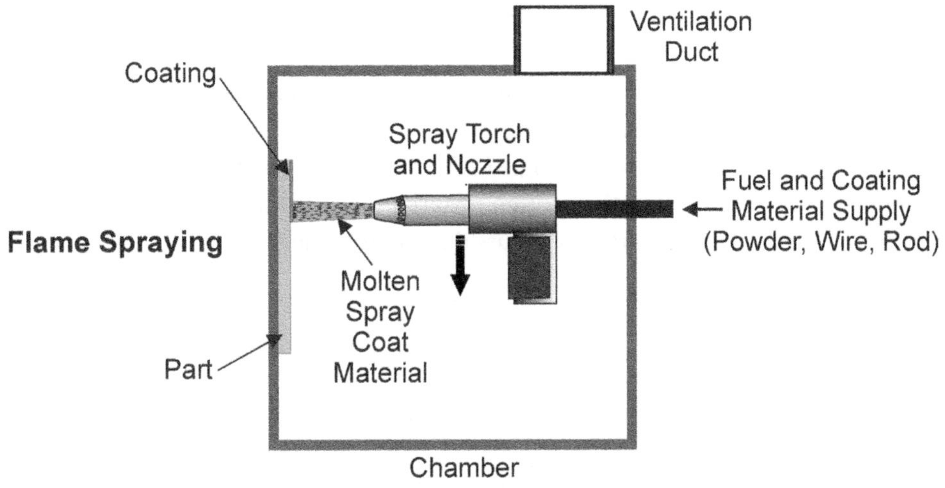

Figure 9.13: Thermal Spraying.

Materials

- Materials coated: ferrous and non-ferrous metals, ceramics, some plastics and carbon fibre composites.
- Coating materials: pure metals, alloy metals, ceramics and composites.

Process Variations

- Flame Spraying: coating material in powder, wire or rod form is heated up to 4000°C using a combustible gas, e.g. acetylene or propane, which ejects the molten droplets at about 35 m/s on to a surface of the part, creating a coating. A wide range of materials can be flame sprayed.
- Electric Arc Spraying: uses an electric arc to melt a fed wire up to temperatures of 7000°C, producing very high deposition rates.

- Plasma Spraying: an electric arc ionises argon gas, melting the coating material (supplied in wire, powder or liquid form) into droplets that are fed into a gas stream and sprayed at the surface of the part. Applicable to a range of coating material types, but smaller part sizes and areas. High temperatures can be achieved depending on melting temperature of coating material, typically 6000–16,000°C (ceramics), but requires the torch to be water cooled. Can be done in air or a vacuum (to limit oxidation).
- High-velocity Oxyfuel Spraying: high velocities (up to 750 m/s) of fuel, typically propane and pure oxygen, melt the coating material (wire or powder form) at temperatures up to 5000°C. This is fed into the gas stream and sprayed on to the surface, providing a dense and well adhered coating.
- Polymer Flame Spraying: powdered form of a thermoplastic is heated up to 200°C and sprayed at high speed to coat the part surface.
- Cold Spraying: very high velocity (up to 1200 m/s) heated compressed gas is used to accelerate the powder form of the coating material towards the surface of the part. Plastic deformation of the powder particles occurs due to high-energy impact with the surface, producing dense coatings.

Economic Considerations

- Deposition rates range from 10 (flame) to 50 kg/h (electric arc).
- Lead times are short, generally.
- Equipment costs vary from low for electric arc to high for plasma.
- Tooling costs are generally low.
- Labour costs are low to moderate. Some skilled labour required for some process variants such as plasma spraying.
- Some process variants can be automated, e.g. use of programmed robots.
- Localised treatment means the process can be rapid and cost effective.
- Applicable to a wide range of production volumes.
- Finishing costs can be high. Post-processing such as machining or grinding typically required.

Typical Applications

- Repair and dimensional restoration of worn machine part surfaces.
- To improve thermal and electrical conductance/resistance, to improve radiation resistance and/or for electromagnetic shielding of parts.
- Food processing machinery.
- Oil drilling components.
- Architectural components.
- Sports equipment.

- Turbine blades.
- Gears.
- Valve seats.
- Tool surfaces.
- Moulds.
- Rocket nozzles.

Design Aspects

- Primarily used to improve corrosion and wear resistance.
- Line of sight process, therefore complex geometries are more difficult to treat uniformly, e.g. re-entrant features.
- Fillet radii should be as generous as possible.
- Features such as grooves may prevent effective coating by disruption of spray stream.
- Holes with large length to diameter ratios will make it more difficult to control the uniformity of coating in these features.
- Sharp corners and feathered edges should be avoided or removed before spraying.
- Typical coating thickness ranges from 0.02 to 5 mm.
- Hardness ranges from 10 (plastics) to 1800 HV (ceramics/carbides).
- Size can be limited by the equipment size, e.g. for plasma spraying in a vacuum chamber (typically <1 m^3), otherwise no limit for portable equipment.

Quality Issues

- Degreasing and grit blasting of surface is typically done as a pre treatment to improve coating adhesion.
- The quality and reliability of the coatings depends on careful control of coating material source and spraying parameters.
- Spray stream should be perpendicular to the surface.
- Multiple passes will provide thicker coatings.
- Dust and fumes generated must be extracted from work area. Excessive noise and light radiation requires personal protective equipment.
- Possible to create both tensile (more common) and compressive residual stresses in the coatings, which limits coating thicknesses.
- Heat from the spraying process has minimal impact on the coated part temperature.
- Parts can also be pre-heated up to 200°C to improve coating adhesion.
- Worn or damaged coatings can be recovered without changing the properties or dimensions of the part.
- Density of coatings varies from almost 100% for plasma spraying in a vacuum and high-velocity oxyfuel spraying, to 90% for flame spraying.

- Coating density and bond strength to the part surface increase with spraying velocity.
- Coating bond strength is mainly related to mechanical interlocking.
- Coating bond strengths of up to 70 MPa can be achieved, although this is low in comparison to other coating processes.
- Oxide content of coating varies from virtually 0% for plasma spraying in a vacuum to 20% for flame spraying.
- Porous surfaces can be sealed using wax (for low service temperatures), silicone and epoxy resins (for higher temperature/high humidity service environments).
- Tolerances are fair. A finish machining process is usually required, e.g. grinding.
- Surface finish is poor generally, although with control, <1 μm Ra possible.

CHAPTER 10

Assembly Systems

10.1 Manual Assembly

Process Description

Manual assembly involves the composition of previously manufactured components and/or sub-assemblies into a complete product or unit of a product, primarily performed by human operators using their inherent dexterity, skill and judgment. The operator may be at a workstation (bench) or be part of a transfer system that moves the product as it is being assembled. Manual assembly can be further assisted by mechanised or automated systems for feeding, handling, fitting and checking operations (Figure 10.1).

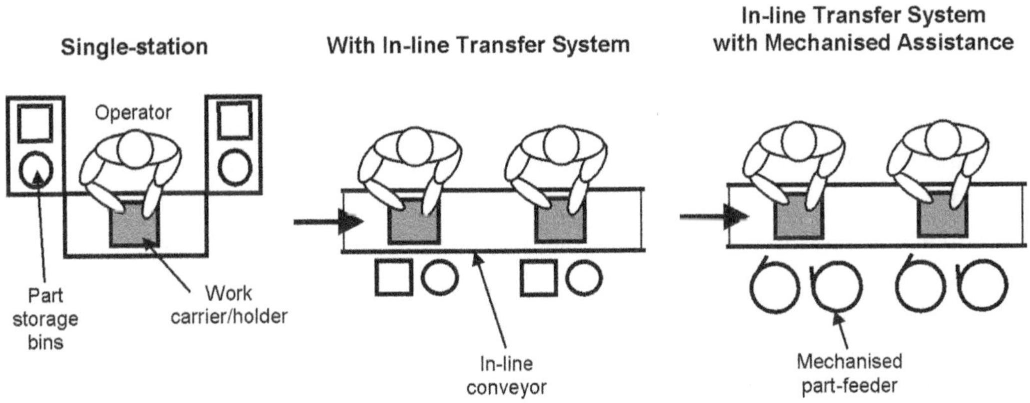

Figure 10.1: Common Manual Assembly System Configurations.

Process Variations

- Feeding: presentation of a component to the handling equipment in the correct orientation by a variety of methods. Parts manually taken from storage bins and then orientated by operator; orientation can be achieved by vibratory/centrifugal bowl feeders, parts already orientated in pallet/magazine/strip form or use of part-feeding escapement mechanisms.
- Handling: bringing components and/or sub-assemblies together such that later composition can occur. Use of hands; simple lifting aids and jigs; and fixtures.

Manufacturing Process Selection Handbook. http://dx.doi.org/10.1016/B978-0-08-099360-7.00010-0

- Fitting: various part placement/location configurations or fastening/joining methods can be utilised, e.g. 'peg in hole', press fit, welding, riveting, adhesive bonding, staking and screwing, using a variety of hand-operated or mechanised/electrical tools.
- Checking: detection of missing, incorrect, misshapen or wrongly orientated components by high-level sensing and checking capabilities of operators and mechanical/electrical aids. Also detection of foreign bodies, part failure and machine inoperation.
- Transfer: typically, the various assembly operations required are carried out at separate stations, usually built up on a work carrier, pallet or holder. Therefore, a system for transferring the partly completed assemblies from workstation to workstation is required. In general, the various types of workstation/transfer system for manual assembly are:
    - Single station: assembly at one workstation or bench where a specific operation or a variety of operations are performed.
    - Continuous: work carrier flows without stopping using conveyors (in-line, rotary dial or carousel), overhead rail or towline.
    - Intermittent: synchronous/indexing (moved with a fixed cycle time) or non-synchronous/free transfer (moved as required or when operation/assembly completed by operator) using in-line or rotary systems.

Economic Considerations

- Production rates are low to moderate, depending on complexity, number and size of component parts. These factors also dictate the degree of mechanised assistance needed.
- Extremely flexible assembly system (many product variants) and therefore most common.
- Lead time is typically days, higher if mechanised assistance devices used.
- Economical for low to moderate production runs. Can be used for one-offs.
- Tooling costs are low to moderate.
- Equipment costs generally low, except where full mechanised assistance exists.
- Direct labour costs are moderate to high. Relatively easy to train operators.

Typical Applications

- Car assembly lines.
- Internal combustion engines.
- Domestic appliances and office equipment.
- Electronic and electrical equipment.
- Machine tools.
- General fabrication.
- Toys, furniture, footware and clothing.

Design Aspects

- Use Design for Assembly (DFA) guidelines/techniques in order to develop assemblies with optimum part-count, improved component geometry for feeding, handling, fitting and checking, and reduce overall assembly costs. See Appendix B – Guidelines for Design for Assembly.
- Use of Poka-Yoke (mistake-proofing) techniques helps reduce operator assembly errors by prescribing component features and/or assembly procedures to aid correct assembly.
- Develop an assembly sequence diagram to optimise the assembly line.
- Overall assembly tolerance must be assessed against component tolerances in stack-up.

Quality Issues

- In general, assembly problems are caused by a number of factors:
    - Components exceeding or being lower than the specified tolerances.
    - Component misalignment and adjustment error.
    - Gross defects (malformed, missing features, wrong lengths, damage in transit, etc.)
    - Foreign matter causing contamination and blockages.
    - Absence of a component due to inefficient feeding or exhausted supply.
    - Incorrect components caused by wrong supply or instructions.
    - Inadequate joining technology.
    - Skill of labour used.
- Manual assembly is not suitable for harsh environments.
- Size and weight of parts to be assembled must be considered for handling safety.
- Operator fatigue, health and relaxation time must be considered, especially for highly repetitive operations.
- Assembly errors increase if components/sub-assemblies are complex, difficult to align, insert, or if there is restricted access for insertion (1% error rate possible for some manual operations).
- Poor quality components can generally be sorted out during the assembly task without difficulty or high loss through the advanced checking capabilities of human operators.
- Repeatable accuracy of component alignment is low to moderate depending on part complexity, typically ±0.5 mm.

10.2 Flexible Assembly

Process Description

Flexible assembly systems use programmable, robotic devices to compose previously manufactured components and/or sub-assemblies into a complete product or unit of a product. A number of transfer mechanisms, feeding devices, robot types and end effectors can be utilised in order to achieve a general assembly system (Figure 10.2).

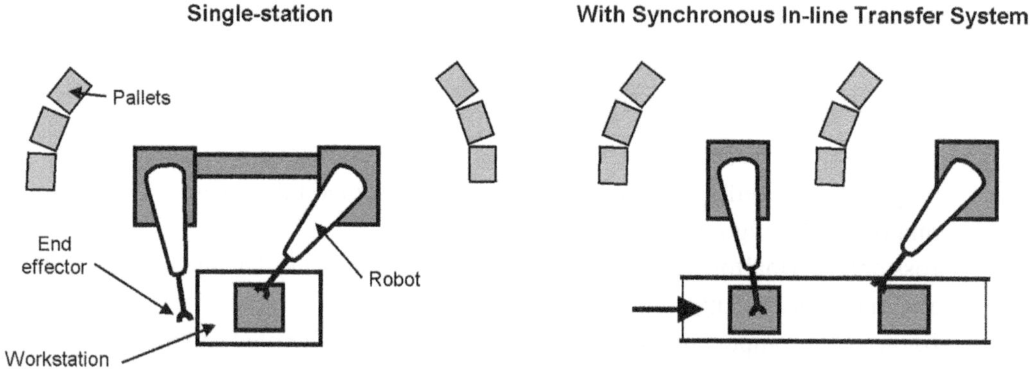

Figure 10.2: Common Flexible Assembly System Configurations.

Process Variations

- Robot types: variety of configurations, loading carrying capacity, working envelope, wrist degrees of freedom and accuracy/repeatability, e.g. revolute, polar, gantry and pendulum.
- End effectors: variety of arrangements depending on part geometry and feeding direction, flexibility, fragility, and overall part size and weight. Either pneumatic, vacuum, electromechanical or electromagnetic actuation of holding/gripping mechanism.
- Feeding: presentation of a component to the robot arm end effector in the correct orientation. Orientation can be achieved by vibratory/centrifugal bowl feeders, by receiving parts already orientated by the supplier in pallet, magazine or by escapement mechanisms for part feeding.
- Handling: bringing components and/or sub-assemblies together such that later composition can occur using robot arm end effectors.
- Fitting: various part placement/location configurations or fastening/joining methods can be utilised, e.g. 'peg in hole', adhesive bonding, staking and screwing.
- Checking: detection of missing, incorrect, misshapen or wrongly orientated components by electronic vision systems, tactile/pressure sensors and proximity sensors.

- Transfer: typically, the various assembly operations required are carried out at separate stations, usually built up on a work carrier, pallet or holder. Therefore, a system for transferring the partly completed assemblies from workstation to workstation is required. In general, the various types of station/transfer system for flexible assembly are:
  - Single/multi-station: assembly at one or more workstations where a specific or, more commonly, a variety of operations are performed. Typically, greater than six components to be assembled require a multi-station arrangement.
  - Synchronous/indexing: moved with a fixed cycle time using in-line, rotary dial or carousel systems.

Economic Considerations

- Only moderate flexibility, despite name.
- Systems can be adapted for the assembly of several different products/variants.
- Production rates are moderate.
- Lead time weeks to months.
- Economical for moderate to high production volumes.
- Tooling costs are high.
- Equipment costs are moderate to very high.
- Direct labour costs are low.
- Programming/teaching of robot operations and movements is complex and lengthy.

Typical Applications

- General assembly, materials handling and transfer of parts and assemblies.
- For hazardous environments (to humans), e.g. radioactive, toxic, dusty and high temperatures.
- Part loading and/or unloading for manufacturing processes, e.g. machining centres, pressure die casting machines and injection moulding machines.
- Spot and MIG welding (see PRIMAs 11.8 and 11.2 respectively).
- Abrasive jet machining.
- Surface finishing, grinding, buffing and spray painting operations.

Design Aspects

- Use Design for Assembly (DFA) guidelines/techniques in order to develop assemblies with optimum part-count, improved component geometry for feeding, handling, fitting and checking, and reduce overall assembly costs. See Appendix B – Guidelines for Design for Assembly.

- Develop an assembly sequence diagram to optimise the assembly line.
- Overall assembly tolerance must be assessed against component tolerances in stack-up.

Quality Issues

- In general, assembly problems are caused by a number of factors:
  - Components exceeding or being lower than the specified tolerances.
  - Component misalignment and adjustment error.
  - Gross defects (malformed, missing features, wrong lengths, damage in transit, etc.)
  - Foreign matter causing contamination and blockages.
  - Absence of a component due to inefficient feeding or exhausted supply.
  - Incorrect components caused by wrong supply or instructions.
  - Inadequate joining technology.
- Approximately 50% of all problems found in automated systems (product defects and downtime) are due to the incoming component quality.
- Robot working envelope must be securely guarded.
- Automated or mechanised systems must be chosen in certain situations, particularly where operator safety is paramount – for example, hazardous or toxic environments, heavy component parts or a high repeatability requirement causing operator fatigue.
- Can use dedicated systems for sterile or clean environment assembly of products.
- Repeatable accuracy of component alignment is high, typically ±0.1 mm.

10.3 Dedicated Assembly

Process Description

Dedicated assembly systems are special purpose, fully mechanised or automated systems for composing previously manufactured components and/or sub-assemblies into a complete product or unit of a product. Typically, a number of workstations comprising automatic part-feeders and fixed work-heads are arranged on an automatically controlled transfer system to compose the product sequentially (Figure 10.3).

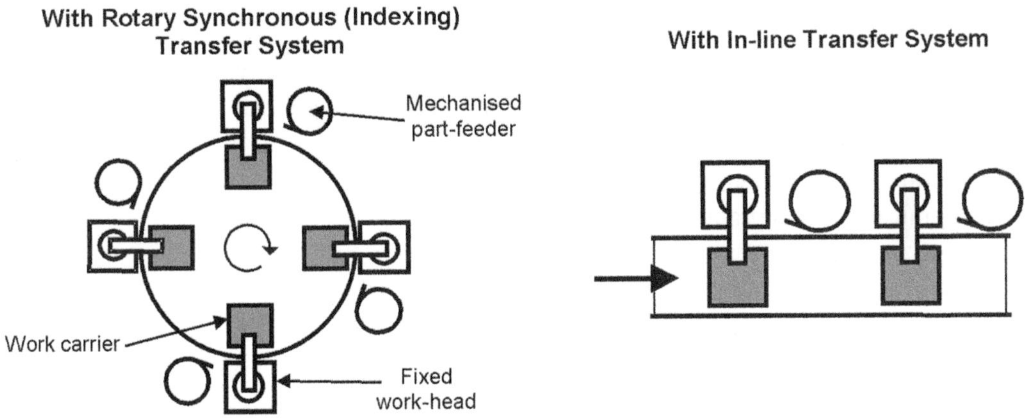

Figure 10.3: Common Dedicated Assembly System Configurations.

Process Variations

- Feeding: presentation of a component to the robot arm end effector in the correct orientation. Orientation can be achieved by vibratory/centrifugal bowl feeders, by receiving parts already orientated by the supplier in pallet, magazine or by escapement mechanisms for part feeding.
- Handling: bringing components and/or sub-assemblies together so that later composition can occur using fixed work-heads and/or pick and place units.
- Fitting: various part placement/location configurations or fastening/joining methods can be utilised, e.g. 'peg in hole', adhesive bonding, staking and screwing.
- Checking: identification of missing, incorrect, misshapen or wrongly orientated components. Also detection of foreign bodies, part failure and machine inoperation. Common technologies include vision systems, tactile/pressure sensors, proximity sensors and 'bed of nails'.

- Transfer: typically, the various assembly operations required are carried out at separate stations, usually built up on a work carrier, pallet or holder. Therefore, a system for transferring the partly completed assemblies from workstation to workstation is required. In general, transfer systems for dedicated assembly are either:
  - Synchronous/indexing: moved with a fixed cycle time.
  - Non-synchronous/free transfer: moved as required or when operation/assembly completed using in-line or rotary systems. Can set up a buffer system using this configuration. Typically, greater than 10 components to be assembled require a free-transfer arrangement.

Economic Considerations

- Almost totally inflexible. Fixed assembly system for one product type typically, except where variants are based on parts missing from original design.
- Production rates are high.
- Lead time typically months.
- Economical for high production volumes.
- Tooling costs are high.
- Equipment costs are high.
- Direct labour costs are very low.

Typical Applications

- Electronic and electrical components and devices.
- Printed circuit-boards.
- Small domestic appliances.
- Medical products.
- Automotive sub-assemblies, e.g. valves, solenoids, relays.
- Office equipment.

Design Aspects

- Use Design for Assembly (DFA) guidelines/techniques in order to develop assemblies with optimum part-count, improved component geometry for feeding, handling, fitting and checking, and reduce overall assembly costs. See Appendix B – Guidelines for Design for Assembly.
- Develop an assembly sequence diagram to optimise the assembly line.
- Overall assembly tolerance must be assessed against component tolerances in stack-up.

Quality Issues

- In general, assembly problems are caused by a number of factors:
  - Components exceeding or being lower than the specified tolerances causing interference or location stability problems.
  - Component misalignment and adjustment error.
  - Gross defects (malformed, missing features, wrong lengths, damage in transit, etc.)
  - Foreign matter causing contamination and blockages.
  - Absence of a component due to inefficient feeding or exhausted supply.
  - Incorrect components caused by wrong supply or instructions.
  - Inadequate joining technology.
- Approximately 50% of all problems found in automated systems (product defects and downtime) are due to the incoming component quality.
- Difficult and expensive to incorporate insensitivity to component variation and faults in assembly systems to reduce this problem. Sensing capabilities are limited in this capacity.
- Automated or mechanised systems must be chosen in certain situations, particularly where operator safety is paramount – for example, hazardous or toxic environments, heavy component parts or a high repeatability requirement causing operator fatigue.
- Can use dedicated systems for sterile or clean environment assembly of products.
- Repeatable accuracy of component alignment is high, typically ±0.1 mm.

CHAPTER 11

Joining Processes

11.1 Tungsten Inert-gas Welding (TIG)

Process Description

An electric arc is automatically generated between the workpiece and a non-consumable tungsten electrode at the joint line. The parent metal is melted and the weld created with or without the addition of a filler rod. Temperatures at the arc can reach 12,000°C. The weld area is shielded with a stable stream of inert gas, usually argon, to prevent oxidation and contamination (Figure 11.1).

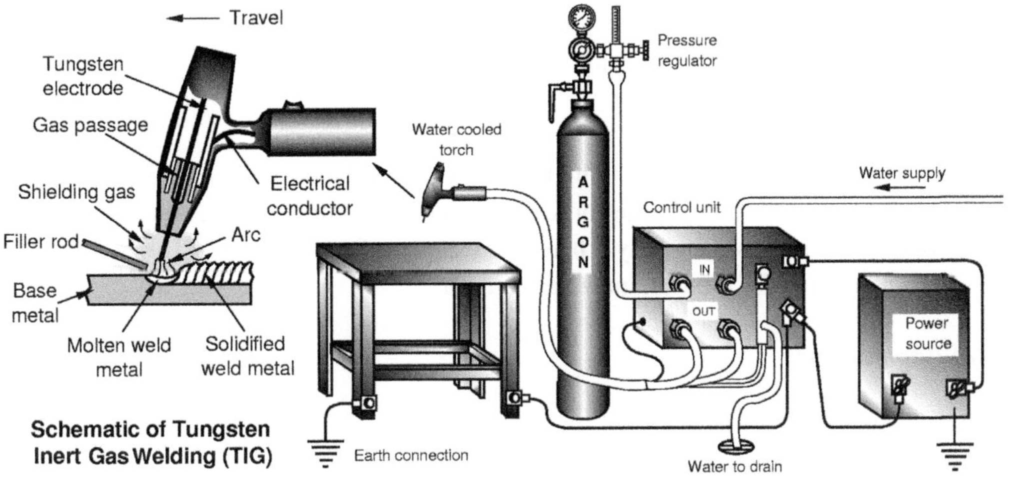

Figure 11.1: Tungsten Inert-gas Welding.

Materials

Most non-ferrous metals (except zinc), commonly aluminium; nickel; magnesium and titanium alloys; copper and stainless steel. Carbon steels, low alloy steels, precious metals and refractory alloys can also be welded. Dissimilar metals are difficult to weld.

Manufacturing Process Selection Handbook. http://dx.doi.org/10.1016/B978-0-08-099360-7.00011-2

Process Variations

- Portable manual or automated a.c. or d.c. systems; a.c. commonly used for welding aluminium and magnesium alloys.
- Pure helium or, more commonly, a helium/argon mix is used as the shielding gas for metals with high thermal conductivity, for example copper, or material thickness greater than 6 mm, giving increased weld rates and penetration.
- Pulsed TIG: excellent for thin sheet or parts with dissimilar thickness (low heat input).
- TIG spot welding: used on lap joints in thin sheets.

Economic Considerations

- Weld rates vary from 0.2 m/min for manual welding to 1.5 m/min for automated systems.
- Automation is suited to long lengths of continuous weld in the same plane.
- Automation is relatively inexpensive if no filler is required, i.e. use of close fitting parts.
- Process is suited to sheet thickness less than 4 mm; heavier gauges become more expensive due to argon cost and decreased production rate. Helium/argon gas expensive but may be viable due to increased production rate.
- Economical for low production runs. Can be used for one-offs.
- Tooling costs are low to moderate.
- Equipment costs are moderate.
- Direct labour costs are moderate to high. Highly skilled labour required for manual welding. Set-up costs can be high for fabrications using automated welding.
- Finishing costs are low, generally. There is no slag produced at the weld area; however, some grinding back of the weld may be required.

Typical Applications

- Chemical plant pipework.
- Nuclear plant fabrications.
- Aerospace structures.
- Sheet-metal fabrication.
- Hardfacing.

Design Aspects

- Design complexity is high.
- Typical joint designs possible using TIG are: butt, lap, fillet and edge. (See Appendix C – Weld Joint Design Configurations.)
- Design joints using minimum amount of weld, i.e. intermittent runs and simple or straight contours, although TIG is suited to automated contour following.

- Design parts to give access to the joint area, for vision, electrodes, filler rods, cleaning, etc.
- Wherever possible horizontal welding should be designed for; however, TIG welding is suited to most welding positions.
- Sufficient edge distances should be designed for. Avoid welds meeting at end of runs.
- Balance the welds around the fabrication's neutral axis where possible.
- Distortion can be reduced by designing symmetry in parts to be welded along weld lines.
- The fabrication sequence should be examined with respect to the above.
- Provision for the escape of gases and vapours in the design is important.
- Minimum sheet thickness = 0.2 mm.
- Maximum thickness commonly:
  - Copper and refractory alloys = 3 mm.
  - Carbon, low alloy and stainless steels, magnesium and nickel alloys = 6 mm.
  - Aluminium and titanium alloys = 15 mm.
- Multiple weld runs required on sheet thickness ≥5 mm.
- Unequal thicknesses are difficult.

Quality Issues

- Clean, high-quality welds with low distortion can be produced.
- Access for weld inspection important, e.g. non-destructive testing (NDT).
- Joint edge and surface preparation is important. Contaminants must be removed from the weld area to avoid porosity and inclusions.
- A heat-affected zone always present. Some stress relieving may be required for restoration of materials' original physical properties.
- Not recommended for site work in wind, where the shielding gas may be gusted.
- Control of arc length is important for uniform weld properties and penetration.
- Need for jigs and fixtures to keep joints rigid during welding and subsequent cooling to reduce distortion on large fabrications.
- Backing strips can be used for avoiding excess penetration, but at added cost and increased set-up times.
- Selection of correct filler rod is important (where required).
- Care is needed to keep filler rod within the shielding gas to prevent oxidation.
- Workpiece and filler rod must be away from the tungsten electrode to prevent contamination, which can cause an unstable arc.
- The shielding gas must be kept on for a second or two to allow tungsten electrode to cool and prevent oxidation.
- Tungsten inclusions can contaminate finished welds.
- Welding variables should be preset and controlled during production.
- Automation reduces the ability to weld mating parts with inherent size and shape variations; however, it does reduce distortion, improves reproduction and produces fewer welding defects.

- 'Weldability' of the material is important and combines many of the basic properties that govern the ease with which a material can be welded and the quality of the finished weld, i.e. porosity and cracking. Material composition (alloying elements, grain structure and impurities) and physical properties (thermal conductivity, specific heat and thermal expansion) are some important attributes that determine weldability.
- Surface finish of weld is excellent.
- Fabrication tolerances are typically ±0.5 mm.

11.2 Metal Inert-gas Welding (MIG)

Process Description

An electric arc is manually created between the workpiece and a consumable wire electrode at the joint line. The parent metal is melted and the weld created with the continuous feed of the wire, which acts as the filler metal. The weld area is shielded with a stable stream of argon or CO_2 to prevent oxidation and contamination (Figure 11.2).

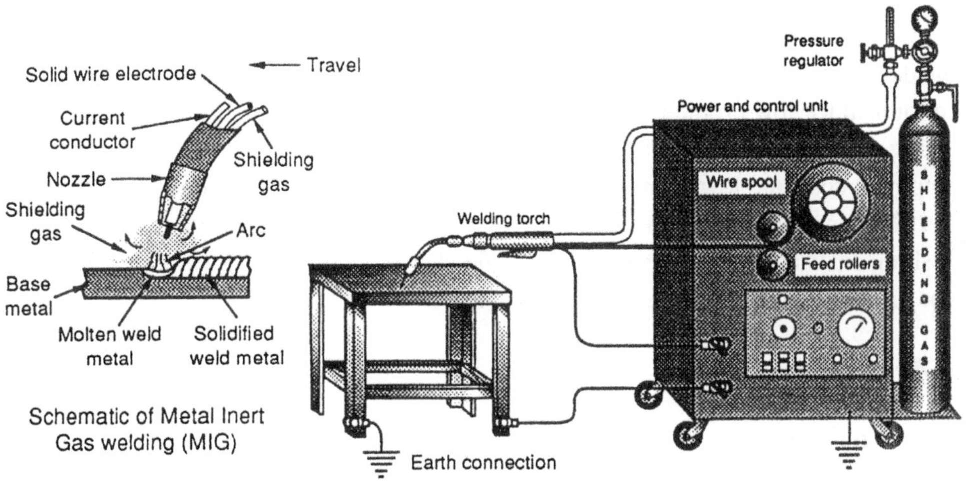

Figure 11.2: Metal Inert-gas Welding.

Materials

Carbon, low alloy and stainless steels. Most non-ferrous metals (except zinc) are also weldable: aluminium, nickel, magnesium and titanium alloys, and copper. Refractory alloys and cast iron can also be welded. Dissimilar metals are difficult to weld.

Process Variations

- Portable semi-automatic (manually operated) or fully automated d.c. systems and robot mounted.
- Three types of metal transfer to the weld area: dip and pulsed transfer use low current for positional welding (vertical, overhead) and thin sheet; spray transfer uses high currents for thick sheet and high deposition rates, typically for horizontal welding.
- Shielding gases: pure CO_2 or argon/CO_2 mix commonly used for carbon and low alloy steels, or a mix of argon/helium also used for nickel alloys and copper. Pure argon is used for aluminium alloys. High chromium steels use an argon/O_2 mix.

- MIG spot welding: used on lap joints.
- Flux Cored Arc Welding (FCAW): uses a wire containing a flux and gas-generating compounds for self-shielding, although flux cored wire is preferred with additional shielding gas for certain conditions. Limited to carbon steels and lower welding rates.

Economic Considerations

- Weld rates from 0.2 m/min for manual welding to 15 m/min for automated set-ups.
- High weld deposition rates with continuous operation reduce production costs.
- Well suited to traversing automated and robotic systems.
- Choice of electrode wire (0.5–1.5 mm diameter) and shielding gas are important cost considerations.
- Economical for low production runs. Can be used for one-offs.
- Tooling costs are low to moderate.
- Equipment costs are low to moderate, depending on degree of automation.
- Direct labour costs are moderate to high. Skill level required is less than TIG.
- Finishing costs are low generally. There is no slag produced at the weld area; however, some grinding back of the weld may be required.

Typical Applications

- General fabrication.
- Structural steelwork.
- Automobile bodywork.
- Hardfacing.

Design Aspects

- All levels of complexity possible.
- Typical joint designs possible using MIG are: butt, lap, fillet and edge. MIG excellent for vertical and overhead welding. (See Appendix C – Weld Joint Design Configurations.)
- Design joints using minimum amount of weld, i.e. intermittent runs and simple or straight contours wherever possible.
- Balance the welds around the fabrication's neutral axis where possible.
- Design parts to give access to the joint area, for vision, electrodes, filler rods, cleaning, etc. MIG good for welds inaccessible by other methods.
- Sufficient edge distances should be designed for and avoid welds meeting at the end of runs.
- Provision for the escape of gases and vapours in the design is important.
- Distortion can be reduced by designing symmetry in parts to be welded along weld lines.

- The fabrication sequence should be examined with respect to the above.
- Minimum sheet thickness = 0.5 mm (6 mm for cast iron).
- Maximum thickness commonly:
  - Carbon, low alloy and stainless steels, cast iron, aluminium, magnesium, nickel, titanium alloys and copper = 80 mm.
  - Refractory alloys = 6 mm.
- Multiple weld runs required on sheet thicknesses ≥5 mm.
- Unequal thicknesses are possible.

Quality Issues

- Clean, high-quality welds with low distortion can be produced.
- Access for weld inspection important, e.g. non-destructive testing (NDT).
- Joint edge and surface preparation is important. Contaminants must be removed from the weld area to avoid porosity and inclusions.
- Shielding gas is chosen to suit parent metal, i.e. it must not react when welding.
- Wire electrode must closely match the composition of the metals being welded.
- The slag created when using a flux cored wire may aid the control of the weld profile and is commonly used for site work (windy conditions where the shielding gas may be gusted or positional welding) and large fillet welds.
- A heat-affected zone is always present. Some stress relieving may be required for restoration of materials' original physical properties.
- Cracking may be experienced when welding high alloy steels.
- Self-adjusting arc length reduces skill level required and increases weld uniformity.
- Backing strips can be used for avoiding excess penetration, but at added cost and increased set-up times.
- Need for jigs and fixtures to keep joints rigid during welding and subsequent cooling to reduce distortion on large fabrications.
- Welding variables should be preset and controlled during production.
- Automation can limit the ability to weld mating parts with large size and shape variations; however, the use of dedicated tooling does reduce distortion, improves reproduction and produces fewer welding defects.
- 'Weldability' of the material is important and combines many of the basic properties that govern the ease with which a material can be welded and the quality of the finished weld, i.e. porosity and cracking. Material composition (alloying elements, grain structure and impurities) and physical properties (thermal conductivity, specific heat and thermal expansion) are some important attributes that determine weldability.
- Surface finish of weld is good.
- Fabrication tolerances are typically ±0.5 mm.

11.3 Manual Metal Arc Welding (MMA)

Process Description

An electric arc is created between a consumable electrode and the workpiece at the joint line. The parent metal is melted and the weld created with the manual feed of the electrode along the weld and downwards as the electrode is being consumed. Simultaneously, a flux on the outside of the electrode melts, covering the weld pool, and generates a gas, shielding it from the atmosphere and preventing oxidation (Figure 11.3).

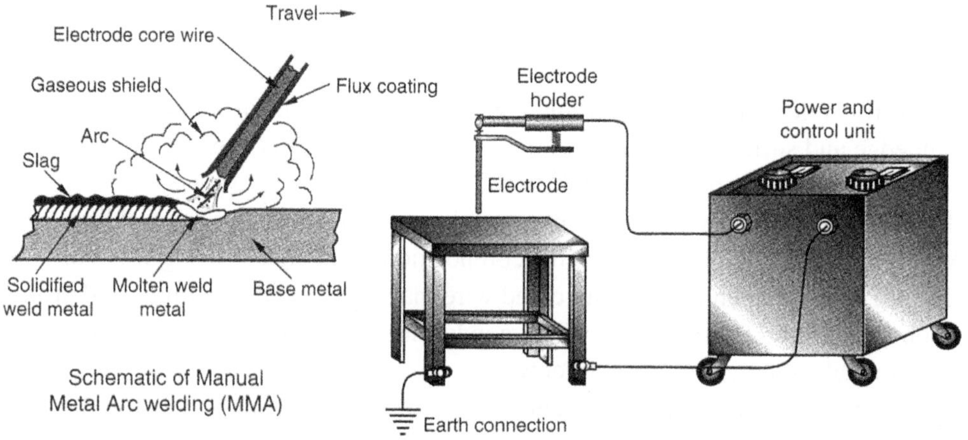

Figure 11.3: Manual Metal Arc Welding.

Materials

Carbon; low alloy and stainless steels; nickel alloys and cast iron, typically. Welding of non-ferrous metals is not recommended, but occasionally performed. Dissimilar metals are difficult to weld.

Process Variations

- Manual d.c. and a.c. sets. Only a few fluxes give stable operation with a.c.
- Large selection of electrode materials with a variety of flux types for the welding of different metals and properties required. Core sizes are between 1.6 and 9.5 mm diameter, and the electrode length is usually 460 mm.
- Stud Arc Welding (SW): for welding pins and stud bolts to structures for subsequent fastening operations. Uses the pin or stud as a consumable electrode to join to the workpiece at one end. Portable semi-automatic or static automated equipment available.

Economic Considerations

- Weld rates up to 0.2 m/min.
- Most flexible of all welding processes.
- Manually performed typically, although some automation possible.
- Can weld a variety of metals by simply changing the electrode.
- A.c. welding requires more power than d.c.
- Suitable for site work. Welding can be performed up to 20 m away from power supply.
- Non-continuous process. Frequent changes of electrode are required.
- Economical for low production runs. Can be used for one-offs.
- Tooling costs are low. Need for jigs and fixtures not as important as other methods and less accuracy required in setting up.
- Equipment costs are low.
- Direct labour costs are high. Skill level required is higher than MIG.
- Finishing costs are high, relative to other welding processes. Slag produced at the weld area, which must be removed during runs, and some grinding back of the weld may be required. Weld spatter often covers the surface, which may need cleaning.

Typical Applications

- Pressure vessels.
- Structural steelwork.
- Shipbuilding.
- Pipework.
- Machine frame fabrication.
- Repair work.
- Hardfacing.

Design Aspects

- All levels of complexity possible.
- Typical joint designs possible using MMA are: butt, lap, fillet and edge in heavier sections. (See Appendix C – Weld Joint Design Configurations.)
- Suitable for all welding positions.
- Design joints using minimum amount of weld, i.e. intermittent runs and simple or straight contours wherever possible.
- Balance the welds around the fabrication's neutral axis.
- Distortion can be reduced by designing symmetry in parts to be welded along weld lines.
- Design parts to give access to the joint area, for vision, electrodes, filler rods, cleaning, etc. MMA excellent for welds inaccessible by other methods.
- Sufficient edge distances should be designed for. Avoid welds meeting at end of runs.

- Provision for the escape of gases and vapours in the design is important.
- The fabrication sequence should be examined with respect to the above.
- Minimum thickness = 1.5 mm (6 mm for cast iron).
- Maximum sheet thickness, commonly for carbon, low alloy and stainless steels, nickel alloys and cast iron = 200 mm.
- Multiple weld runs required on sheet thicknesses ≥10 mm.
- Unequal thicknesses are difficult.

Quality Issues

- Moderate- to high-quality welds with moderate but acceptable levels of distortion can be produced.
- Quality and consistency of weld is related to skill of welder to maintain correct arc length and burn-off rate.
- Access for weld inspection important, e.g. non-destructive testing (NDT).
- Joint edge and surface preparation is important. Contaminants must be removed from the weld area to avoid porosity and inclusions after each pass.
- A heat-affected zone is always present. Some stress relieving may be required for restoration of materials' original physical properties.
- Need for jigs and fixtures to keep joints rigid during welding and subsequent cooling to reduce distortion on large fabrications.
- Backing strips can be used to avoid excess penetration, but at added cost and increased set-up times.
- Can alter composition of weld by addition of alloying elements in the electrode. Addition of deoxidants in the flux minimises carbon loss, which reduces weld strength.
- Electrodes must be dry and free from oil and grease to prevent weld contamination.
- Low hydrogen electrodes should be used when welding high carbon steels to reduce chance of hydrogen cracking.
- The protective slag can help the weld to keep its shape during positional welding.
- Weld is ideally left to cool to room temperature before the slag is removed.
- When the electrode's length is reduced to approximately 50 mm it should be replaced.
- Welding current should be maintained during welding with a stable power supply.
- Arc deflection can sometimes occur with d.c. supplies, especially in magnetised metals. The workpiece may need demagnetising or the return cable repositioned.
- Pre-heating of workpiece can reduce porosity and hydrogen cracking.
- 'Weldability' of the material is important and combines many of the basic properties that govern the ease with which a material can be welded and the quality of the finished weld, i.e. porosity and cracking. Material composition (alloying elements, grain structure and impurities) and physical properties (thermal conductivity, specific heat and thermal expansion) are some important attributes that determine weldability.
- Surface finish of weld is fair to good. Weld spatter often covers the surface.
- Fabrication tolerances are typically ±1 mm.

11.4 Submerged Arc Welding (SAW)

Process Description

A blanket of flux is fed from a hopper in advance of an electric arc created between a consumable electrode wire and the workpiece at the joint line. The arc melts the parent metal and the wire creates the weld as it is automatically fed downwards and traversed along the weld, or the work is moved under the welding head. The flux shields the weld pool from the atmosphere, preventing oxidation. Any flux that is not used is recycled (Figure 11.4).

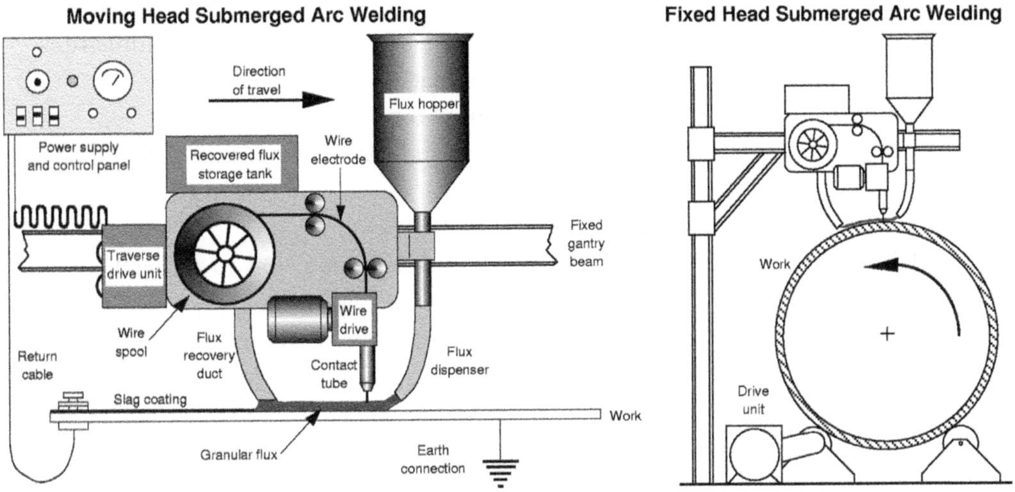

Figure 11.4: Submerged Arc Welding.

Materials

- Carbon; low alloy and stainless steels; and some nickel alloys.
- Dissimilar metals are difficult to weld.

Process Variations

- Self-contained, mainly automated a.c. or d.c systems with up to three welding heads.
- Can have portable traversing welding unit using a wheeled buggy (for long welds on ships' deck plates, for example), self-propelled traversing unit on a gantry or moving head type (for shorter weld lengths) and fixed head where the work rotates under the welding unit (for pressure vessels).
- Copper-coated electrode wire can be solid or tubular. Tubular is used to supply the weld with additional alloying elements. Wire sizes range from 0.8 to 9.5 mm in diameter.
- Can use a strip electrode for surfacing to improve corrosion resistance (pressure vessels) or for hardfacing parts subject to wear (bulk materials handling chute).

- Fluxes available in powdered or granulated form, either neutral or basic. Neutral fluxes used for low carbon steel and basic fluxes for higher carbon steels.
- Bulk welding: uses an iron powder placed in the joint gap in advance of the flux and electrode to increase deposition rates.
- For thin sections can use a flux-coated electrode wire.

Economic Considerations

- Highest weld deposition rate of all arc welding processes.
- Speeds range from 0.1 to 5 m/min.
- Economic for straight, continuous welds on thick plate using single or multiple runs.
- High power consumption offset by high productivity.
- Economical for low production runs. Can be used for one-offs.
- Tooling costs are low to moderate. Need for jigs and fixtures important for accurate joint alignment.
- Equipment costs are moderate to high.
- Direct labour costs are low to moderate. Skill level required is low to moderate.
- Flux handling costs can be high.
- Finishing costs are moderate to high. Slag produced at the weld area needs to be removed.

Typical Applications

- Ships.
- Bridges.
- Pressure vessels.
- Structural steelwork.
- Pipework.

Design Aspects

- Design complexity is limited.
- Typical joint designs possible using SAW are butt and fillet in heavier sections. (See Appendix C – Weld Joint Design Configurations.)
- Suitable for horizontal welding, but can perform vertical welding with special copper side plates to retain flux and mould the weld pool.
- Welds should be designed with straight runs.
- Minimum sheet thickness = 5 mm (6 mm for nickel alloys).
- Maximum sheet thickness commonly:
  - Carbon, low alloy and stainless steels = 300 mm.
  - Nickel alloys = 20 mm.
- Multiple weld runs required on sheet thicknesses ≥40 mm.
- Unequal thicknesses are very difficult.

Quality Issues

- High-quality welds can be produced with low levels of distortion due to fast welding rates.
- Good weld uniformity and properties, although on large deposit welds a coarse grain structure is formed, giving inferior weld toughness.
- Access for weld inspection important, e.g. non-destructive testing (NDT).
- Large weld beads can cause cracking. Weld penetration can be controlled by using a backing strip when using high currents.
- Joint edge and surface preparation is important. Contaminants must be removed from the weld area to avoid porosity and inclusions on each pass.
- A heat-affected zone is always present. Some stress relieving may be required for restoration of materials' original physical properties.
- Can alter composition of weld by addition of alloying elements in the electrode.
- Flux must be clean and free from moisture to prevent weld contamination.
- Weld is ideally left to cool to room temperature to allow the slag to peel off.
- Welding variables are automatically controlled. Monitoring of welding voltage is used to control arc length through varying the wire feed rate, improving weld quality.
- Pre-heating of workpiece can reduce porosity and hydrogen cracking, especially on high carbon steels.
- 'Weldability' of the material is important and combines many of the basic properties that govern the ease with which a material can be welded and the quality of the finished weld, i.e. porosity and cracking. Material composition (alloying elements, grain structure and impurities) and physical properties (thermal conductivity, specific heat and thermal expansion) are some important attributes that determine weldability.
- Surface finish of weld is good.
- Fabrication tolerances are typically ±2 mm.

11.5 Electron Beam Welding (EBW)

Process Description

A controlled high-intensity beam of electrons (0.5–1 mm diameter) is directed to the joint area of the work (anode) by an electron gun (cathode) where fusion of the base material takes place. The operation takes place in a vacuum and typically the work is traversed under the electron beam (Figure 11.5).

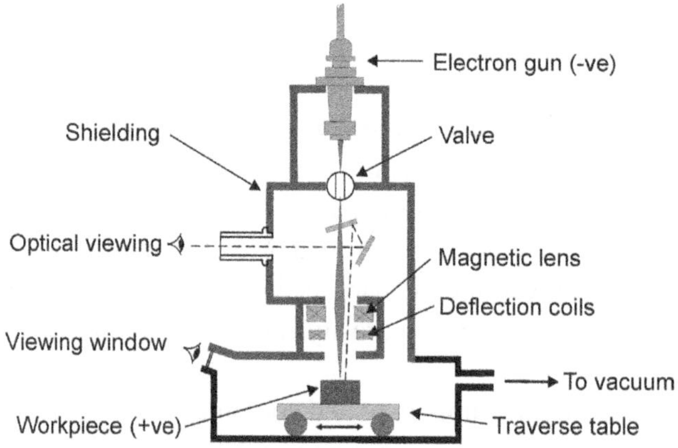

Figure 11.5: Electron Beam Welding.

Materials

- Most metals and combinations of metals are weldable, including low to high carbon and alloy steels, aluminium, titanium, copper, refractory and precious metals.
- Copper alloys and stainless steel difficult to weld. Cast iron, lead or zinc alloys are not weldable.
- Metals that experience gas evolution or vaporisation on welding are difficult.

Process Variations

- High-vacuum (most common), semi-vacuum and atmospheric (out-of-vacuum) equipment available, depending on type of work, size and location.
- Semi-vacuum set-up is used for transportable equipment. Only the area to be welded is surrounded by a vacuum using suction cups.
- Joint advanced under beam for high-vacuum EBW, but for short weld lengths the beam can be moved along the joint using magnetic coils, rather than the work under the beam on a traversing system.

- Electron Beam Machining (EBM; see PRIMA 7.3): an electron gun is used to generate heat, evaporating the workpiece surface for fusion.
- The electron beam process can also be used for cutting, profiling, slotting and surface hardening, using the same equipment by varying process parameters.

Economic Considerations

- Weld rates range from 0.2 to 2.5 m/min.
- Production rates range from 10 to 100/h using high-vacuum equipment.
- Lead times can be several weeks.
- Set-up times can be short, but the time to create a vacuum in the chamber at each loading cycle is an important consideration.
- High flexibility. Possible to perform many operations on the same machine by varying process parameters.
- Full automation of process possible and gives best results.
- Economical for low to moderate production runs.
- Material utilisation is excellent.
- High power consumption.
- Tooling costs are very high.
- Equipment costs are very high.
- Direct labour varies depending on level of automation.
- No finishing needed, typically.

Typical Applications

- Aerospace assemblies (turbine vanes, filters, high-pressure pump bodies).
- Automotive assemblies (crankshaft, gears, valves, bearings).
- Machine parts.
- Instrumentation devices.
- Pipes.
- Reactor shells.
- Hermetic sealing of assemblies.
- Medical implants.
- Bimetallic saw blades.
- Repair work.

Design Aspects

- Typical joint designs possible using EBW are: butt, fillet and lap. (See Appendix C – Weld Joint Design Configurations.)

- Horizontal welding position is the most suitable.
- Path to joint area from the electron beam gun must be a straight line.
- Beam and joint must be aligned precisely.
- Depth to width ratio can exceed 20:1.
- Balance the welds around the fabrication's neutral axis.
- Size limited by vacuum chamber dimensions unless semi-vacuum equipment is used. Maximum height of work in a chamber is 1.2 m typically.
- Possible to weld thin and delicate sections due to no mechanical processing forces.
- Maximum thickness (dependent on vacuum integrity):
  - Aluminum and magnesium alloys = 450 mm.
  - Carbon, low alloy and stainless steels = 300 mm.
  - Copper alloys = 100 mm.
- Minimum thickness = 0.05 mm.
- Single pass maximum = 75 mm.
- Highly dissimilar thicknesses commonly welded.

Quality Issues

- High-quality welds possible with little or no distortion.
- No flux or filler used.
- Integrity of vacuum important. Beam dispersion occurs due to electron collision with air molecules.
- Out-of-vacuum systems must overcome atmospheric pressures at weld area.
- Beams can be generated up to 700 mm from workpiece surface for high-vacuum systems; reduces to less than 40 mm for out-of-vacuum.
- Precise alignment of work required and held using jigs and fixtures.
- Hazardous X-rays are produced during processing, which requires lead shielding.
- Vacuum removes gases from weld area, e.g. hydrogen to minimise hydrogen embrittlement in hardened steels.
- Localised thermal stresses lead to a very small heat-affected zone. Distortion of thin parts may occur.
- Surface finish is excellent.
- Fabrication tolerances are a function of the accuracy of the component parts and the assembly/jigging method. Joint gaps less than 0.1 mm required. Therefore abutment faces should be machined to close tolerances.

11.6 Laser Beam Welding (LBW)

Process Description

Heat for fusion at the weld area is provided by a controlled laser. Focusing of the laser is performed by mirrors or lenses and filler wire material can be supplied for multiple passes (Figure 11.6).

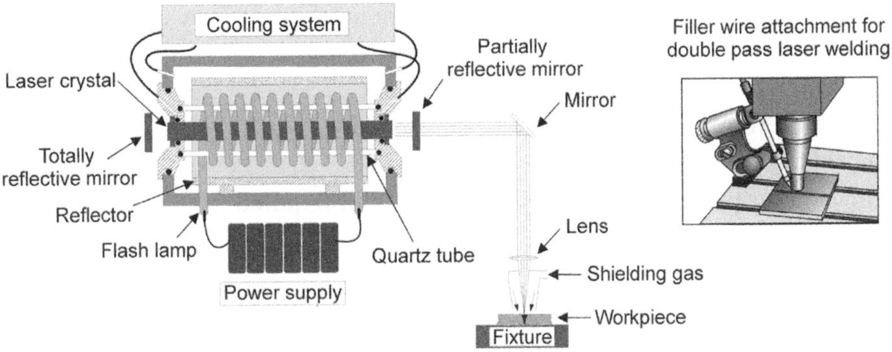

Figure 11.6: Laser Beam Welding.

Materials

- Dependent on thermal diffusivity and to a lesser extent the optical characteristics of material, rather than chemical composition, electrical conductivity or hardness.
- Stainless steel and carbon steels, typically.
- Aluminium alloys and alloy steels are difficult to weld. Not used for cast iron.

Process Variations

- Many types of laser are available, used for different applications. Common laser types available are CO_2, Nd:YAG, Nd:glass, ruby and excimer. Depending on economics of process, pulsed and continuous wave modes are used.
- Shielding gas such as argon sometimes employed to reduce oxidation.
- Laser beam machines can also be used for cutting, surface hardening, machining (LBM; see PRIMA 7.4), drilling, blanking, engraving and trimming, by varying the power density.
- Laser beam spot and seam welding can also be performed on the same equipment.
- Laser soldering: provides very precise heat source for precision work.

Economic Considerations

- Weld rates range from 0.25 to 13 m/min for thin sheet.
- Production rates are moderate.
- High power consumption.
- Lead times can be short, typically weeks.
- Set-up times are short.
- Material utilisation is excellent.
- High degree of automation possible.
- Possible to perform many operations on the same machine by varying process parameters.
- Economical for low to moderate production runs.
- Tooling costs are very high.
- Equipment costs are high.
- Direct labour costs are medium. Some skilled labour required depending on degree of automation.

Typical Applications

- Structural sections.
- Transmission casings.
- Hermetic sealing (pressure vessels, pumps).
- Transformer lamination stacks.
- Instrumentation devices.
- Electronics fabrication.
- Medical implants.

Design Aspects

- Laser can be directed, shaped and focused by reflective optics, permitting high spatial freedom in two dimensions. Horizontal welding position is the most suitable.
- Typical joint designs using LBW: lap, butt and fillet. (See Appendix C – Weld Joint Design Configurations.)
- Mostly for horizontal welding.
- Balance the welds around the fabrication's neutral axis.
- Path to joint area from the laser must be a straight line. Laser beam and joint must be aligned precisely.
- Intimate contact of joint faces is required.
- Filler rod rarely utilised, but for thick sheets or requiring multi-pass welds, a wire-feed filler attachment can be used.
- Minimal work holding fixtures are required.

- Minimum thickness = 0.1 mm.
- Maximum thickness = 20 mm.
- Multiple weld runs required on sheet thickness ≥13 mm.
- Dissimilar thicknesses are difficult.

Quality Issues

- Difficulty of material processing is dictated by how close the material's boiling and vaporisation points are.
- Localised thermal stresses lead to a very small heat-affected zone. Distortion of thin parts may occur.
- No cutting forces, so simple fixtures can be used.
- Inert gas shielding, argon commonly, is employed to reduce oxidation.
- Control of the pulse duration is important to minimise the heat-affected zone, depth and size of molten metal pool surrounding the weld area.
- The reflectivity of the workpiece surface is important. Dull and unpolished surfaces are preferred and cleaning prior to welding is recommended.
- Hole wall geometry can be irregular. Deep holes can cause beam divergence.
- Surface finish good.
- Fabrication tolerances are a function of the accuracy of the component parts and the assembly/jigging method.

11.7 Plasma Arc Welding (PAW)

Process Description

A plasma column is created by constricting an ionised gas through a water-cooled nozzle, reaching temperatures of around 20,000°C. The plasma column flows around a non-consumable tungsten electrode, which provides the electrical current for the arc. The plasma provides the energy for melting and fusion of the base materials and filler rod when used (Figure 11.7).

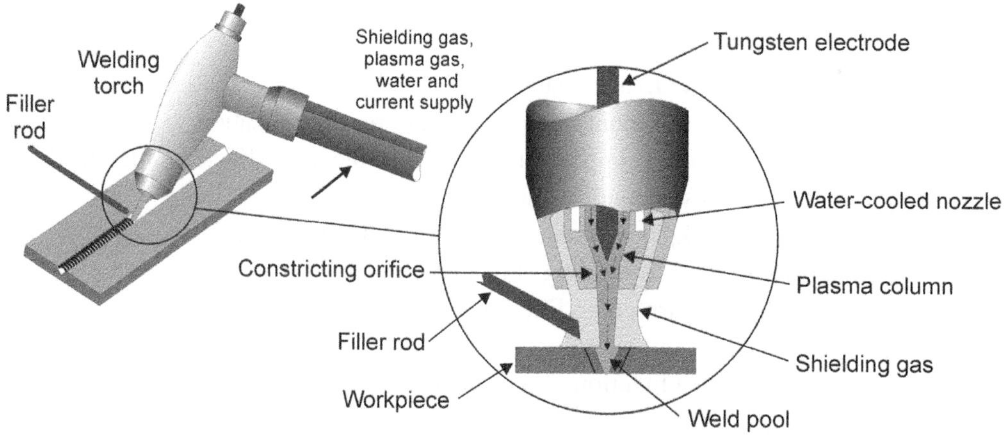

Figure 11.7: Plasma Arc Welding.

Materials

- Most electrically conductive materials.
- Commonly stainless steels; aluminium; copper and nickel alloys; refractory and precious metals.
- Not cast iron, magnesium, lead or zinc alloys.

Process Variations

- Portable manual or automated a.c. or d.c. systems: d.c. system most common.
- There are two modes of operation used for welding:
  - Melt-in fusion for reduced distortion uses low currents.
  - Keyhole fusion at higher currents for full penetration on thick materials.
- Choice of gas and their proportions important for two modes of operation:
  - Plasma gas: argon or argon–hydrogen mix.
  - Shielding gas: argon or argon–hydrogen mix. Also helium or helium–argon mix used.
- Plasma arc cutting: for cutting, slotting and profiling materials up to about 40 mm thickness using the keyholing mode of operation.

- Plasma arc spraying: melting of solid feedstock (e.g. powder, wire or rod) and propelling the molten material on to a substrate to alter its surface properties, such as wear resistance or oxidation protection.
- Filler rod sizes are between 1.6 and 3.2 mm diameter typically.

Economic Considerations

- Weld rates vary from 0.4 m/min for manual welding to 3 m/min for automated systems.
- Alternative to TIG for high automation potential using keyhole mode.
- Welding circuit and system more complex than TIG. Additional controls needed for plasma arc and filters, and deionisers for cooling water mean more frequent maintenance and additional costs.
- Economical for low production runs. Can be used for one-offs.
- Tooling costs are low to moderate.
- Equipment costs are generally high.
- Direct labour costs are moderate.
- Finishing costs are low.

Typical Applications

- Engine components.
- Sheet-metal fabrication.
- Domestic appliances.
- Instrumentation devices.
- Pipes.

Design Aspects

- Design complexity is high.
- Typical joint designs possible using PAW are: butt, lap, fillet and edge. (See Appendix C – Weld Joint Design Configurations.)
- Design joints using minimum amount of weld, i.e. intermittent runs and simple or straight contours wherever possible.
- Balance the welds around the fabrication's neutral axis.
- Distortion can be reduced by designing symmetry in parts to be welded along weld lines.
- The fabrication sequence should be examined with respect to the above.
- Design parts to give access to the joint area, for vision, filler rods, cleaning, etc.
- Sufficient edge distances should be designed for. Avoid welds meeting at end of runs.
- Mostly for horizontal welding, but can also perform vertical welding using higher shielding gas flow rates.
- Filler can be added to the leading edge of the weld pool using a rod, but not necessary for thin sections.

- Minimum sheet thickness = 0.05 mm.
- Maximum thickness commonly:
  - Aluminium = 3 mm.
  - Copper and refractory metals = 6 mm.
  - Steels = 10 mm.
  - Titanium alloys = 13 mm.
  - Nickel = 15 mm.
- Multiple weld runs required on sheet thickness ≥10 mm.
- Unequal thicknesses are difficult.

Quality Issues

- High-quality welds possible with little or no distortion.
- Provides good penetration control and arc stability.
- Access for weld inspection important, e.g. non-destructive testing (NDT).
- Tungsten inclusions from electrode are not present in welds, unlike TIG.
- Joint edge and surface preparation is important. Contaminants must be removed from the weld area to avoid porosity and inclusions.
- A heat-affected zone is always present. Some stress relieving may be required for restoration of materials' original physical properties.
- Not recommended for site work in wind, where the shielding gas may be gusted.
- Need for jigs and fixtures to keep joints rigid during welding and subsequent cooling to reduce distortion on large fabrications.
- Care is needed to keep filler rod within the shielding gas to prevent oxidation.
- Tungsten inclusions can contaminate finished welds.
- The nozzle is used to increase the temperature gradient in the arc, concentrating the heat and making the arc less sensitive to arc length changes in manual welding.
- Plasma arc is very delicate and orifice alignment with tungsten electrode crucial for correct operation.
- Important process variables for consistency in manual welding are welding speed, plasma gas flow rate, current and torch angle.
- 'Weldability' of the material is important and combines many of the basic properties that govern the ease with which a material can be welded and the quality of the finished weld, i.e. porosity and cracking. Material composition (alloying elements, grain structure and impurities) and physical properties (thermal conductivity, specific heat and thermal expansion) are some important attributes that determine weldability.
- Surface finish of weld is excellent.
- Fabrication tolerances are a function of the accuracy of the component parts and the assembly/jigging method, but typically ±0.25 mm.

11.8 Resistance Welding

Process Description

Covers a range of welding processes that use the resistance to electrical current between two materials to generate sufficient heat for fusion. A number of processes use a timed or continuous passage of electric current at the contacting surfaces of the two parts to be joined to generate heat locally, fusing them together and creating the weld with the addition of pressure provided by current supplying electrodes or platens (Figure 11.8).

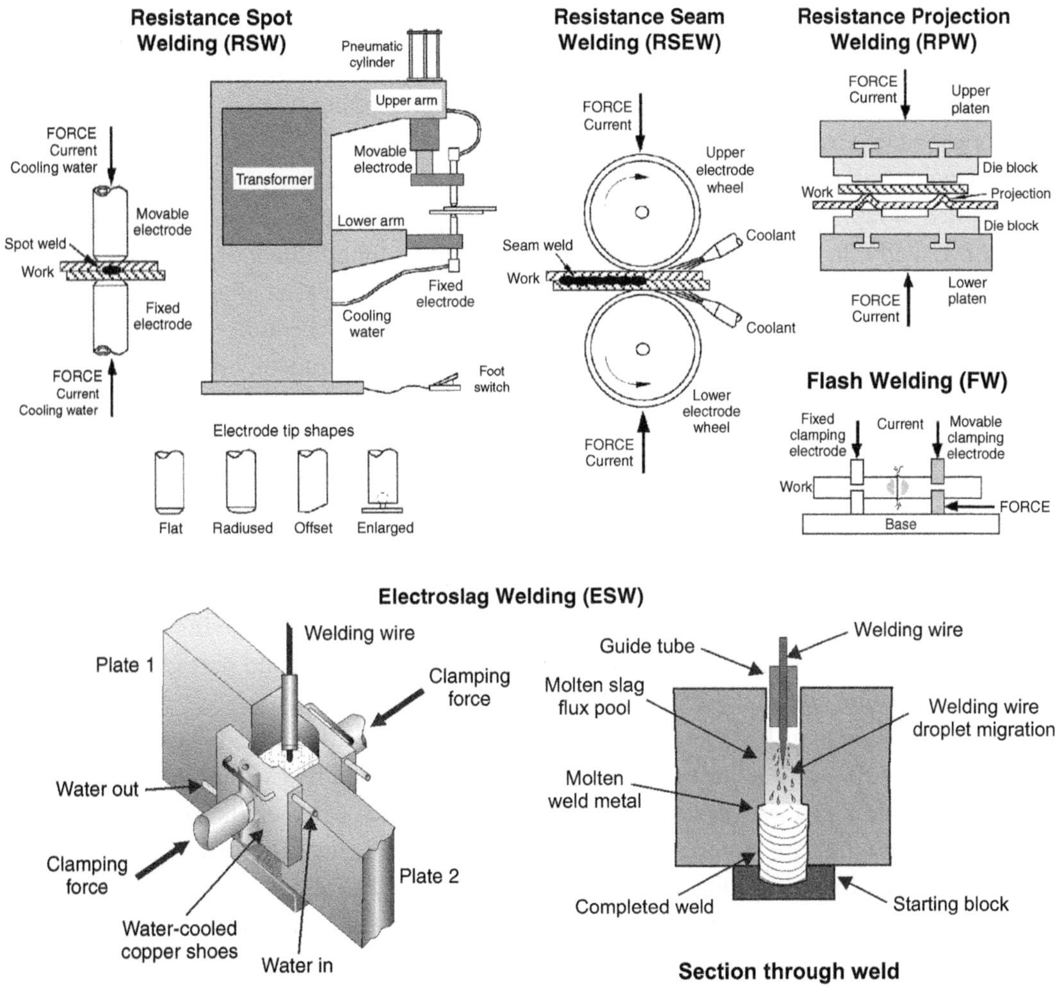

Figure 11.8: Resistance Welding.

Materials

- Low carbon steels, commonly; however, almost any material combination can be welded using conventional resistance welding techniques. Not recommended for cast iron, low-melting-point metals and high carbon steels.
- Electroslag Welding (ESW) is used to weld carbon and low alloy steels, typically. Nickel, copper and stainless steel less common.

Process Variations

- Resistance Spot Welding (RSW): uses two water-cooled copper alloy electrodes of various shapes to form a joint on lapped sheet metal. Can be manual portable (gun), single- or multi-spot semi-automatic, automatic floor standing (rocker arm or press) or robot-mounted as an end effector.
- Resistance Seam Welding (RSEW): uses two driven copper alloy wheels. Current is supplied in rapid pulses, creating a series of overlapping spot welds that are pressure tight. Usually floor standing equipment, either circular, longitudinal or universal types.
- Resistance Projection Welding (RPW): a component and sheet metal are clamped between current-carrying platens. Localised welding takes place at the projections on the component(s) at the contact area. Usually floor standing equipment, either single- or multi-projection press type.
- Upset resistance welding: electrical resistance between two abutting surfaces and additional pressure used to create butt welds on small pipe assemblies, rings and strips.
- Percussion resistance welding: rapid discharge of electrical current and then percussion pressure for welding rods or tubes to sheet metal.
- Flash Welding (FW): parts are accurately aligned at their ends and clamped by the electrodes. The current is applied and the ends brought together, removing the high spots at the contact area deoxidising the joint (known as flashing). Second part is the application of pressure, effectively forging the weld.
- Electroslag Welding (ESW): the joint is effectively 'cast' within joint edges within a gap of about 20–50 mm. An electric arc is used initially to heat a flux within water-cooled copper moulding shoes spanning the joint area. Resistance between the consumable electrode and the base material is then used to generate the heat for fusion. The weld pool is shielded by the molten flux as welding progresses up the joint.
- A variant of ESW is Electrogas Welding (EGW). However, the process does not use electrical resistance as a heat source, but a gas-shielded arc, therefore the molten flux pool above the weld is not necessary. Used for thick sections of carbon steel.

Economic Considerations

- Full automation and integration with component assembly is relatively easy.
- High production rates possible due to short weld times, e.g. RSW = 20 spots/min, RSEW = 30 m/min, FW = 3 s/10 mm^2 area.
- Automation readily achievable using all processes.
- No filler metals or fluxes required (except ESW).
- Little or no post-welding heat treatment is required.
- Minimal joint preparation needed.
- Economical for low production runs. Can be used for one-offs.
- Tooling costs are low to moderate.
- Equipment costs are low to moderate.
- Direct labour costs are low. Skilled operators are not required.
- Finishing costs are very low. Cleaning of welds not necessary typically, except with FW, which requires machining or grinding to remove excess material.
- High deposition rates for ESW, but can still be slow.

Typical Applications

- RSW: car bodies, aircraft structures, light structural fabrications and domestic appliances.
- RSEW: fuel tanks, cans and radiators.
- RPW: reinforcing rings, captive nuts, pins and studs to sheet metal, wire mesh.
- FW: for joining parts of uniform cross-section, such as bar, rods and tubes, and occasionally sheet metal.
- ESW: joining structural sections of buildings and bridges such as columns, machine frames and on-site fabrication.

Design Aspects

- Typical joint designs are: lap (RSW and RSEW); edge (RSEW); butt (FW and ESW); attachments (PW). (See Appendix C – Weld Joint Design Configurations.)
- Access to joint area is important.
- Can be used for joints inaccessible by other methods or where welded components are closely situated.
- Spot weld should have a diameter between four and eight times the material thickness.
- Can process some coated sheet metals (except ESW).

- Same end cross-sections are required for FW.
- For RSW, RSEW and PW:
  - Minimum sheet thickness = 0.3 mm.
  - Maximum sheet thickness commonly = 6 mm.
  - Mild steel sheet up to 20 mm thick has been spot and seam welded but requires high currents and expensive equipment.
- For FW, sizes range from 0.2-mm-thick sheet to sections up to 0.1 m^2 in area.
- Unequal thicknesses are possible with RSW and RSEW (up to 3:1 thickness ratio).
- ESW applied to sheet thicknesses of same order from 25 mm up to 500 mm using several guide tubes and electrodes in one pass, but down to 75 mm for a single set. Vertical welds can restrict design freedom in ESW.

Quality Issues

- Clean, high-quality welds with very low distortion can be produced. Although a heat-affected zone is always created, it can be small.
- Coarse grain structures may be created in ESW due to high heat input and slow cooling.
- Surface preparation is important to remove any contaminants from the weld area, such as oxide layers, paint, and thick films of grease and oil. Resistance welding of aluminium requires special surface preparation.
- Welding variables for spot, seam and projection welding should be preset and controlled during production. These include: current, timing and pressure (where necessary).
- Electrodes or platens must efficiently transfer pressure to the weld, conduct and concentrate the current, and remove heat away from the weld area, therefore maintenance should be performed at regular intervals.
- Spot, seam and projection welds can act as corrosion traps.
- RSW, RSEW and PW welds can be difficult to inspect. Destructive testing should be intermittently performed to monitor weld quality.
- Depression left behind in RSW and RSEW serves to prevent cavities or cracks due to contraction of the cooling metal.
- Possibility of galvanic corrosion when resistance welding some dissimilar metals.
- High strength welds are produced by FW. Always leaves a ridge at the joint area that must be removed.
- 'Weldability' of the material is important and combines many of the basic properties that govern the ease with which a material can be welded and the quality of the finished weld, i.e. porosity and cracking. Material composition (alloying elements, grain structure and impurities) and physical properties (thermal conductivity, specific heat and thermal expansion) are some important attributes that determine weldability.

- Surface finish of the welds is fair to good for RSW, RSEW, FW and PW. Excellent for ESW.
- No weld spatter and no arc flash (except ESW initially).
- Alignment of parts to give good contact at the joint area is important for consistent weld quality.
- Repeatability is typically ±0.5 to ±1 mm for robot RSW.
- Axes alignment total tolerance for FW between 0.1 and 0.25 mm.

11.9 Solid-state Welding

Process Description

A range of methods utilising heat, pressure and/or high energy to plastically deform the material at the joint area in order to create a solid-phase mechanical bond (Figure 11.9).

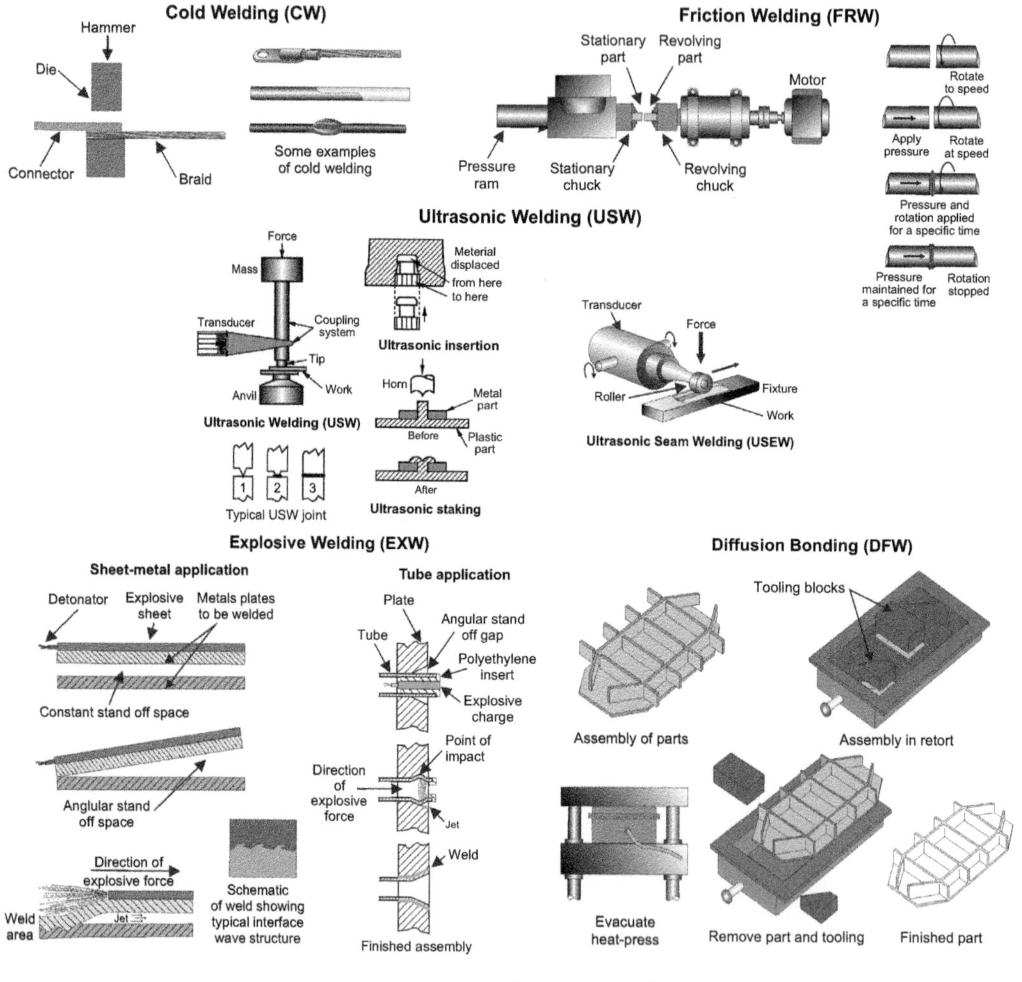

Figure 11.9: Solid-state Welding.

Materials

- Cold Welding (CW): ductile metals such as carbon steels, aluminium, copper and precious metals.
- Friction Welding (FRW): can weld many material types and dissimilar metals effectively, including aluminium to steel. Also thermoplastics and refractory metals.
- Ultrasonic Welding (USW): can be used for most ductile metals, such as aluminium and copper alloys, carbon steels and precious metals, and some thermoplastics. Can bond dissimilar materials readily.
- Explosive Welding (EXW): carbon steels, aluminium, copper and titanium alloys. Welds dissimilar metals effectively.
- Diffusion (welding) Bonding (DFW): stainless steel, aluminium, low alloy steels, titanium and precious metals. Occasionally, copper and magnesium alloys are bonded.

Process Variations

- CW: process is performed at room temperature using high forces to create substantial deformation (up to 95%) in the parts to be joined. Surfaces require degreasing and scratch-brushing for good bonding characteristics.
- Cold pressure spot welding: for sheet-metal fabrication using suitably shaped indenting tools.
- Forge welding: the material is heated in a forge or oxyacetylene ring burners. Hand tools and anvil used to hammer together the hot material to form a solid-state weld. Commonly associated with the blacksmiths' trade, and used for decorative and architectural work.
- Thermocompression bonding: performed at low temperatures and pressures for bonding wires to electrical circuit-boards.
- USW: hardened probe introduces a small static pressure and oscillating vibrations at the joint face, disrupting surface oxides and raising the temperature through friction and pressure to create a bond. Can also perform spot welding using similar equipment.
- Ultrasonic Seam Welding (USEW): ultrasonic vibrations imparted through a roller traversing the joint line.
- Ultrasonic soldering: uses an ultrasonic probe to provide localised heating through high-frequency oscillations. Eliminates the need for a flux, but requires pre-tinning of surfaces.
- Ultrasonic insertion: for introducing metal inserts into plastic parts for subsequent fastening operations.
- Ultrasonic staking: for light assembly work in plastics.
- FRW: the two parts to be welded, one stationary and one rotating at high speed (up to 3000 rpm), have their joint surfaces brought into contact. Axial pressure and frictional heat at the interface create a solid-state weld on discontinuation of rotation and on cooling.

- Friction stir welding: uses the frictional heat to soften the material at the joint area using a wear-resistant rotating tool.
- EXW: uses explosive charge to supply energy for a cladding sheet metal to strike the base sheet metal, causing plastic flow and a solid-state bond. Bond strength is obtained from the characteristic wavy interlocking at the joint face. Can also be used for tube applications.
- DFW: the surfaces of the parts to be joined are brought together under moderate loads and temperatures in a controlled inert atmosphere or vacuum. Localised plastic deformation and atomic interdiffusion occurs at the joint interface, creating the bond after a period of time.
- Superplastic diffusion bonding: combination of DFW with superplastic forming to produce complex fabrications (see PRIMA 4.7).

Economic Considerations

- Production rates vary: high for CW and FW (30 s cycle time), moderate for USW, and low for EXW and DFW.
- Lead times low typically.
- Material utilisation excellent. No scrap generated.
- High degree of automation possible with many processes (except EXW).
- No filler materials needed.
- Economical for low production runs. Can be used for one-offs.
- Tooling costs are low to moderate.
- Equipment costs are low (CW, EXW) to high (USW, FRW, DFW).
- Direct labour costs are low to moderate. Some skilled labour maybe required.
- Finishing costs are low. Cleaning of welds not necessary typically, except with FRW, which requires machining or grinding to remove excess material.

Typical Applications

- CW: welding caps to tubes, electrical terminations and cable joining.
- USW: for sheet-metal fabrication, joining plastics, electrical equipment and light assembly work.
- FRW: for welding hub-ends to axle casings, welding valves stems to heads and gear assemblies.
- EXW: used mainly for cladding, or bonding one plate to another, to improve corrosion resistance in the process industry, for marine parts and joining large pipes in the petrochemical industry.
- DFW: for joining high-strength materials in the aerospace and nuclear industries; honeycomb structures; biomedical implants and metal laminates for electrical devices.

Design Aspects

- Typical joint designs are: lap (CW, USW, USEW, EXW, DFW); edge (USEW); butt (CW, FRW, ESW); T-joint (DFW); flange (EXW). (See Appendix C – Weld Joint Design Configurations.)
- Access to joint area is important.
- Unequal thicknesses are possible with CW, USW, EXW and DFW.
- CW: thicknesses range from 5 to 20 mm.
- USW: thicknesses range from 0.1 to 3 mm.
- EXW: thicknesses range from 20 to 500 mm and maximum surface area = 20 m^2.
- FRW: diameters range from 2 to 150 mm and maximum surface area = 0.02 m^2. Parts must have rotational symmetry.
- DFW: thicknesses range from 0.5 to 20 mm.

Quality Issues

- Little or no deformation takes place (except EXW).
- No weld spatter and no arc flash.
- Alignment of parts is crucial for consistent weld quality.
- Parts must be able to withstand high forces and torques to create bond over long period of time.
- Safety concerns for EXW include explosives handling, noise and provision for controlled explosion.
- Welds as strong as base material in many cases.
- Surface preparation is important to remove any contaminants from the weld area, such as oxide layers, paint and thick films of grease and oil.
- Possibility of galvanic corrosion when welding some material combinations.
- Surface finish of the welds is good.
- Fabrication tolerances vary from close for DFW; moderate for FRW, CW and USW; and low dimensional accuracy for EXW.

11.10 Thermit Welding (TW)

Process Description

A charge of iron oxide and aluminium powder is ignited in a crucible. The alumino-thermic reaction produces molten steel and alumina slag. On reaching the required temperature, a magnesite thimble melts and allows the molten steel to be tapped off to the mould surrounding the pre-heated joint area. On cooling, a cast joint is created (Figure 11.10).

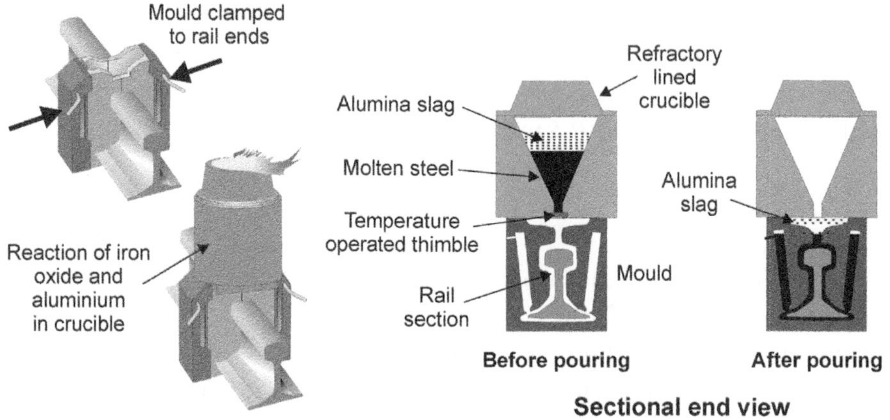

Figure 11.10: Thermit Welding.

Materials

Carbon and low alloys steels, and cast iron only.

Process Variations

- Moulds can be refractory sand or carbon.
- Can be used to repair broken areas of structural sections using special moulds.

Economic Considerations

- Production rates very low. Cycle times typically 1 hour.
- Lead time a few days.
- Twenty percent of welding metal lost in runners and risers.
- Scrap material cannot be recycled directly.
- Economical for low production runs. Can be used for one-offs.
- Manual operation only.
- Tooling costs are low to moderate.

- Equipment costs are low to moderate.
- Direct labour costs are moderate to high. Some labour involved.
- Finishing costs are moderate. Excess metal around joint not always removed, but gates and risers must be ground off.

Typical Applications

- Site welding of rails to form continuous lengths.
- Joining heavy structural sections and low-loaded structural joints.
- Machine frame fabrication.
- Shipbuilding.
- Joining thick cables.
- Concrete reinforcement steel bars.
- Repair work.

Design Aspects

- The cross-section of the parts to be joined can be complex, otherwise limited design freedom.
- Joint gaps typically 20–80 mm.
- Butt joint design possible only (see Appendix C – Weld Joint Design Configurations).
- Minimum sheet thickness = 10 mm.
- Maximum thickness = 1,000 mm.

Quality Issues

- Weld quality is fair.
- The cast joint typically has inferior properties to that of the base material.
- Pre-heating times range from 1 to 7 minutes depending on section thickness. Small section thicknesses may not require pre-heating.
- Joint area must be cleaned thoroughly.
- Joint edges must be aligned with a suitable gap depending on section size.
- Alloying elements can be added to the charge to match the physical properties of materials to be joined.
- Exothermic chemical reaction has safety concerns, and proper precautions and ventilation are necessary.
- Surface finish is poor to fair.
- Fabrication tolerances are a function of the accuracy of the component parts (hot rolled sections usually, which have poor dimensional accuracy) and the clamping/jigging method used, but typically ±1.5 mm.

11.11 Gas Welding (GW)

Process Description

High-pressure gaseous fuel and oxygen are supplied by a torch through a nozzle where combustion takes place providing a controllable flame. The high temperature generated (greater than 3000°C) is sufficient to melt the base metal at the joint area. Shielding from the atmosphere is performed by the outer flame. Filler metal can be supplied to the weld pool if needed (Figure 11.11).

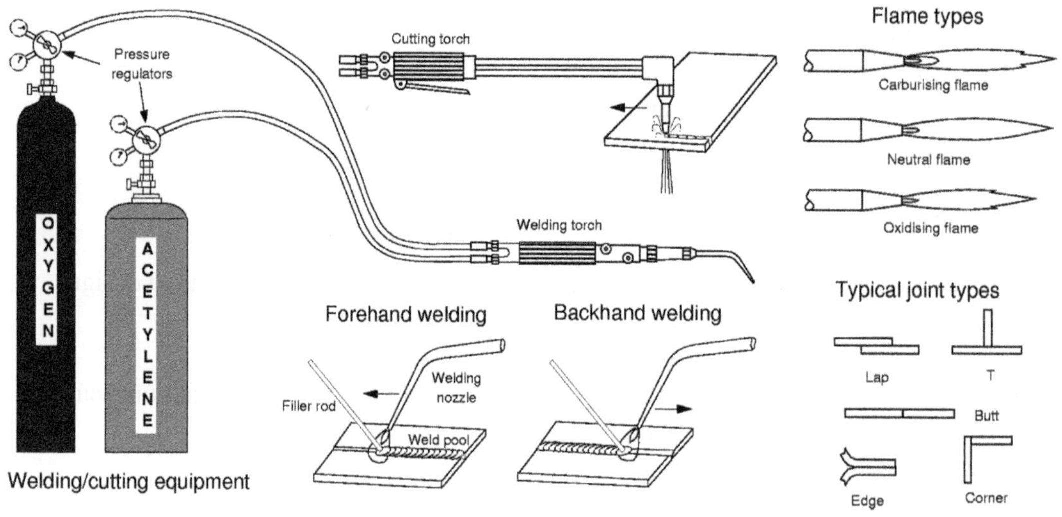

Figure 11.11: Gas Welding.

Materials

- Commonly ferrous alloys: low carbon, low alloy and stainless steels, and cast iron.
- Also, nickel, copper and aluminium alloys, and some low-melting-point metals (zinc, lead and precious metals).
- Refractory metals cannot be welded.

Process Variations

- Commonly manually operated, portable and self-contained welding sets.
- Can use forehand or backhand welding procedures.
- Gas fuel commonly used is acetylene for most welding applications and materials, known as oxyacetylene welding.
- Hydrogen, propane, butane and natural gas are used for low-temperature brazing and welding small and thin parts.

- Air can be used instead of oxygen for brazing, soldering and welding lead sheet.
- Flux may be necessary for welding metals other than ferrous alloys.
- By regulating the oxygen flow, three types of flame can be produced:
  - Carburising: for flame hardening, brazing, welding nickel alloys and high carbon steels.
  - Neutral: for most welding operations.
  - Oxidising: used for welding copper, brass and bronze.
- Braze welding: base metal is pre-heated with an oxyacetylene or oxypropane gas torch at the joint area. Brazing filler metal, usually supplied in rod form, and a flux are applied to joint area, where the filler becomes molten and fills the joint gap through capillary action. Although no fusion takes place, very high temperatures are required, typically 700°C. Some finishing may be necessary to clean flux residue and excess braze.
- Pressure gas welding: heat from oxyacetylene burner is used to melt ends of the parts to be joined and then applied pressure creates the weld.
- Gas cutting: an oxyacetylene or oxypropane flame from a specially designed nozzle is used to pre-heat the parent metal and an additional high-pressure oxygen supply effectively cuts the metal by oxidising it. Can perform straight cuts or profiles (when automated) in plate over 500 mm thick.

Economic Considerations

- Weld rates very low, typically 0.1 m/min.
- Lead times very short.
- Very flexible process. Same equipment can be used for welding, cutting and several heat treatment processes.
- Economical for very low production runs. Can be used for one-offs.
- Automation is not practical for most situations.
- Tooling costs are low to moderate. Little tooling required, and jigs and fixtures are simple for manual operation.
- Equipment costs are low to moderate.
- Direct labour costs are moderate. Skilled operators may be required.
- Finishing costs are low to moderate. No slag produced, but cleaning may be required.

Typical Applications

- Sheet-metal fabrication.
- Ventilation ducts.
- Small-diameter pipe welding.
- Repair work.

Design Aspects

- Moderate levels of complexity possible. Capability to weld parts with large size and shape variations.
- Typical joint designs possible using gas welding are: butt, fillet, lap and edge, in thin sheet. (See Appendix C – Weld Joint Design Configurations.)
- All welding positions possible.
- Design joints using minimum amount of weld, i.e. intermittent runs and simple or straight contours wherever possible.
- Balance the welds around the fabrication's neutral axis.
- Distortion can be reduced by designing symmetry in parts to be welded along weld lines.
- The fabrication sequence should be examined with respect to the above.
- Sufficient edge distances should be designed for, and welds meeting at the end of runs should be avoided.
- Minimum sheet thickness commonly:
  - Carbon steel = 0.5 mm
  - Cast iron = 3 mm
- Maximum sheet thickness commonly:
  - Carbon steel and cast iron = 30 mm
  - Low alloy steel, stainless steel, nickel and aluminium alloys = 3 mm
- Multiple weld runs required on sheet thicknesses ≥4 mm.
- Unequal thicknesses are possible.

Quality Issues

- Good-quality welds with moderate but acceptable levels of distortion can be produced. Repeatability can be a problem.
- Access for weld inspection important.
- Attention to adequate jigs and fixtures, when welding thin sheet is recommended to avoid excessive distortion of parts, by providing good fit-up and to take heat away from the surrounding metal.
- Heat-affected zone always created. Some stress relieving may be required for restoration of materials' original physical properties.
- Surface preparation is important to remove any contaminants from the weld area, such as oxide layers, paint, and thick films of grease and oil.
- Gas flow rates should be preset and regulated during production. Even-gas mix gives the neutral flame most commonly used for welding. Even-heating of joint area required for consistent results.
- Shielding integrity at the weld area not as high as arc welding methods, and some oxidation and atmospheric attack may occur.

- 'Weldability' of the material is important and combines many of the basic properties that govern the ease with which a material can be welded and the quality of the finished weld, i.e. porosity and cracking. Material composition (alloying elements, grain structure and impurities) and physical properties (thermal conductivity, specific heat and thermal expansion) are some important attributes that determine weldability.
- Surface finish of weld is fair to good.
- Fabrication tolerances are typically ±1 mm.

11.12 Brazing

Process Description

Heat is applied to the parts to be joined, which melts a manually fed or pre-placed filler braze metal (which has a melting temperature ≥450°C) into the joint by capillary action. A flux is usually applied to facilitate 'wetting' of the joint, prevent oxidation, remove oxides and reduce fuming (Figure 11.12).

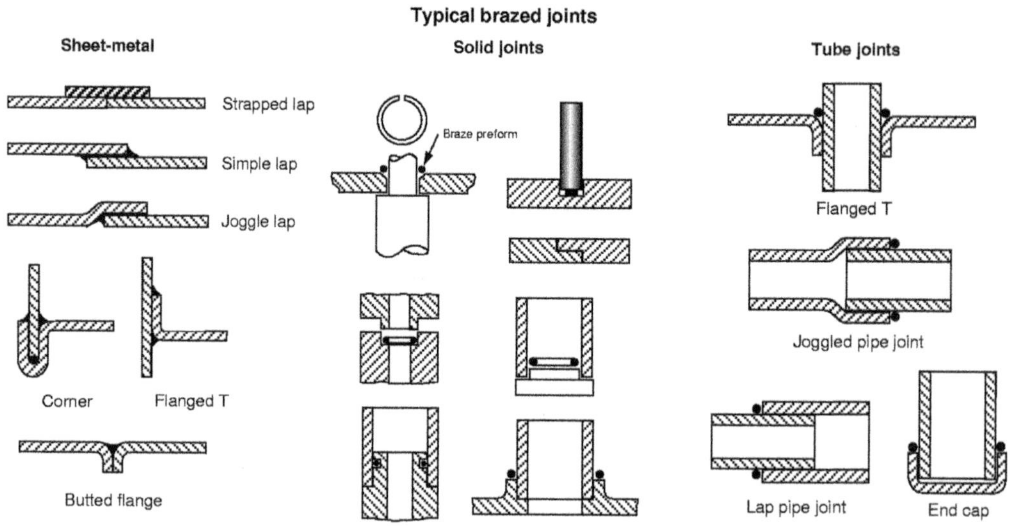

Figure 11.12: Typical Brazed Joints.

Materials

Almost any metal and combination of metals can be brazed. Aluminium difficult due to oxide layer.

Process Variations

- Gas brazing: neutral or carburising oxyfuel flame is used to heat the parts. Can be manual torch brazing (TB) for small production runs or automated with a fixed burner (ATB).
- Induction Brazing (IB): components are placed in a magnetic field surrounding an inductor carrying a high-frequency current, giving uniform heating.
- Resistance Brazing (RB): high electric resistance at joint surfaces causes heating for brazing. Not recommended for brazing dissimilar metals.
- Dip Brazing (DB): parts immersed to a certain depth in a bath of molten chemical or brazing alloy covered with molten flux. Commonly used for brazing aluminium.

- Furnace Brazing (FB): heating takes place in carburising/inert atmosphere or a vacuum. The filler metal is pre-placed at the joint and no additional flux is needed. Large batches of parts of varying sizes and joint types can be brazed simultaneously. Good for parts that may distort using localised heating methods and dissimilar metals.
- Infrared Brazing (IRB): uses quartz–iodine incandescent lamps as heat energy. For joining pipes typically.
- Diffusion Brazing (DFB): braze filler actually diffuses into the base metal, creating a new alloy at the joint interface. Gives a strong bond of equal strength to that of the base metal.
- Braze welding: base metal is pre-heated with an oxyacetylene or oxypropane gas torch at the joint area. Brazing filler metal, usually supplied in rod form, and a flux are applied to joint area, where the filler becomes molten and fills the joint gap through capillary action (see PRIMA 7.11).
- Filler metal can be in preforms, wire, foil, coatings, slugs and pastes in a variety of metal alloys; commonly the alloys are based on copper, silver, nickel and aluminium.
- Flux types: borax, borates, fluoroborates, alkali fluorides and alkali chlorides (for brazing aluminium and its alloys) in powder, paste or liquid form.

Economic Considerations

- High production rates are possible using FB and IB, but low with TB.
- Cycle times vary. Long for FB and DFB, short for TB.
- Very flexible process.
- Large fabrications may be better suited to welding than brazing.
- Economical for very low production volumes. Can be used for one-offs.
- Tooling costs are low. Little tooling required.
- Equipment costs vary depending on process and degree of automation. Low for TB, high for FB.
- Direct labour costs are low to moderate. Cost of joint preparation can be high.
- Finishing costs are moderate. Cleaning of the parts to remove corrosive flux residues is critical.

Typical Applications

- Machine parts.
- Pipework.
- Bicycle frames.
- Repair work.
- Cutting tool inserts.

Design Aspects

- All levels of complexity.
- Joints should be designed to operate in shear or compression, not tension.
- Typical joint designs using brazing: lap and scarf in thin joints with large contact areas or a combination of lap and fillet. Fillets can help to distribute stresses at the joint. Butt joints are possible but can cause stress concentrators in bending.
- Lap joints should have a length to thickness ratio of between three and four times that of the thinnest part for optimum strength.
- Joints should be designed to give a clearance between the mating parts of, typically, 0.02–0.2 mm depending on the process to be used and the material to be joined (can be zero for some process/material combinations). The clearance directly affects joint strength. If the clearance is too great the joint will lose a considerable amount of strength.
- Tolerances on mating parts should maintain the joint clearances recommended.
- Parts in the assembly should be arranged to promote capillary action by gravity.
- Machine marks should be in line with the flow of solder.
- Joint strength is between that of the base and filler metals in a well-designed joint.
- Vertical brazing should integrate chamfers on parts to create reservoirs.
- Jigs and fixtures should be used only on parts where self-locating mechanisms (staking, press fits, knurls and spot welds) are not practical. If jigs and fixtures are used they should support the joint as far from the joint area as possible, have minimum contact and have low thermal mass.
- Provision for the escape of gases and vapours in the joint design is important.
- Metals with a melting temperature less than 650°C cannot be brazed.
- Minimum sheet thickness = 0.1 mm.
- Maximum thickness = 50 mm.
- Unequal thicknesses are possible, but sudden changes in section can create stress concentrators.
- Dissimilar metals can cause thermal stresses on cooling.

Quality Issues

- Good-quality joints with very low distortion are produced.
- Virtually a stress-free joint is created with proper control of cooling.
- Choice of filler metal important in order to avoid joint embrittlement. Possibility of galvanic corrosion.
- A limited amount of inter-alloying takes place between the filler metal and the part metal; however, excessive alloying can reduce joint strength. Control of the time and temperature of the applied heat is important with respect to this.
- Subsequent heating of assembly after brazing could melt the filler metal again.

- Filler metal selection is based upon the metals to be brazed, process to be used and its economics, and the operating temperature of the finished assembly.
- Surface preparation is important to remove any contaminants from the joint area, such as oxide layers, paint, and thick films of grease and oil; and promote wetting. Pickling and degreasing commonly performed before brazing of parts.
- Smooth surfaces are preferred to rough ones. Sand-blasted surfaces are not recommended as they tend to reduce joint strength. Abrading the joint area using emery cloth is acceptable.
- Correct clearance, temperature gradients and effective use of gravity promote flow of braze filler through capillary action.
- Flux residues after the joint has been made must be removed to avoid corrosion.
- Surface finish of brazed joints is good.
- Fabrication tolerances are a function of the accuracy of the component parts and the assembly/jigging method.

11.13 Soldering

Process Description

Heat is applied to the parts to be joined, which melts a manually fed or pre-placed filler solder metal (which has a melting temperature <450°C) into the joint by capillary action. A flux is usually applied to facilitate 'wetting' of the joint, prevent oxidation, remove oxides and reduce fuming (Figure 11.13).

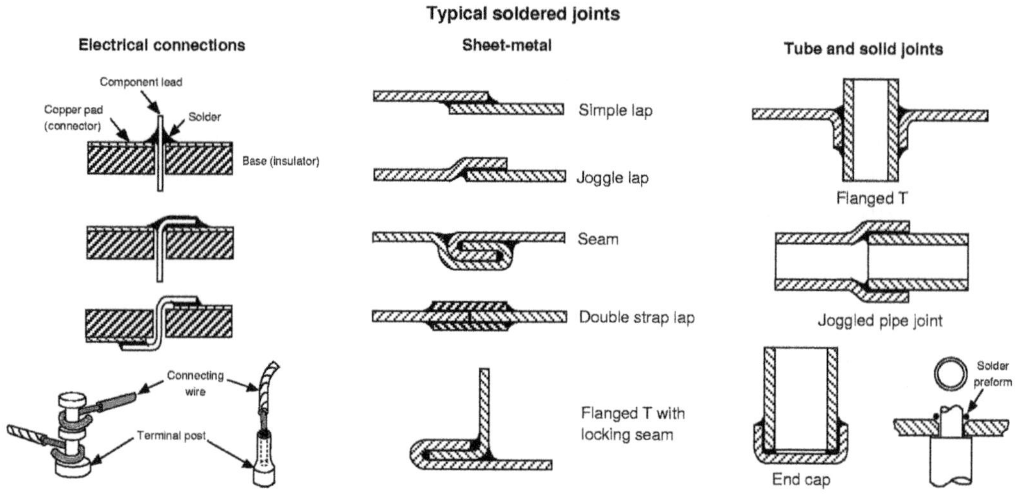

Figure 11.13: Typical Soldered Joints.

Materials

- Most metals and combinations of metals can be soldered with the correct selection of filler metal, heating process and flux. Commonly, copper; tin; mild and low alloy steels; nickel and precious metals are soldered. Some ceramics can be soldered.
- Magnesium, titanium, cast iron and high carbon or alloy steels are not recommended.

Process Variations

- Gas soldering: air-fuel flame is used to heat the parts. Can be manually performed with a torch (TS) for small production runs or automated (ATS) with a fixed burner for greater economy.
- Furnace Soldering (FS): uniform heating takes place in an inert atmosphere or vacuum.
- Induction Soldering (IS): components are placed in a magnetic field surrounding an inductor carrying a high-frequency current, giving uniform heating.

- Resistance Soldering (RS): high electric resistance at joint surfaces causes heating for brazing. Not recommended for brazing dissimilar metals.
- Dip Soldering (DS): assemblies immersed to a certain depth in bath of molten solder. Can require extensive jigging and fixtures.
- Wave Soldering (WS): similar to dip soldering, but the solder is raised to the joint area on a wave. Used extensively for soldering electronic components to printed circuit-boards.
- Contact or Iron Soldering (INS): uses an electrically heated iron or hot plate. Most common soldering process used for general electrical and sheet-steel work.
- Infrared Soldering (IRS): heat application through directed spot of infrared radiation. Used for small precision work and difficult to reach joints.
- Laser soldering: provides very precise heat source for precision work, but at high cost.
- Ultrasonic soldering: uses an ultrasonic probe to provide localised heating through high-frequency oscillations. Eliminates the need for a flux, but requires pre-tinning of surfaces.
- Filler metal (solder) can be in preforms, wire, foil, coatings, slugs and pastes in a variety of metal alloys, commonly: tin–lead, tin–zinc, lead–silver, zinc–aluminium and cadmium–silver. The selection is based upon the metals to be soldered.
- Flux types: either corrosive (rosin, muriatic acid, metal chlorides) or non-corrosive (aniline phosphate), in powder, paste or liquid form.

Economic Considerations

- High production rates are possible for WS.
- Very flexible process.
- Economical for very low production runs. Can be used for one-offs.
- Tooling costs are low. Little tooling required.
- Equipment costs vary depending on degree of automation.
- Direct labour costs are low to moderate. Cost of joint preparation can be high.
- Finishing costs are moderate. Cleaning of the parts to remove corrosive flux residues is critical.

Typical Applications

- Electrical connections.
- Printed circuit-boards.
- Light sheet-metal fabrication.
- Pipes and plumbing.
- Automobile radiators.
- Precision joining.
- Jewellery.
- Food handling equipment.

Design Aspects

- Design complexity is high, but low load capacity joints.
- Most common joint is the lap with large contact areas or a combination of lap and fillet. Fillets joints are predominantly used in electrical connections.
- Can be used to provide electrical or thermal conductivity or provide pressure-tight joints.
- Joints should be designed to operate in shear and not tension. Additional mechanical fastening is recommended on highly stressed joints.
- Joints should be designed to give a clearance between the mating parts of 0.08–0.15 mm.
- The clearance directly affects joint strength. If the clearance is too great the joint will lose a considerable amount of strength.
- Tolerances on mating parts should maintain the joint clearances recommended.
- On lap joints the length of lap should be between three and four times that of the thinnest part for optimum strength.
- Parts in the assembly should be arranged to promote capillary action by gravity.
- Machine marks should be in line with the flow of solder.
- Design joints using minimum amount of solder.
- Jigs and fixtures should be used only on parts where self-locating mechanisms, i.e. seaming, staking, knurls, bending or punch marks, are not practical.
- If jigs and fixtures are used they should support the joint as far from the joint as possible, have minimum contact with the parts to be soldered and have low thermal mass.
- Soldered joints in electronic printed circuit-boards should be spaced more than 0.8 mm apart.
- Provision for the escape of gases and vapours in the design is important with vent-holes.
- Minimum sheet thickness = 0.1 mm.
- Maximum thickness commonly = 6 mm.
- Unequal thicknesses are possible but may create unequal joint expansion.
- Dissimilar metals can cause thermal stresses at the joint on cooling, due to different expansion coefficients.

Quality Issues

- Virtually stress- and distortion-free joints can be produced.
- Coating metals with tin improves solderability.
- Coatings should be used on parts to protect the parent metal prior to soldering, classed as: protective, fusible, soluble, non-soluble and stop-off coatings.
- Control of the time and temperature of the applied heat is important.
- Contamination-free environment important for electronics soldering.
- Subsequent operations should have a lower processing temperature than the solder melting temperature.

- Heat sinks should be used when soldering heat-sensitive components, especially in electronics manufacture.
- Jigs and fixtures should be used to maintain joint location during solder cooling for delicate assemblies.
- Choice of solder important in order to avoid possibility of galvanic corrosion.
- Surface preparation is important to remove any contaminants from the joint area, such as oxide layers, paint, and thick films of grease and oil; and promote wetting. Degreasing and pickling of the parts to be soldered is recommended.
- Smooth surfaces are preferred to rough ones. Abrading the joint area using emery cloth is acceptable.
- Correct clearance, temperature gradients and use of effective use of gravity promote flow of solder metal through capillary action.
- Flux residues after the joint has been made must be removed to avoid corrosion.
- Surface finish of soldered joints is excellent.
- Fabrication tolerances are a function of the accuracy of the component parts and the assembly/jigging method.

11.14 Thermoplastic Welding

Process Description

Joint edges are heated using hot gas from a hand-held torch, causing the thermoplastic material to soften. A consumable thermoplastic filler rod of the same composition as the base material is used to fill the joint and create the bond with additional pressure from the filler rod at the joint area (Figure 11.14).

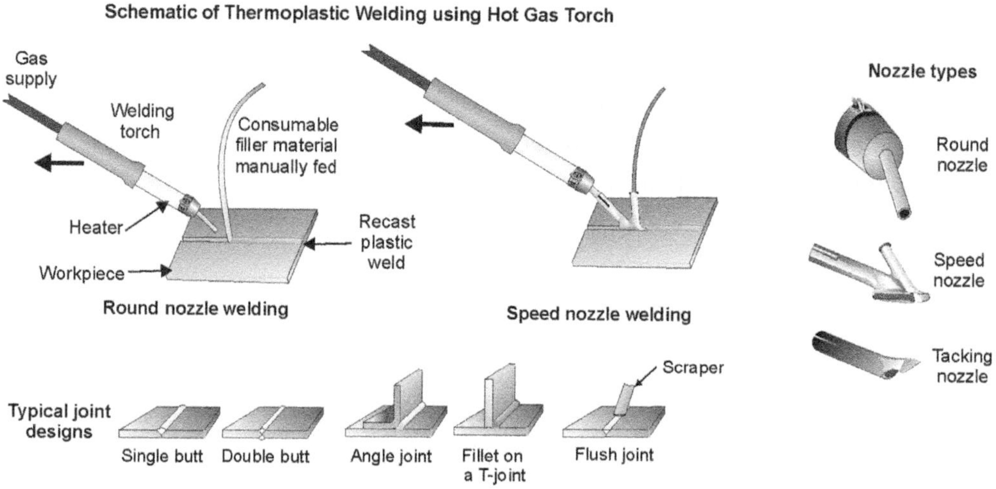

Figure 11.14: Thermoplastic Welding.

Materials

Only thermoplastic materials.

Process Variations

- Hot gas can be either nitrogen or air depending on thermoplastic to be joined. Nitrogen minimises oxidation of some thermoplastic materials.
- Various nozzle types for normal welding, speed welding and tacking.
- Other thermoplastic welding techniques available:
  - Spin welding: similar to Friction Welding (FRW), where the two parts to be joined, one stationary and one rotating at speed, have their joint surfaces brought into contact. Axial pressure and frictional heat at the interface create a solid-state weld on discontinuation of rotation and on cooling (see PRIMA 11.9).

- Ultrasonic Welding (USW): hardened probe introduces a small static pressure and oscillating vibrations at the joint face, disrupting surface oxides and raising the temperature through friction and pressure to create a bond. Can also perform spot welding using similar equipment (see PRIMA 11.9).
- Hot plate welding: electrically heated platens are used to soften base material at the joint and a bond is created with additional pressure, giving good joint strength.

Economic Considerations

- Production rates are very low.
- Weld rates typically less than 1.5 m/min.
- Lead time typically hours.
- Manual operation typically using transportable equipment.
- Automation possible using a trolley system traversing over joint.
- Economical for low production runs. Can be used for one-offs.
- Tooling costs are low.
- Equipment costs are generally low.
- Direct labour costs are moderate to high. Some skill needed by operator.
- Finishing costs are low. Scraping the joint flush may be required for aesthetic reasons.
- Other thermoplastic welding techniques have a moderate to high production rate, are applicable to large volumes, have a moderate to high equipment cost and are more readily automated.

Typical Applications

- Joining plastic pipes.
- Ducts.
- Containers.
- Repair work.

Design Aspects

- Moderate levels of complexity possible.
- Typical joint designs possible using hot gas welding are: butt, lap and fillet, in thin sheet.
- Horizontal welding position only.
- Parts to be joined must be in contact.
- Minimum overlap for lap joints = 13 mm.
- Minimum sheet thickness = 2 mm.
- Maximum sheet thickness = 8 mm.
- Multiple weld runs required on sheet thicknesses ≥5 mm.

Quality Issues

- Filler rods must be the same thermoplastic as base material.
- The force from the filler rod is applied to encourage mixing of softened material and must be consistent through the operation.
- Joints are weakened by incomplete softening, oxidation and thermal degradation of plastic material.
- Process variables are hot gas temperature, pressure (either from filler rod or fixtures) and speed of welding.
- Hot gas needs excess moisture and contaminants removed using filters.
- Weld strength is between 50% and 100% of base material.
- Recast plastic filler at the joint can be made flush with base material using a scraper.
- Tack welding of parts to be joined should be performed before welding commences.
- Use of additional fixtures is advised for large parts, also to provide additional pressure to aid joint formation.
- Surface finish of weld is fair to good.
- Fabrication tolerances are typically ±0.5 mm.

11.15 Adhesive Bonding

Process Description

Joining of similar or dissimilar materials (adherend) by the application of a natural or synthetic substance (adhesive) to their mating surfaces, which subsequently cures to form a bond (Figure 11.15).

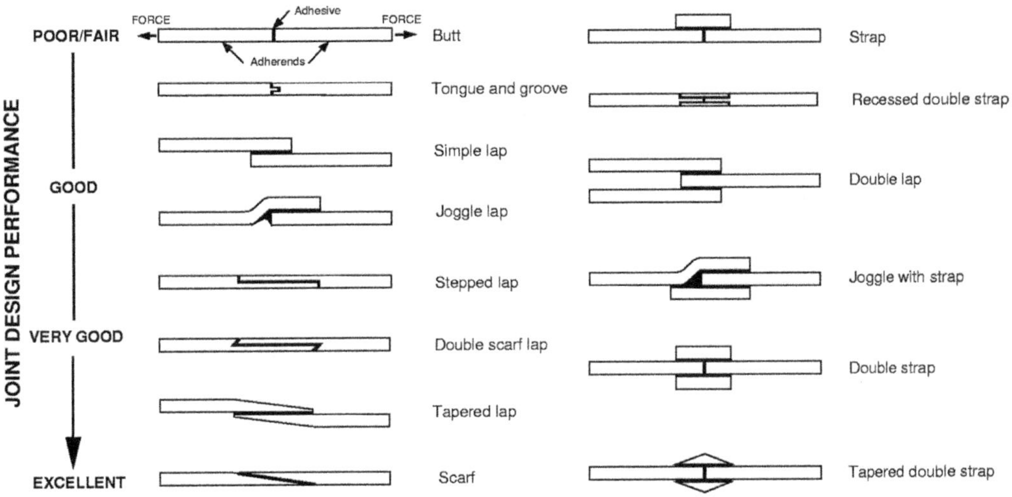

Figure 11.15: Typical Adhesive Bonded Joints.

Materials

- Most materials can be bonded with the correct selection of adhesive, surface preparation and joint design. Metals, plastics, composites, wood, glass, paper, leather and ceramics are bonded commonly.
- Can join dissimilar materials readily with proper adhesive selection, even materials with marked differences in coefficient of linear expansion, strength and thickness.

Process Variations

- Adhesives available in many forms: liquids, emulsions, gels, pastes, films, tapes, powder, rods and granules.
- Curing mechanisms: heat, pressure, time, chemical catalyst, ultraviolet light, vulcanisation or reactivation, or a combination of these.
- Various additives: catalysts, hardeners, accelerators and inhibitors to alter curing characteristics; silver metal flakes for electrical conduction, and aluminium oxide to improve thermal conduction.

- Adhesives can be applied manually or automatically by: brushing, spreading, spraying, roll coating, placed using a backing strip or dispensed from a nozzle.
- Many types of adhesive are available:
  - Natural animal (beeswax, casein), vegetable (gum, wax, dextrin, starch) and mineral (amber, paraffin, asphalt) based glues. Commonly low-strength applications such as paper, cardboard (packaging) and wood.
  - Epoxy resins: typically uses a two-part resin and hardener or single part cured by heat for large structural applications.
  - Anaerobics: set in the absence of atmospheric oxygen. Commonly known as thread-locking compounds and used for locating and sealing closely mated machined parts such as bearings and threads.
  - Cyanoacrylates: better known as super glues and use the presence of surface moisture as the hardening catalyst. Create good bonds when used for assembling small plastic, rubber and most metal parts.
  - Hot melts: thermoplastic resin bonds as it cools. Used for low-load situations.
  - Phenolics: based on phenol formaldehyde thermosetting resins, two-part cold or heat and pressure cured. More expensive than most adhesives, but give strong bonds for structural applications and good environmental resistance.
  - Plastisols: based on Polyvinyl Chloride (PVC) and use heat to cure. For larger parts such as furniture and automotive panels.
  - Polyurethanes: similar to epoxies. Fast-acting adhesive for low-temperature applications and low loads. Footwear commonly uses this type of adhesive.
  - Solvent-borne rubber adhesives: rubber compounds in a solvent where upon the solvent evaporates to cure the rubber; for minimal load applications.
  - Toughened adhesives: acrylic- or epoxy-based adhesives cured by a number of methods and can withstand high shock loads and high loads in large structures.
  - Tapes: pressure-sensitive adhesives on a backing strip for light loading applications such as packaging, automotive trim, cable secure and craft work.
  - Emulsions: based on Poly Vinyl Acetate (PVA), highly versatile, suitable for cold bonding of plastic laminates, wood, plywood, paper, cardboard, cork and concrete.
  - Polyimides: requires very high curing temperatures and pressures. Used in electronics and aerospace industries. High temperature capability.

Economic Considerations

- High production rates are possible.
- Lead time hours, typically, but weeks if automated.
- Time for curing heavily dictates achievable production rate: tapes are instant, cyanoacrylates take several seconds, anaerobics can take 15–30 minutes, epoxy resins may take 2–24 hours, although this can be reduced using catalysts.

- The viscosity of the adhesive must be suitable for the mixing and dispersion method chosen in production.
- Very flexible process.
- Simplifies the assembly process and therefore can reduce costs.
- Can replace or complement conventional joining methods such as welding and mechanical fasteners.
- Very little waste produced. Liquid adhesives require accurate metering to avoid excess.
- Economical for low production runs. Can be used for one-offs.
- Tooling costs are low to medium. Jigs and fixtures recommended during curing procedure to maintain position of assembled parts can be costly.
- Equipment costs are generally low.
- Direct labour costs are low to moderate. Cost of joint preparation can be high.
- Finishing costs are low. Little or no finishing required except removal of excess adhesive in some situations.

Typical Applications

- Building and structural applications.
- Electrical, electronic, automotive, marine and aerospace assemblies.
- Packaging and stationary.
- Furniture and footwear.
- Craft and decorative work.

Design Aspects

- All levels of complexity.
- Can be used where other forms of joining are not possible or practical.
- Joints should be designed to operate in shear, not tension or compression.
- Adhesives have relatively low strength and additional mechanical fixing is recommended on highly stressed joints to avoid peeling.
- Most common joint is the lap or variations on the lap, for example the tapered lap and scarf (preferred). Can also incorporate straps and self-locating mechanisms. Butt joints are not recommended on thin sections.
- A loaded lap joint tends to produce high stresses at the ends of the joints due to the slight eccentricity of the force line. Excessive joint overlap also increases the stress concentrations at the joint ends.
- For lap joints, the length of lap should be approximately 2.5 times that of the thinnest part for optimum strength. Increasing the width of the lap, adhesive thickness or increasing the stiffness of the parts to be joined can improve joint strength.

- Adhesive selection should also be based on: joint type and loading, curing mechanism and operating conditions.
- Can aid weight minimisation in critical applications or where other joining methods are not suitable or where access to joint area is limited.
- Inherent fluid sealing and insulation capabilities (electricity, heat and sound).
- Life prediction at operating temperature should be assessed.
- Adequate space should be provided for the adhesive at the joint ($\approx$0.05 mm optimum clearance).
- Adhesives can be used to provide electrical, sound and heat insulation.
- Can provide a barrier to prevent galvanic corrosion between dissimilar metals or to create a pressure-tight seal.
- Design joints using minimum amount of adhesive and provide for uniform thin layers.
- Jigs and fixtures should be used to maintain joint location during adhesive curing.
- Provision for the escape of gases and vapours in the design is important.
- Minimum sheet thickness = 0.05 mm.
- Maximum sheet thickness commonly = 50 mm.
- Unequal thicknesses are commonly bonded.

Quality Issues

- Excellent quality joints with little or no distortion.
- Residual stresses may be problematic with long curing time adhesives in combination with poor surface condition of base material, but otherwise not problematic.
- Dissimilar materials can cause residual stresses on cooling due to different expansion coefficients, especially if heat is used in the curing process.
- Problems are encountered with materials that are prone to solvent attack, stress cracking, water migration or low surface energy.
- Problems may be encountered in bonding materials that have surface oxides, loose surface layers or that are plated or painted (delamination may occur from the base material).
- Stress distribution over the joint area is more uniform than other joining techniques.
- Joint fatigue resistance is improved due to inherent damping properties of adhesives to absorb shocks and vibrations.
- Heat-sensitive materials can be joined without any change of base material properties.
- Adhesives generally have a short shelf-life.
- Optimum joint strength may not be immediate following assembly.
- Various adhesives can operate in temperatures up to approximately 250°C.
- Control of surface preparation, adhesive preparation, assembly environment and curing procedure is important for consistent joint quality.

- In surface preparation it is important to remove any contaminants from the joint area, such as oxide layers, paint, and thick films of grease and oil; to aid 'wetting' of the joint. Mechanical abrasion (grit blasting, abrasive cloth), solvent degreasing, chemical etching, anodising or surface primers may be necessary depending on the base materials to be joined.
- Adhesive almost invisible after assembly. Joint surface free of irregular shapes and contours as produced by mechanical fastening techniques and welding.
- Joint inspection is difficult after assembly and non-destructive testing (NDT) techniques are currently inadequate. Quality control should include intermittent testing of joint strength from samples taken from the production line.
- Quality control of adhesive mix is also important.
- Consideration of joint permanence is important for maintenance purposes. Bonded structures are not easily dismantled.
- Joint strength may deteriorate with time and severe environmental conditions (ultraviolet, radiation, chemicals, humidity and water) can greatly reduce joint integrity.
- Flammability and toxicity of adhesives can present problems to the operator. Fume extraction facilities may be required and safety procedures for chemical spillage need to be observed.
- Rough surfaces are preferred to smooth ones to provide surface-locking mechanisms.
- Fabrication tolerances are a function of the accuracy of the component parts and the assembly/jigging method during curing time.

11.16 Mechanical Fastening

Process Description

A mechanical fastening system is a separate device or integral component feature that will position and hold two or more components in a desired relationship to each other. The joining of parts by mechanical fastening systems can be generally classified as (see Figure 11.16):

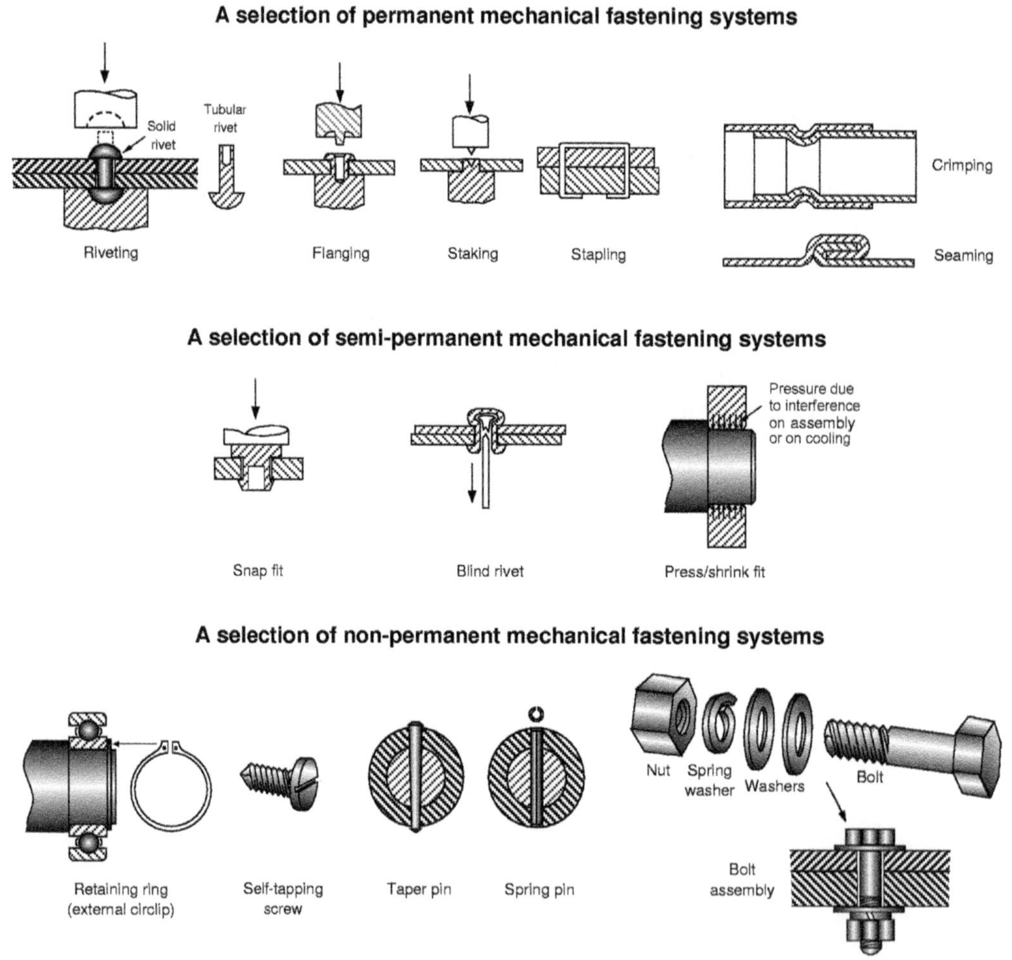

Figure 11.16: Mechanical Fastening.

- Permanent: can only be separated by causing irreparable damage to the base material, functional element or characteristic of the components joined, for example surface integrity. A permanent joint is intended for a situation where it is unlikely that a joint will be dismantled under any servicing situation.
- Semi-permanent: can be dismantled on a limited number of occasions, but may result in loss or damage to the fastening system and/or base material. Separation may require an additional process, for example plastic deformation. A semi-permanent joint can be used when disassembly is not performed as part of regular servicing, but for some other need.
- Non-permanent: can be separated without special measures or damage to the fastening system and/or base material. A non-permanent joint is suited to situations where regular dismantling is required, for example at scheduled maintenance intervals.

Materials

- Can join most materials and combinations of materials using various processes. Metals, plastics, ceramics and wood are commonly joined.
- Fastening elements made from most metal alloys such as ferrous (steel most common), copper, nickel, aluminium and titanium, depending on strength of joint and environmental requirements. Use of plastics for fastening methods common for low loading conditions.
- Variety of coatings available for metal fasteners to improve corrosion resistance, commonly: zinc (electroplated and hot-dip), cadmium, chromate, phosphate and bluing.

Process Variations

- Permanent fastening systems:
  - Riveting: used to create a closed mechanical element spanning an assembly. The rivet is located through a previously created hole through the materials to be joined and then the rivet shank is plastically deformed (either hot or cold) on one side, typically. Used for joining sheet materials of varying type and thickness by solid, tubular (both semi-tubular and eyelet), split, compression and explosive types.
  - Flanging: the plastic deformation of an amount of excess material exposed on one component to locate and hold it to an adjacent face of another component. Readily lends itself to full automation. Deformation can be performed through direct pressure, rotary or vibratory tool movement.
  - Staking: similar to flanging, but plastic deformation is localised to where the components are closely assembled through a punch mark in the centre of a protrusion. Location of the parts is by friction and pressure at their interface. Low joint strengths.
  - Stapling: joins materials using U-shaped staples fed on strips to the head of a semi-automatic tool. Can join dissimilar materials of thin section and no hole prior to the operation is needed.

    - Stitching: similar to stapling, but the stitching is made by the machine itself into a U-shaped form.
    - Crimping: a pressure-tight joint is created on thin section assembled components by localised plastic deformation at dimple points, by swaging or shrinkage. Also notching, which shears and bends the same portion of the assembled parts to maintain location.
    - Seaming: creation of a pressure-tight joint in sheet-metal assemblies by hooking together two sheets through multiple bends and pressing down the joint area. Joint strength and integrity can be further improved by soldering, adhesive bonding or brazing.
    - Nailing: uses the friction between a nail and the pierced materials to maintain location of the parts. Typically used for joining wood to wood, or wood to masonry.
- Semi-permanent fastening systems:
    - Snap fits: integral features of the components to be joined are typically hooked tabs that lock into notches on the adjacent part to be assembled with the application of a modest force. Commonly used for large-volume production of plastic assemblies. Require special design attention to determine deflections and dimensional clearances.
    - Press fits: use of the negative difference in dimensions (or interference) on the components to impart an interface pressure through the force for assembly.
    - Shrink fits: use of the negative difference in component dimensions to impart an interface pressure on assembly by heating one component (usually the external), causing expansion, and then allowing it to cool and contract in situ.
    - Blind rivets: located in a previously created hole in the assembly from a single direction using a special tool. The tool retracts a headed pin from the rivet body, deforming it enough to hold the components. The head is left inside the rivet body on joint completion. Used for thin sheet material fabrication.
- Non-permanent fastening systems:
    - Retaining rings: provide a removable shoulder within a groove of a bore or on the surface of a shaft to locate and lock components assembled to it. Presented either axially, radially or pushed into the groove using special tools. Self-locking, circlip, E-clip and wire-formed types available for various applications. Made from spring steel, typically.
    - Self-tapping screws: for assembling thin sheet material by passing a large pitch screw through previously created holes in the parts. Also self-drilling and thread-forming types for soft materials.
    - Quick release mechanisms: for rapid securing and release of parts, e.g. doors, access panels, tooling jigs and fixtures. Various types available, such as clips, locks, latches, cams, clamps and quarter-turn fastening systems.
    - Pins: for locating and retaining collars, hubs, gears and wheels on shafts, or to act as pivots in machinery or stops. Various types available, such as taper, spring, grooved, split and cotter.

- Tapered and gib-head keys: for locating and holding gears, wheels and hubs on shafts through friction.
- Magnetic devices: for locating or holding items such as doors and work holders for machine tools. Can be permanent type, mechanically or electrically actuated. Parts must be ferrous, nickel or cobalt based if direct magnetic attraction is required.
- Threaded fastening systems: include a number of standard thread forms and pitches. Variety of drive types (hexagonal head, socket head, slotted head); washers (plain, spring, double coil, toothed locking, crinkle, tab); nuts (plain, thin, nyloc, castle nut); locking mechanisms (split pin, lock plate, wiring); and bolt, screw, stud and set-screw configurations.
- Anchor and rag bolts: used for fixing structural sections and fabrications to concrete.
- Threaded inserts: for use in brittle and flexible materials such as ceramics and plastics. Can be moulded or cast in situ or inserted in previously threaded holes. Also Helicoil wire thread inserts for protecting and strengthening previously tapped threads.
- Collets: for locating gears, hubs and wheels on shafts through friction mechanisms. Various types, such as expanding, taper and Morse.
- Zips, studs, buttons, plastic tie-wraps, wire and Velcro are all very useful non-permanent fastening systems that have, from time to time, been used in engineering assemblies, particularly the last three.

- All mechanical fastening systems can be manually or semi-automatically performed during assembly or installation; however, not all fastening systems readily lend themselves to full automation.

Economic Considerations

- High production rates are possible depending on the fastening system and degree of automation. Also dependent on time to 'open' and 'close' fastening system.
- Economical for very low production runs.
- All production quantities viable.
- Regular use of same fastening system type on an assembly is more cost effective than the use of many different types.
- A smaller number of large fasteners may be more economical than many small ones.
- Consideration must be given to fastener replacement costs for maintenance or service requirements.
- Tooling costs range from low to moderate depending on degree of automation.
- Equipment costs are low.
- Direct labour costs are low to moderate.
- Cost and skill of joint preparation can be high.
- Finishing costs are very low. Usually no finishing is required.
- Little or no scrap, except where hole generation is concerned.

Typical Applications

- Structures for buildings and bridges.
- Automotive, aerospace, electrical and marine assemblies.
- Domestic and office appliances.
- Machine tools.
- Pipework and ducting.
- Furniture.
- Clothing.

Design Aspects

- Applicable to all levels of design complexity.
- Identification of possible failure modes (tension, shear, bearing, fatigue) and calculation of stresses in the fastener at the design stage are recommended in joints subjected to high static, impact and/or fluctuating loads.
- Examination of the stresses in the joint area under the fastener is important to determine the load-bearing capability and stiffness of the parts to be joined.
- Use of recommended torque values for bolted connections is critical for obtaining correct pre-loads and should be indicated on assembly drawings.
- Differentials in thermal expansion must be taken into consideration when using a fastener of different material to that of the base material.
- Provide for anti-vibration mechanisms in the fastening system where necessary, e.g. Nyloc, lock nuts in combination with split pins, spring washers.
- The damping characteristics of the assembled product must be considered when using a specific fastening system with fluctuating loads.
- Can incorporate pressure-tight seals with most bolted joints, e.g. gaskets.
- Try to use standard fastener sizes, lengths and common fastening systems for a product.
- Keep the number of fasteners to a minimum for economic reasons.
- Design for the easy disassembly and maintenance of non-permanent fasteners, i.e. provide enough space for spanners, sockets and screwdrivers.
- Avoid placing fasteners too close to the edge of parts or too close to each other because of assembly difficulty and reduced strength capacity, i.e. pull-out and rupture.
- Maximum operating temperatures of mechanical fastenings is approximately 700°C using nickel–chromium steel bolts.
- When joining plastics it is good practice to use metal-threaded inserts or plastic fasteners.
- Minimum section thickness = 0.25 mm.
- Maximum section thickness typically = 200 mm.
- Unequal section thicknesses commonly joined.

Quality Issues

- Galvanic corrosion between dissimilar metals requires careful consideration, e.g. aluminium and steel.
- There is a risk of damage to joined parts or fasteners when using permanent systems or non-permanent fasteners that have been disassembled many times.
- Stress relaxation can cause the joint to loosen over time (especially in high-temperature operating conditions over long periods). Subsequent re-torquing is recommended at regular intervals. This should be written into the service requirements for critical applications.
- High-temperature applications in combination with harsh environments accelerate creep and fatigue failure.
- Rolled threads on bolts and screws preferred over machined threads due to improved strength and surface integrity.
- Variations in flatness and squareness of abutment faces in assemblies can affect joint rigidity, corrosion resistance and sealing integrity.
- Variations in tolerances and accumulations of tolerances can result in mismatched parts and cause high assembly stresses. Dissimilar materials will also cause additional stresses, if reactions to the assembly environment result in unequal size changes.
- Variation in bolt pre-load is dependent on degree of automation of torquing method and frictional conditions at the component interfaces. Both should be controlled wherever possible.
- Lubricants and plate finishes on fasteners can help reduce torque required and improve corrosion resistance.
- Hydrogen embrittlement in electroplated steel fasteners can be problematic and accelerates failure.
- Stress concentrations in fastener and joint designs should be minimised by incorporating radii, gradual section changes and recesses.
- Hole size and preparation (where required) is important. Holes can act as stress concentrations. Fatigue life can be improved by inducing compressive residual stresses in the hole, e.g. by caulking.
- Reliability of joint and consistency of operation improved with automation, generally. Can be highly reliant on operator skill where automation not feasible.
- Fabrication tolerances are a function of the accuracy of the component parts and the fastening system used.

CHAPTER 12

Component Costing

12.1 Introduction

For financial control and successful marketing it is necessary to have cost targets and realisations throughout the product introduction process. Product cost is virtually always a prime element in decision-making in the manufacturing industry. The main problem in product introduction is the provision of reliable cost information in the early stages of the design process, for the comparison of alternative conceptual designs and assessment of the myriad of ways in which a product may be structured during concept development.

Cost estimates are needed to determine the viability of projects and to minimise project and product costs. The inadequate nature of the historical standard costing methods and cost-estimating practices found in most companies has been highlighted by researchers over a number of years [1–4]. One signal that emerges from all workers is that it is crucial to reject uneconomic designs early, for it is not often possible to reduce costs productively once production has commenced, largely due to the high cost of change at this stage in the product life cycle. Hence, costing is best utilised at the stage in the design process when rough designs for a component have been prepared.

The aim of the component costing method presented here is to highlight expensive and difficult-to-manufacture designs, thus indicating areas that will benefit from further attention before the design has been completed. Benefits of the method include:

- Lower component costs.
- Systematic component costing.
- Identification of feasible manufacturing processes.
- Rapid comparison of alternative designs and competitor products.
- Reduced engineering change.
- Shorter development time and reduced time to market.
- Education and training.

The costing method described is ideally applicable to team-based applications, both manually and in the form of computer software. The initial work was primarily designed to cater for components found in the light engineering, aerospace and automotive business sectors. Other cost models and approaches useful to the reader can be found in Refs [5–9].

Manufacturing Process Selection Handbook. http://dx.doi.org/10.1016/B978-0-08-099360-7.00012-4

In order to produce a practical and widely applicable method for designers with the capability to provide feedback on the technological and economic consequences of component design decisions, it was considered useful to develop a sample model that is widely applicable to a number of different manufacturing processes. In addition, the model was designed such that appropriate manufacturing processes and equipment requirements can be specified early in the product introduction process. Recognising the problem that the relationship between a design and its manufacturing feasibility and cost is not easily amenable to precise scientific formulation, the model has come out of knowledge engineering work in a number of user companies and those specialising in particular manufacturing processes.

The model is logically based on material volume and processing considerations. The process cost is determined using a basic processing cost (the cost of producing an ideal design for that process) and design-dependent relative cost coefficients (which enable any component design to be compared with the ideal). Material costs are calculated taking into account the transformation of material to yield the final form.

Thus, a single process model for manufacturing cost, M_i, can be formulated as:

$$M_i = V \cdot C_{mt} + R_c \cdot P_c \tag{12.1}$$

where

V = volume of material required in order to produce the component;
C_{mt} = cost of the material per unit volume in the required form;
P_c = basic processing cost for an ideal design of the component by a specific process;
R_c = relative cost coefficient assigned to a component design (taking account of shape complexity, suitability of material for processing, section dimensions, tolerances and surface finish).

The initial hypothesis can be expanded to allow for secondary processing and thus the model can take the general form:

$$M_i = V \cdot C_{mt} + (R_{c_1} \cdot P_{c_1} + R_{c_2} \cdot P_{c_2} + \ldots + R_{c_n} \cdot P_{c_n})$$

$$M_i = V \cdot C_{mt} + \sum_{i=1}^{n} (R_{c_i} \cdot P_{c_i}) \tag{12.2}$$

where n = number of operations required to achieve the finished component.

In order for such a formulation to be used in practice it is necessary to define relationships enabling the determination of the quantities P_c and R_c for design–process combinations. In practice, it has been found that Equation (12.1) is the form preferred by industry. This is based on the need to work in the early stages of the design process with incomplete component data and without the necessity for detailing the sequence of manufacturing operations.

The approach has been to build the secondary processing requirements into the relative cost coefficient.

12.2 Basic Processing Cost (P_c)

In order to represent the basic processing cost of an ideal design for a particular process, it is first necessary to identify the factors on which it is dependent. These factors include:

- Equipment costs including installation.
- Operating costs (labour, number of shifts worked, supervision and overheads, etc.).
- Processing times.
- Tooling costs.
- Component demand.

The above variables are taken account of in the calculation of P_c using the simple equation:

$$P_c = \alpha \cdot T + \frac{\beta}{N} \tag{12.3}$$

where

α = cost of setting up and operating a specific process, including plant, labour, supervision and overheads, per second;
β = process-specific total tooling cost for an ideal design;
T = process time in seconds for processing an ideal design of component by a specific process;
N = total production quantity per annum.

Values for α and β are based on expertise from companies specialising in producing components in specific technological areas. Using these process-specific values in Equation (12.3), it is possible to produce comparative cost curves for any process.

Data for P_c against annual production quantity, N, is illustrated in Figures 12.1–12.5 for several main process groups (casting and moulding, forming, machining, continuous extrusion and chemical milling) covering 20 individual manufacturing processes. While the data presented might be adequate in most cases, the methodology was devised with the idea that users would develop their own data for the process they would wish to consider. Such an approach has many benefits to a business, including ownership of the data and a confidence in the results produced. The values of P_c represent the minimum likely costs associated with a particular manufacturing process at a given annual production quantity. In this way, it is possible to indicate the lowest likely cost for a component associated with a particular manufacturing process route assuming an ideal design for the process, one-shift working and a 2-year payback on investment.

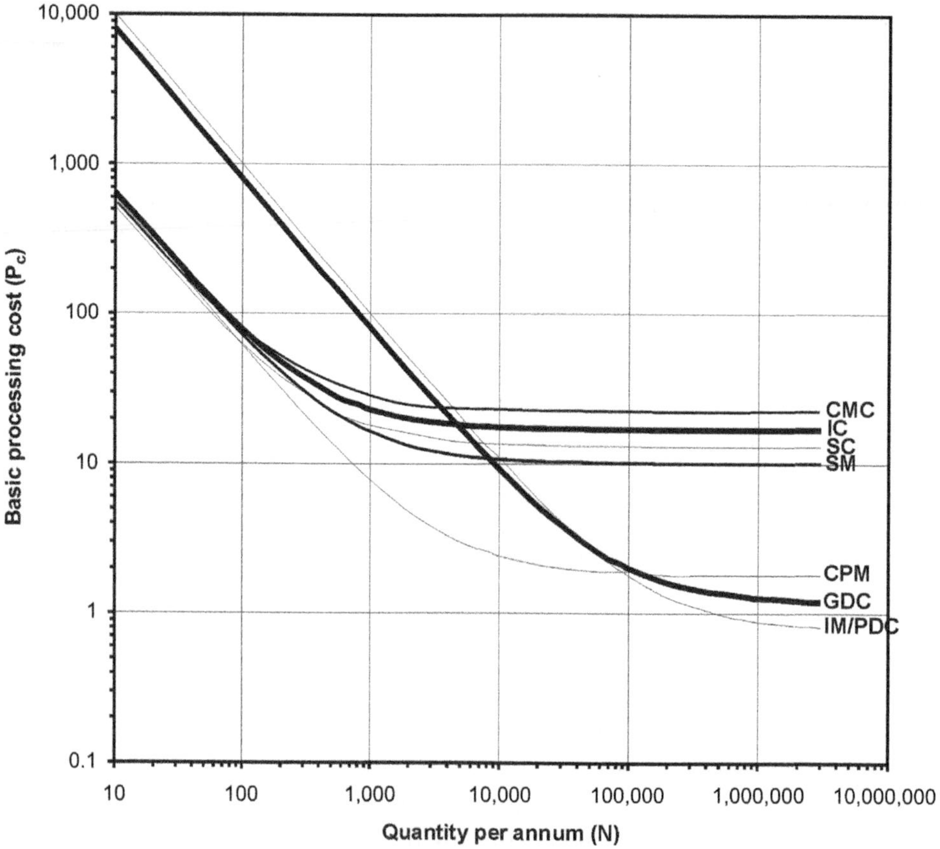

Figure 12.1: Basic Processing Cost (P_c) against Annual Production Quantity (N) for Casting and Moulding Processes.

A process key for the figures is provided below:

AM	Automatic Machining
CCEM	Cold Continuous Extrusion (Metals)
CDF	Closed-die Forging
CEP	Continuous Extrusion (Plastics)
CF	Cold Forming
CH	Cold Heading
CM2.5	Chemical Milling (2.5 mm depth)
CM5	Chemical Milling (5 mm depth)
CMC	Ceramic Mould Casting
CNC	Computer Numerical Controlled Machining
CPM	Compression Moulding
GDC	Gravity Die Casting

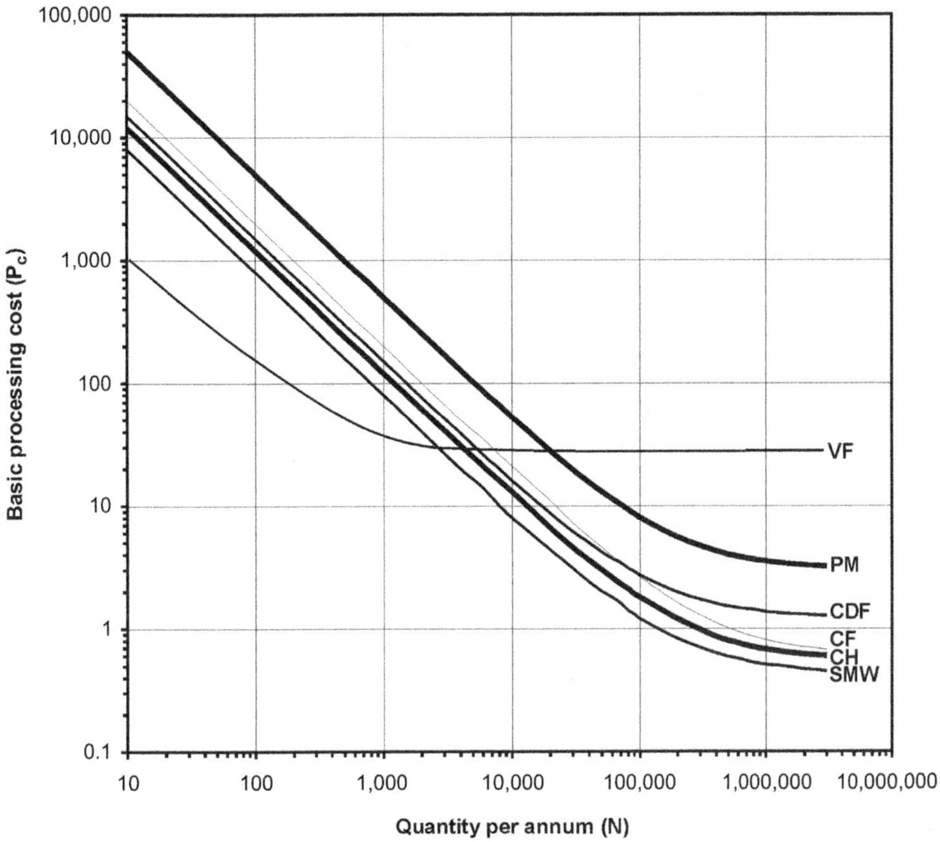

Figure 12.2: Basic Processing Cost (P_c) against Annual Production Quantity (N) for Forming Processes.

HCEM	Hot Continuous Extrusion (Metals)
IC	Investment Casting
IM	Injection Moulding
MM	Manual Machining
PDC	Pressure Die Casting
PM	Powder Metallurgy
SM	Shell Moulding
SC	Sand Casting
SMW	Sheet-metal Work
VF	Vacuum Forming

Having defined P_c, it is necessary to determine the design-dependent factors. The variables, shape complexity, tolerances, etc. modify the relationship between the curves. The relative cost coefficient R_c in Equation (12.1) is one way in which these variables can be expressed.

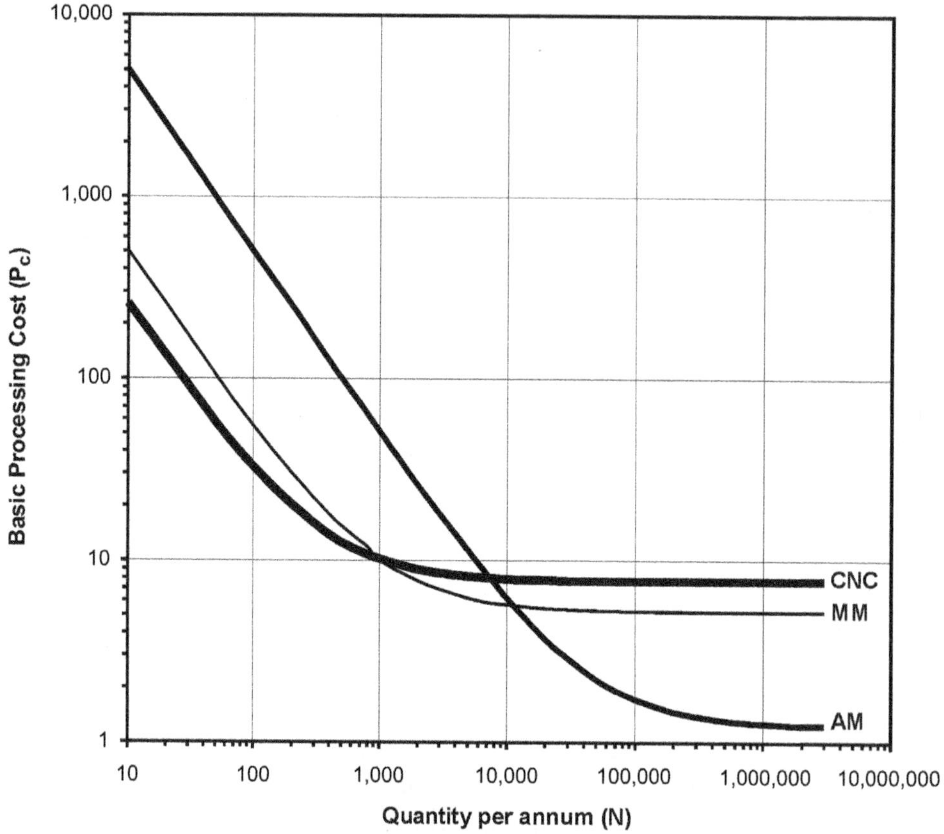

Figure 12.3: Basic Processing Cost (P_c) against Annual Production Quantity (N) for Machining Processes.

12.3 Relative Cost Coefficient (R_c)

This coefficient will determine how much more expensive it will be to produce a component with more demanding features than the 'ideal design'. The characteristics that we have assumed to influence the relative cost coefficient, R_c, are given below:

$$R_c = \phi(C_{mp}, C_c, C_s, C_t, C_f)$$

where

C_{mp} = relative cost associated with material-process suitability;
C_c = relative cost associated with producing components of different geometrical complexity;
C_s = relative cost associated with size considerations and achieving component section reductions/thickness;

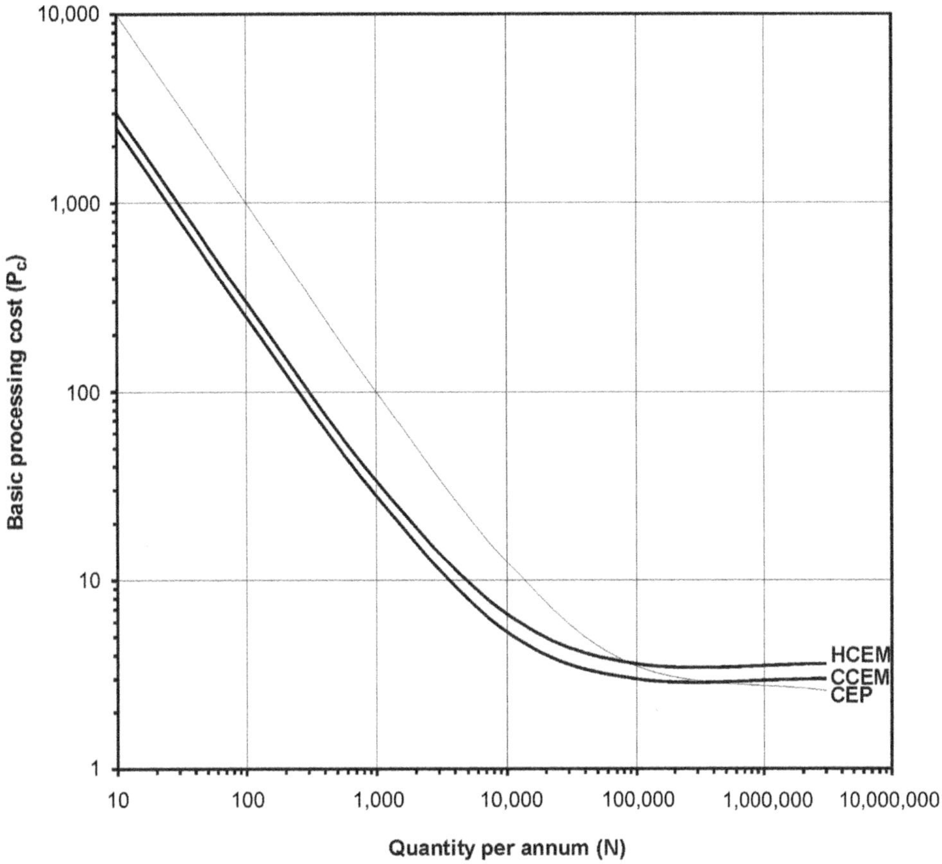

Figure 12.4: Basic Processing Cost (P_c) against Annual Production Quantity (N) for Continuous Extrusion Processes.

C_t = relative cost associated with obtaining a specified tolerance;
C_f = relative cost associated with obtaining a specified surface finish.

Analysis of the influence of the above quantities and discussions with experts led to the idea that these could be combined as shown below:

$$R_c = C_{mp}^a \cdot C_c^b \cdot C_s^c \cdot C_t^d \cdot C_f^e \tag{12.4}$$

where *a*, *b*, *c*, *d* and *e* are weighting exponents.

However, it was found that the knowledge could be structured to enable each of the exponents to be assigned the value of unity. Therefore, the relative cost coefficient can be represented by the formula:

$$R_c = C_{mp} \cdot C_c \cdot C_s \cdot C_{ft} \tag{12.5}$$

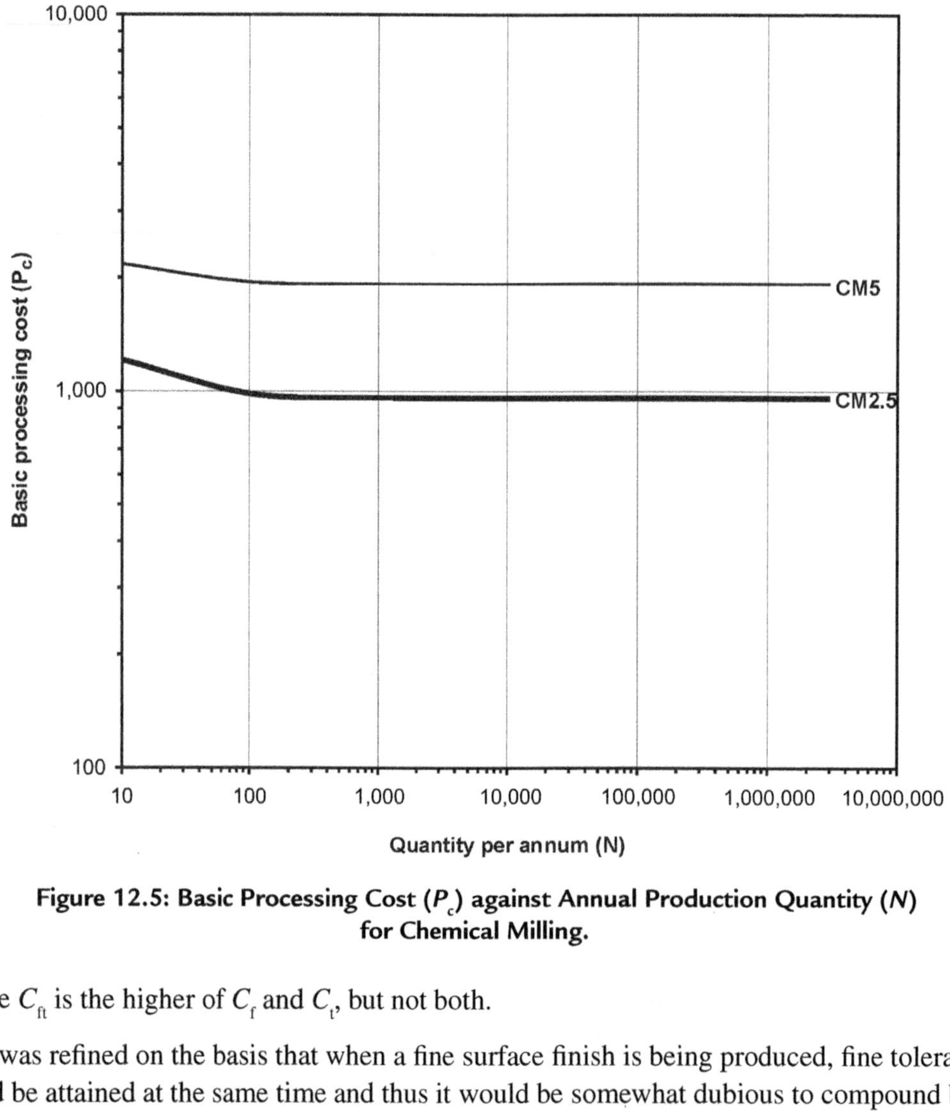

Figure 12.5: Basic Processing Cost (P_c) against Annual Production Quantity (N) for Chemical Milling.

where C_{ft} is the higher of C_f and C_t, but not both.

This was refined on the basis that when a fine surface finish is being produced, fine tolerances could be attained at the same time and thus it would be somewhat dubious to compound both relative cost coefficients.

Knowledge engineering indicated that Equation (12.5) was the most appropriate combination of coefficients at the present stage of development. The method of comparison and accumulation of costs was shown to be analogous to those methodologies employed by experts in the field of cost engineering/estimating. For the ideal design, each of these coefficients is unity, but as the component design moves away from this state, then one or more of the coefficients may increase in magnitude, thus changing the manufacturing cost term, M_i, in Equation (12.1).

Figure 12.6 shows how the cost curves for P_c are modified according to the model proposed as a component design shifts away from the ideal for that process. As the design

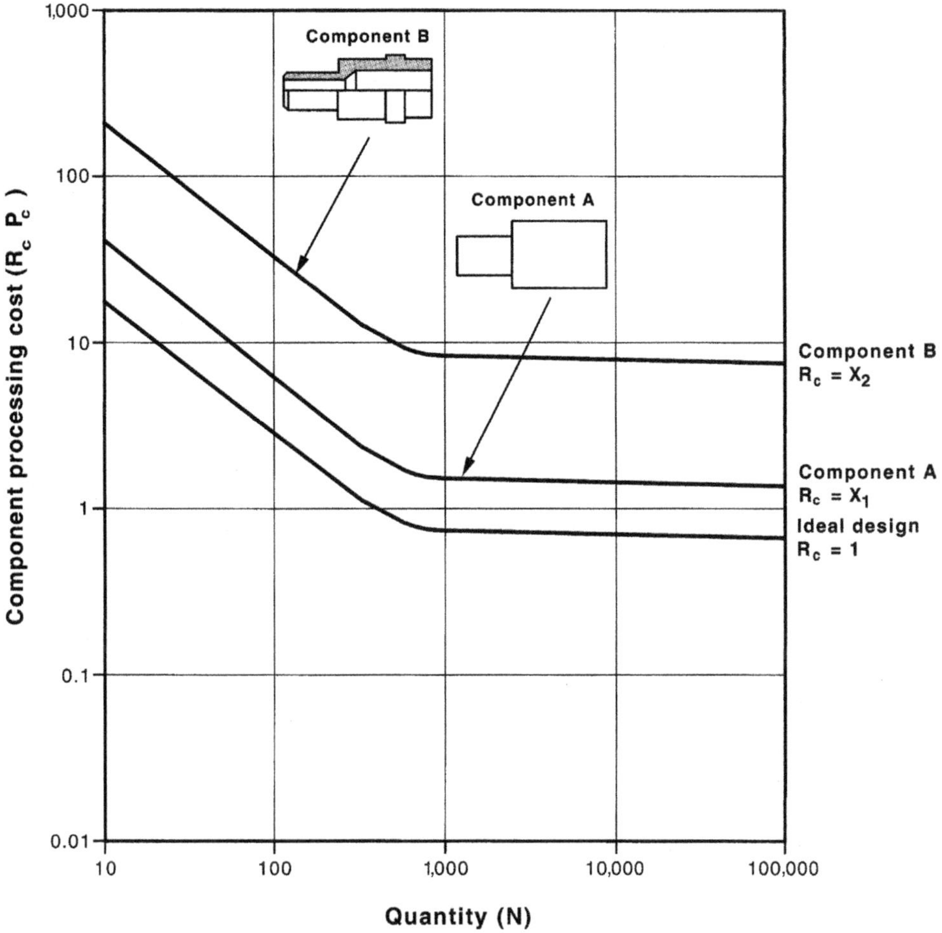

Figure 12.6: Relative Manufacturing Cost versus Quantity Curves.

becomes more difficult to process, because of material types or geometrical features for example, its cost curve progresses up the cost axis, as illustrated, moving from Design A to B in the figure.

Data for the above relative cost coefficients C_{mp}, C_c, C_s, C_t and C_f are provided in the following sections.

12.3.1 Material to Process Suitability (C_{mp})

In Table 12.1 the C_{mp} data indicates the suitability of using various materials with different processes. Clearly some combinations are inappropriate and C_{mp} values only appear at nodes currently considered to be technologically and economically feasible.

Table 12.1: Relative Cost Data for Material Processing Suitability (C_{mp})

	Process																					
Material	**AM**	**CCEM**	**CDF**	**CEP**	**CF**	**CH**	**C.M2.5**	**CM5**	**CMC**	**CNC**	**CPM**	**GDC**	**HCEM**	**IC**	**IM**	**MM**	**PDC**	**PM**	**SM**	**SC**	**SMW**	**VF**
Cast iron	1.2						1	1	1	1.2				1		1.2		1.6	1	1		
Low carbon steel	1.4	1.3	1		1.3	1.3	1	1	1.2	1.4			1.3	1		1.4		1.2	1.2	1.2	1.2	
Alloy steel	2.5	2	2		2	2	1	1	1.3	2.5			2	1		2.5		1.1	1.3	1.3	1.5	
Stainless steel	4	2	2		2	2	1	1	1.5	4			2	1		4		1.1	1.5	1.5	1.5	
Copper alloy	1.1	1.1	1		1	1			1	1.1			1	1		1.1	3	1	1	1	1	
Aluminium alloy	1	1.1	1		1	1			1	1		1.5	1.1	1		1	1.5	1	1	1	1	
Zinc alloy	1.1	1	1		1	1			1	1.1		1.2	1	1		1.1	1.2	1	1	1	1	
Thermoplastic	1.1			1						1.1	1.2				1	1.1						1
Thermoset	1.2			1.2						1.2	1				1	1.2						
Elastomer	1.1			1.5						1.1	1.5				1.5	1.1						

12.3.2 Shape Complexity (C_c)

Figures 12.7 and 12.8 present a shape classification scheme for the determination of coefficient C_c. The first step is to read the supporting notes and complexity definitions provided in Figure 12.7.

Notes - Geometrical Considerations

The shape complexity index is obtained by using a feature-based classification system which enables the important design/manufacturing issues to be taken into account. Firstly, determine the shape category:

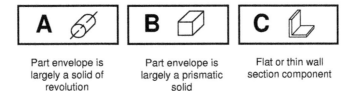

A	B	C
Part envelope is largely a solid of revolution	Part envelope is largely a prismatic solid	Flat or thin wall section component

Within the above classes, components are divided into five bands of complexity.

Note that the classification process should reflect the finished form of the component and the features listed in the tables should be used to aid the selection of the appropriate band. Always determine the classification by working from the left-hand side of the table.

Notes - Shape Complexity Definitions

- **Basic Features** - Straightforward processing where the operation can be carried out without a change of setting or the need of complex tooling. Parts are usually uniform in cross section.
- **Secondary Features** - As above, but where additional processing is necessary or more complex tooling is required.
- **Multi-axis Features** - Parts require to be processed in more than a single axis/set-up.
- **Non-uniform Features** - Parts require the development of more complex processing techniques/set-up.
- **Complex Forms** - Parts need dedicated tooling and the development of specialised processing techniques.
- **Single Axis** - This is usually the axis along the component's largest dimension, however, in the case of cylindrical or disc-shaped components, it is more convenient to consider the axis of revolution as the primary axis.
- **Through Features** - Features which run along, across or through a component from one end or side to the other.
- **Important** - If the component falls into more than one category, always choose the one that gives the highest value of C_c.

Figure 12.7: Notes on Shape Classification used in the Determination of C_c.

(A) Part Envelope is Largely a Solid of Revolution

Single/Primary Axis		Secondary Axes: Straight line features parallel and/or perpendicular to primary axis		Complex Forms
Basic rotational features only	Regular secondary/ repetitive features	Internal	Internal and/or external features	Irregular and/or complex forms
A 1	A 2	A 3	A 4	A 5
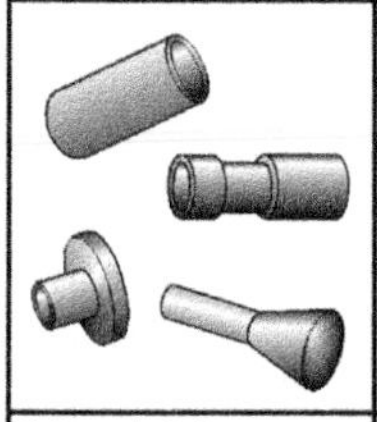	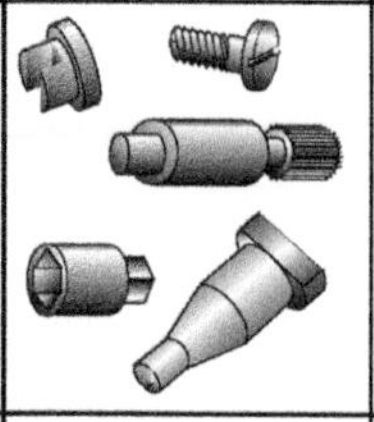	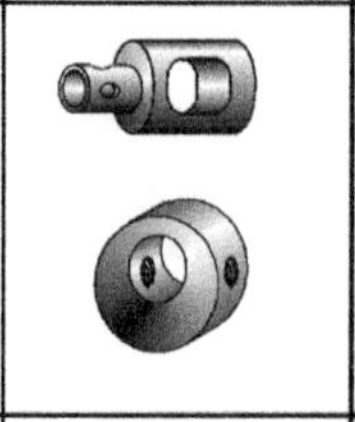		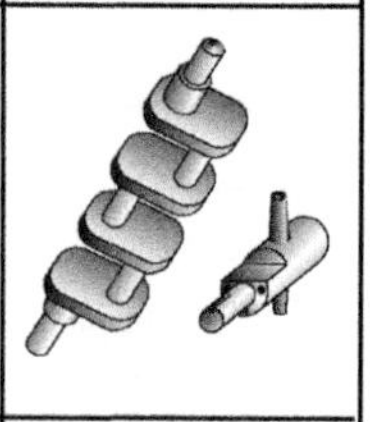
Category Includes: Rotationally symmetrical/ grooves, undercuts, steps, chamfers, tapers and holes along primary axis/centre line.	Internal/external threads, knurling and simple contours through flats/splines/keyways on/around the primary axis/centre line.	Holes/threads/ counterbores and other internal features not on the primary axis .	Projections, complex features, blind flats, splines, keyways on secondary axes.	Complex contoured surfaces,and /or series of features which are not represented in previous categories.

(B) Part Envelope is Largely a Rectangular or Cubic Prism

Single Axis/Plane		Multiple Axes		Complex Forms
Basic features only	Regular secondary/ repetitive features	Orthogonal/straight-line-based features	Simple curved features on a single plane	Irregular and/or contoured forms
B 1	B 2	B 3	B 4	B 5
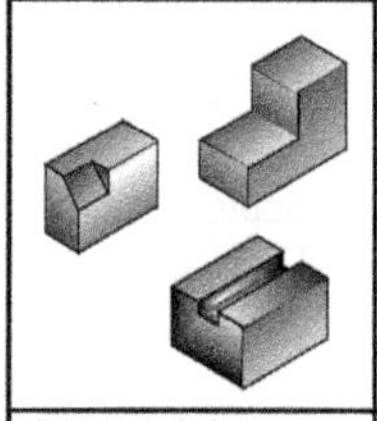	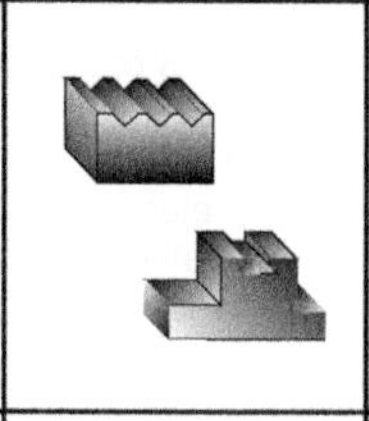	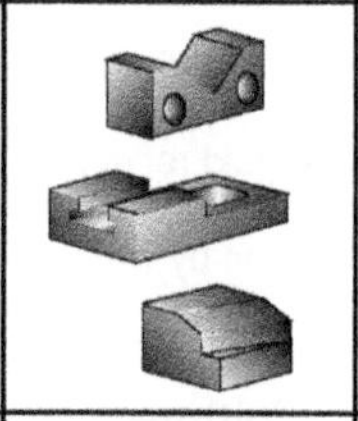	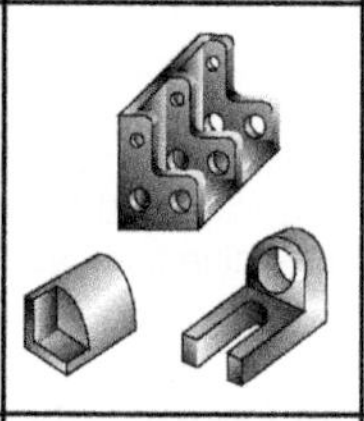	
Category Includes: Through steps, chamfers and grooves/channels/slots and holes/threads on a single axis.	Regular through features, T-slots and racks/plain gear sections etc. Repetitive holes/threads/counter bores on a single plane.	Regular orthogonal/straight line based pockets and/or projections on one or more axis. Angled holes/threads/ counter bores.	Curves on internal and/or external surfaces.	Complex 3-D contoured surfaces/geometries which cannot be assigned to previous categories.

(C) Flat Or Thin Wall Section Components

Single Axis	Secondary/Repetitive Regular Features		Regular Forms	Complex Forms
Basic features only	Uniform section/ wall thickness	Non-uniform section/ wall thickness	Cup, cone and box-type parts	Non-uniform and/or contoured forms
C 1	C 2	C 3	C 4	C 5
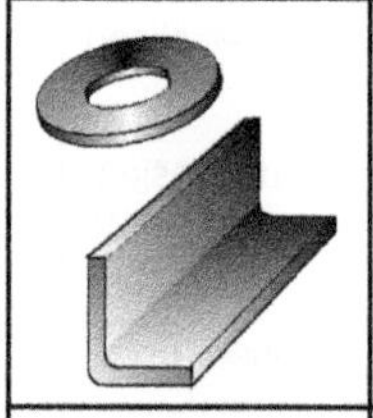		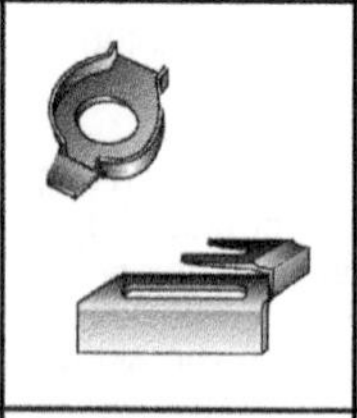		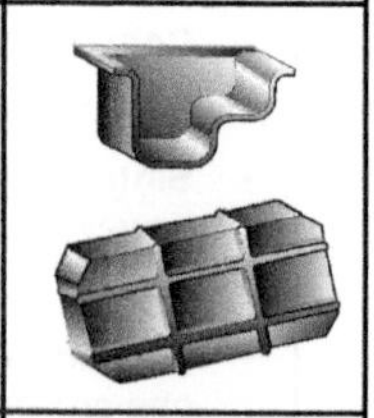
Category Includes: Blanks, washers, simple bends, forms and through features on or parallel to primary axis.	Plain cogs/gears, multiple or continuous bends and forms.	Component section changes not made up of multiple bends or forms. Steps, tapers and blind features.	Components may involve changes in section thickness.	Complex or irregular features or series of features which are not represented in previous categories.

Figure 12.8: Shape Classification Categories used in the Determination of C_c.

There are three basic shape categories for components:

- Solid of revolution.
- Prismatic solid.
- Flat or thin wall section.

These three fundamental shape categories can be subdivided into five bands of complexity as shown pictorially in Figure 12.8. The shape class must reflect the finished form of the component and the features listed in the tables should be used as an aid to the selection of the appropriate value of C_c from Figures 12.9, 12.10 or 12.11 for classification categories 'A', 'B' or 'C', respectively. Determination of shape complexity is important. Failure to classify the geometry properly may affect the final component cost result, M_i, quite significantly, and studying the shape complexity definitions is crucial in this connection.

12.3.3 Section Coefficient (C_s)

Figures 12.12, 12.13 and 12.14 show the relative cost consequences of producing specific section/wall thicknesses for the sample set of processes. Data required are the maximum dimension where the section acts, and the specified section size. Using a process outside its normal size domain can result in an additional cost. In the situation where a process is being considered for a component that is longer or smaller than the size range given in the macro (length, area, volume or weight), the estimate of cost produced is likely to be a lower bound value. Note that for Chemical Milling (CM2.5 and CM5), $C_s = 1$, as this penalty is taken account of in the formulation of the basic processing cost, P_c.

12.3.4 Tolerance (C_t) and Surface Finish (C_f) Coefficients

The sample data on the affects of tolerance (C_t) and surface finish (C_f) can be found in Figures 12.15–12.17 and 12.18–12.20, respectively. These indicate the relative cost consequences of achieving specific tolerance and surface finish levels for the various manufacturing processes. The process of analysis is:

1. Determine the most important tolerance values.
2. Identify the tolerance band on the C_t table.
3. Count the number of tolerances in the same band.
4. Identify the number of planes on which the critical values lie.
5. Select the appropriate C_t index from the table.

If the tolerance falls to the left of the thick grey line, a final machining, lapping, honing, polishing or grinding process is necessary to achieve the tolerance. This is already taken into account in the indices shown. Only the tightest tolerance required should be used, even if it only occurs on one plane. Included in the graph are separate lines for the number of

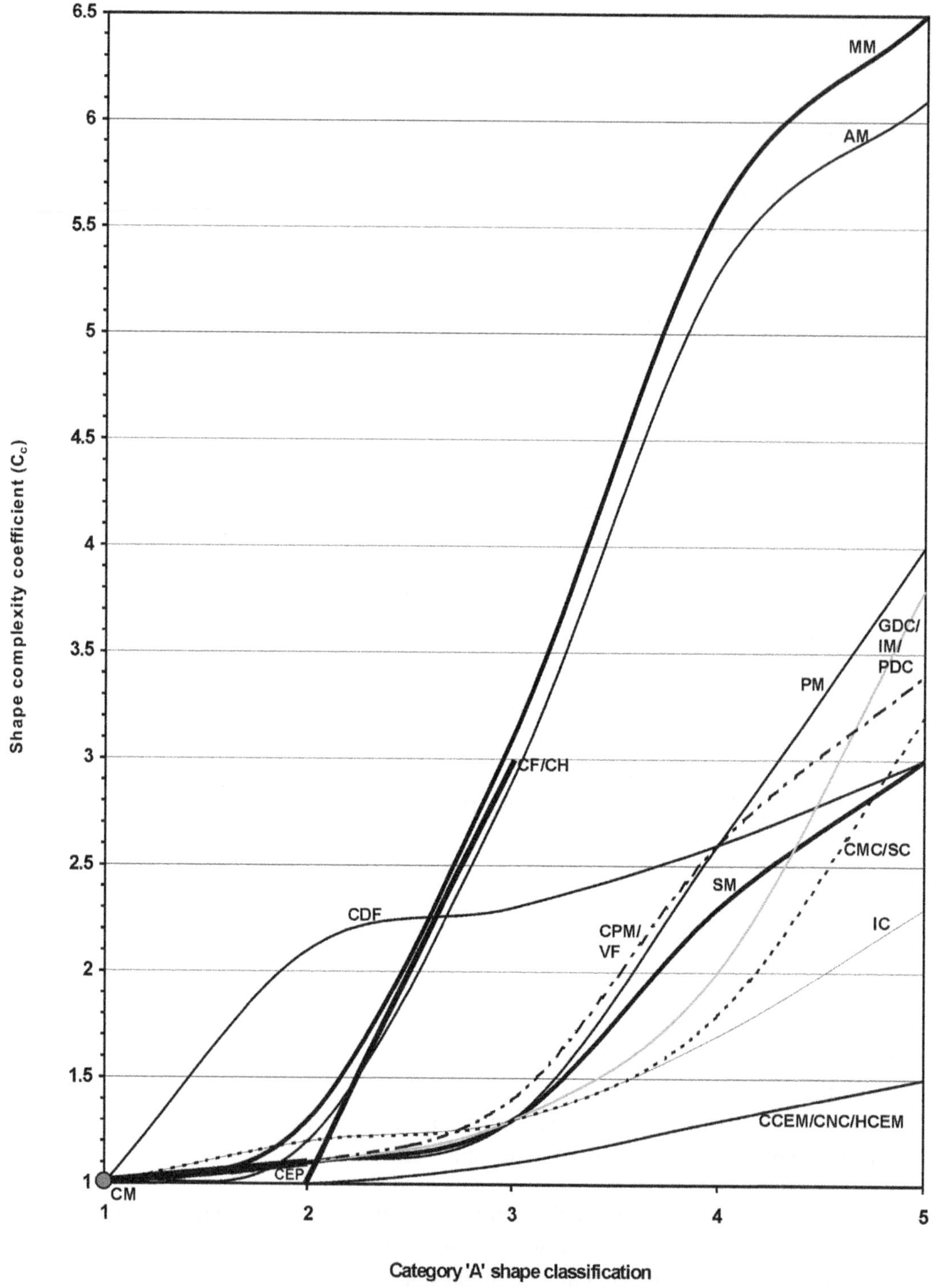

Figure 12.9: Determination of Shape Complexity Coefficient (C_c) – Category 'A' Shape Classification.

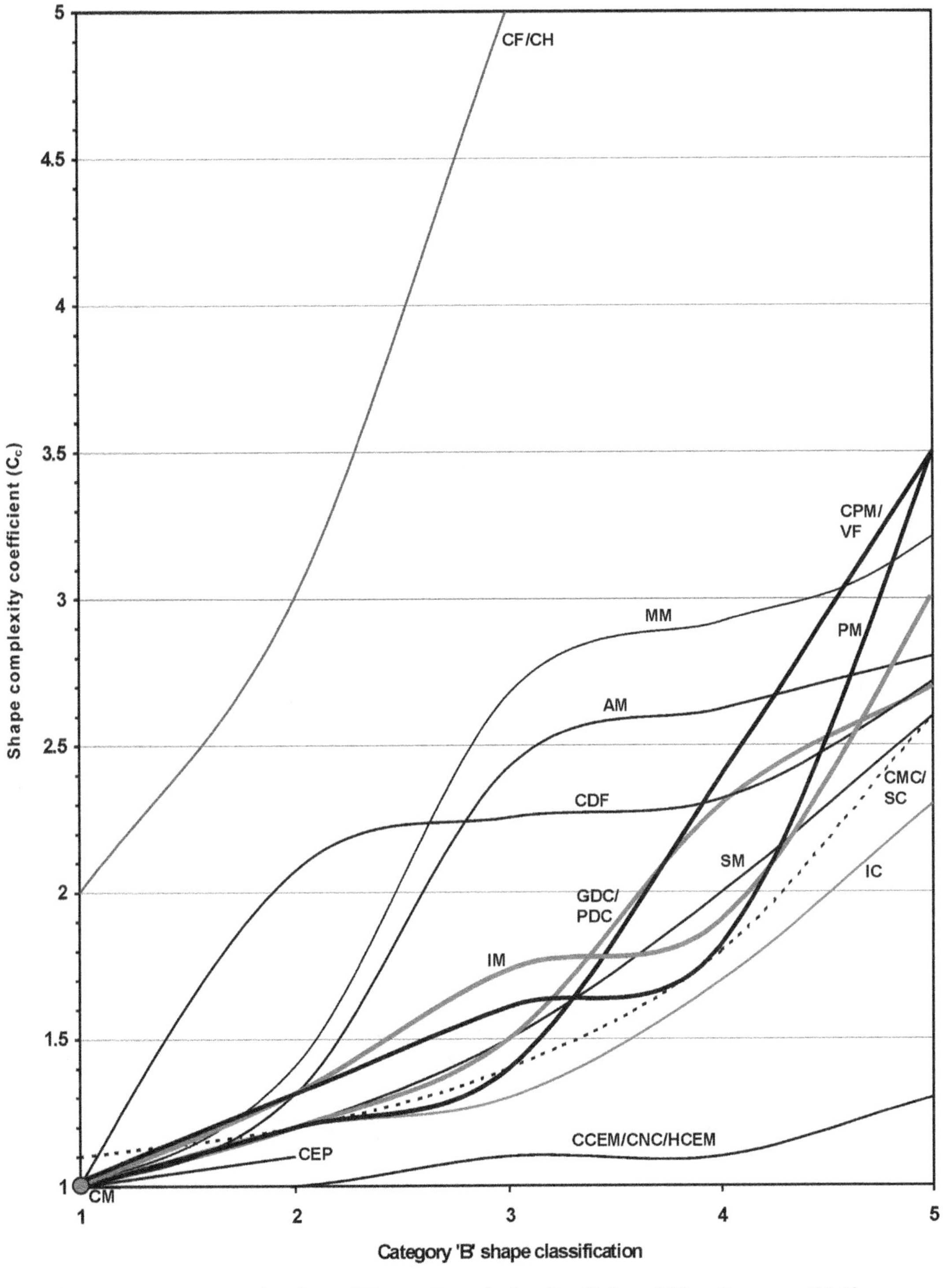

Figure 12.10: Determination of Shape Complexity Coefficient (C_c) – Category 'B' Shape Classification.

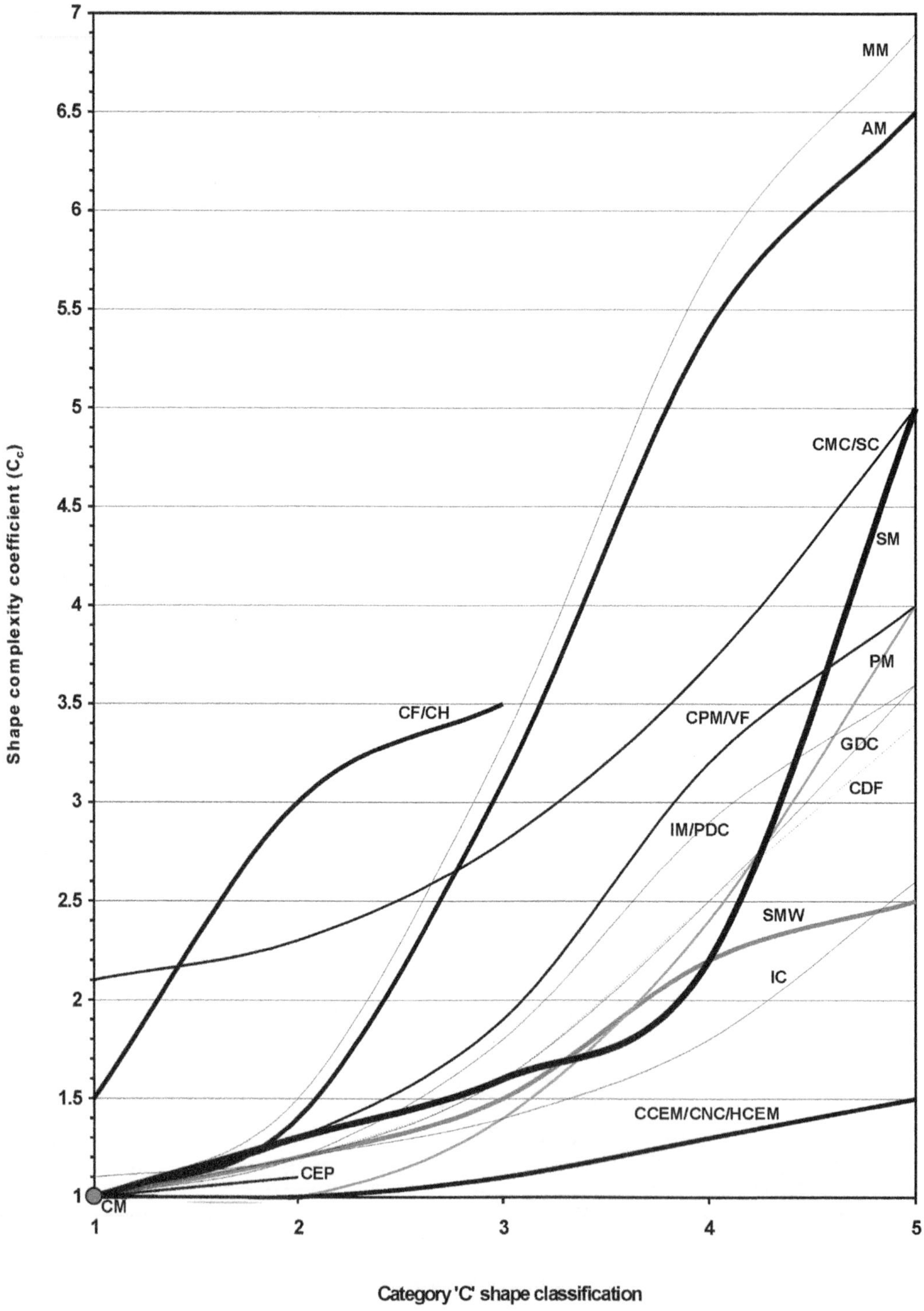

Figure 12.11: Determination of Shape Complexity Coefficient (C_c) – Category 'C' Shape Classification.

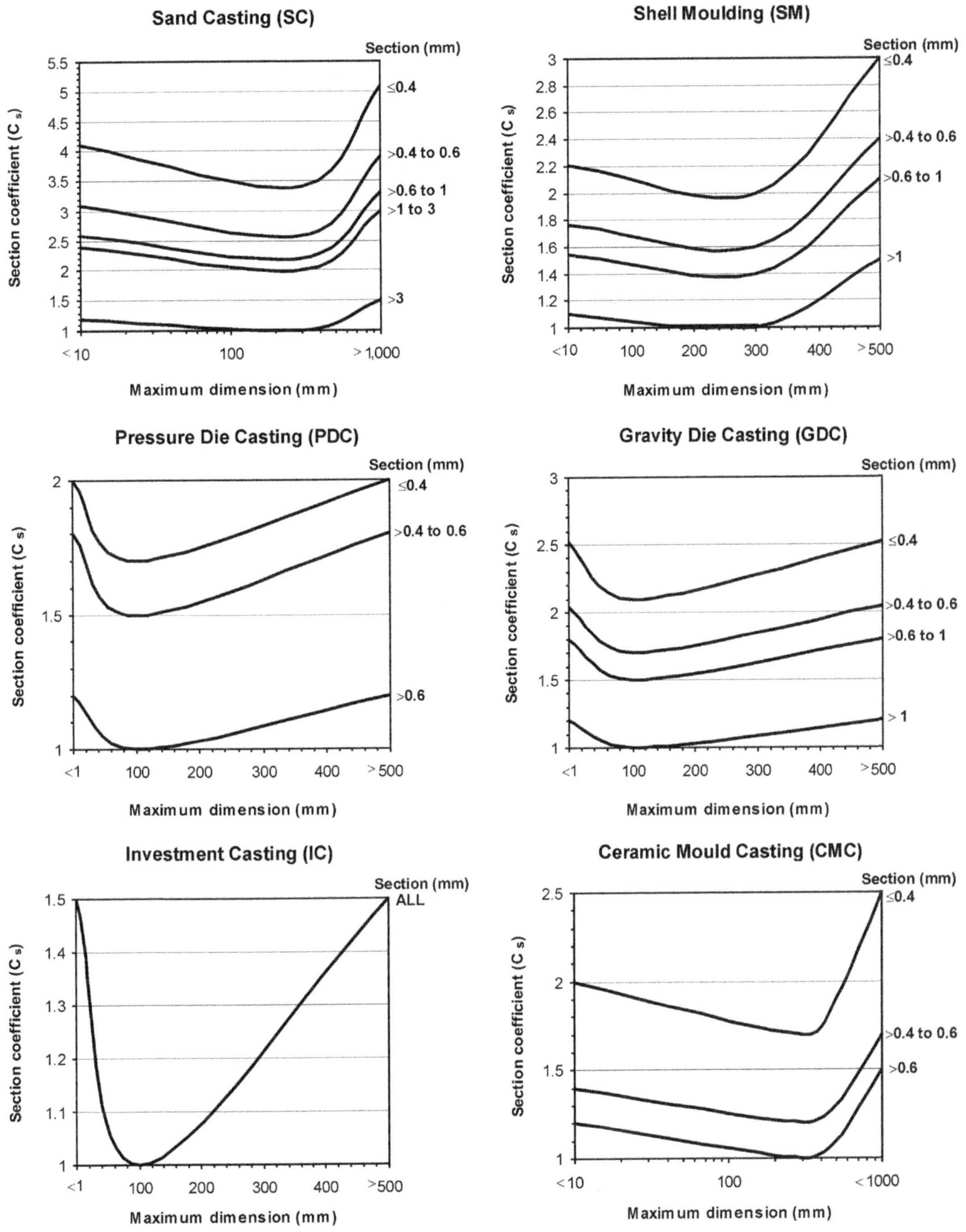

Figure 12.12: Chart Used for the Determination of the Section Coefficient (C_s) for Casting Processes.

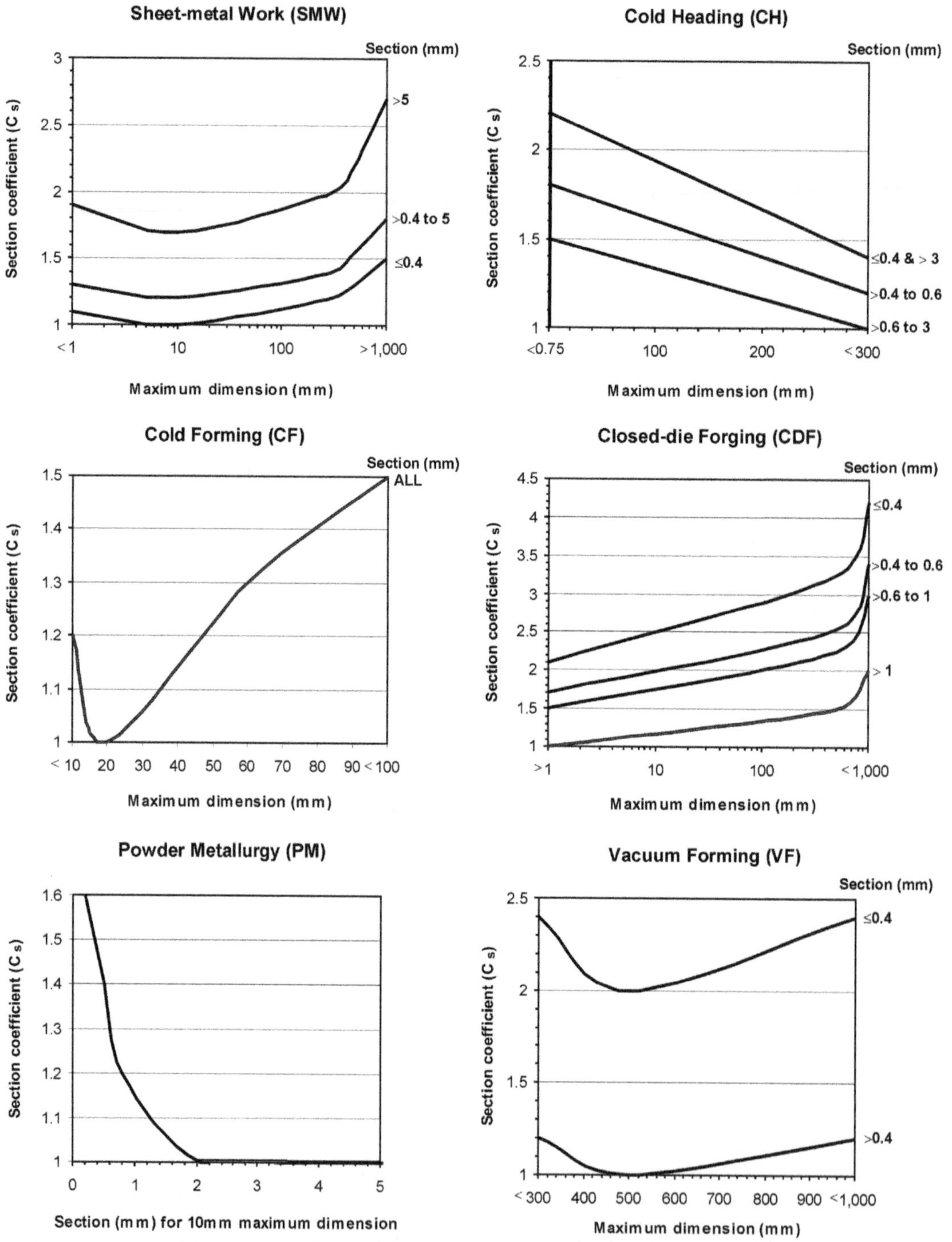

Figure 12.13: Chart Used for the Determination of the Section Coefficient (C_s) for Forming Processes.

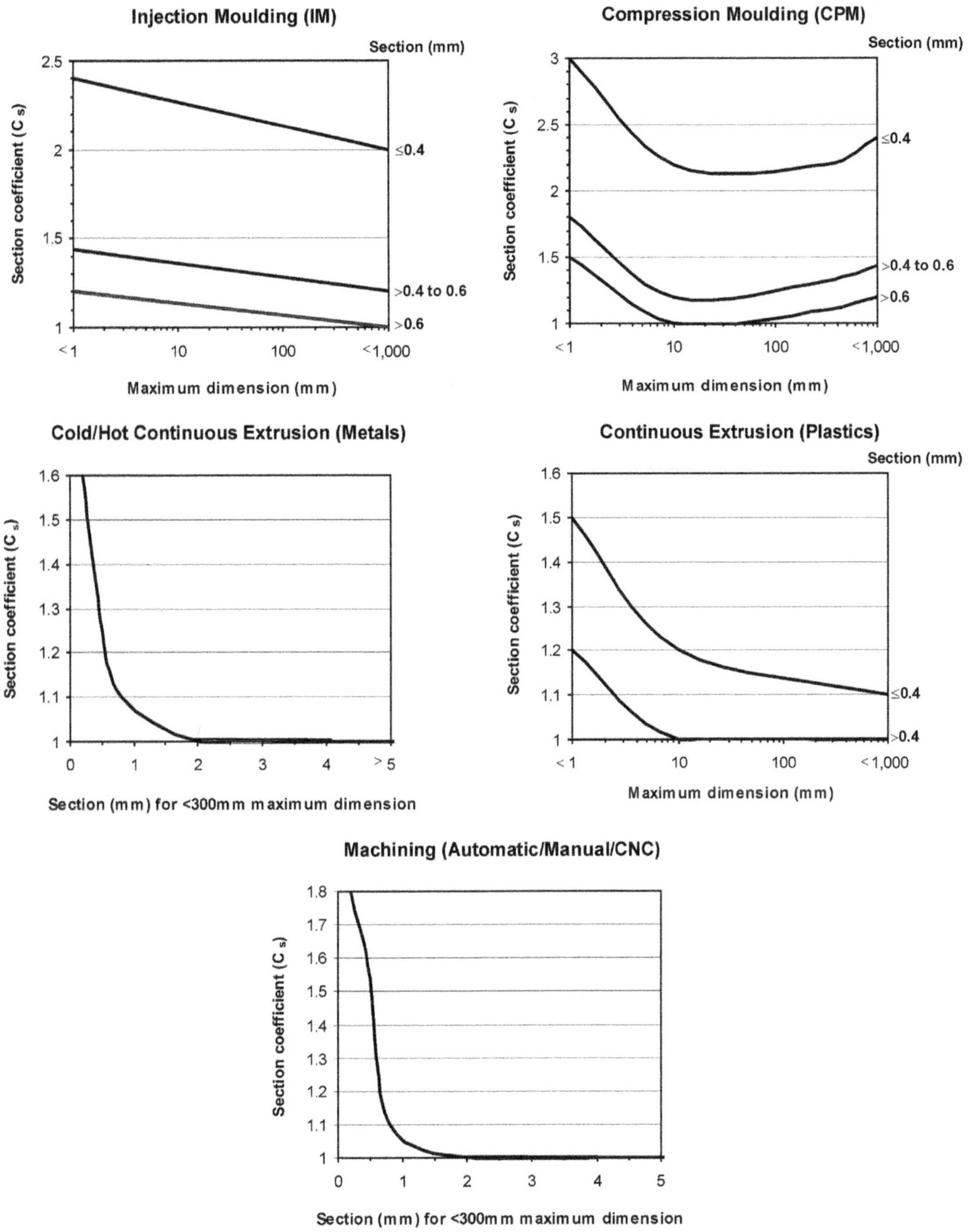

Figure 12.14: Chart Used for the Determination of the Section Coefficient (C_s) for Plastic Moulding, Continuous Extrusion and Machining Processes.

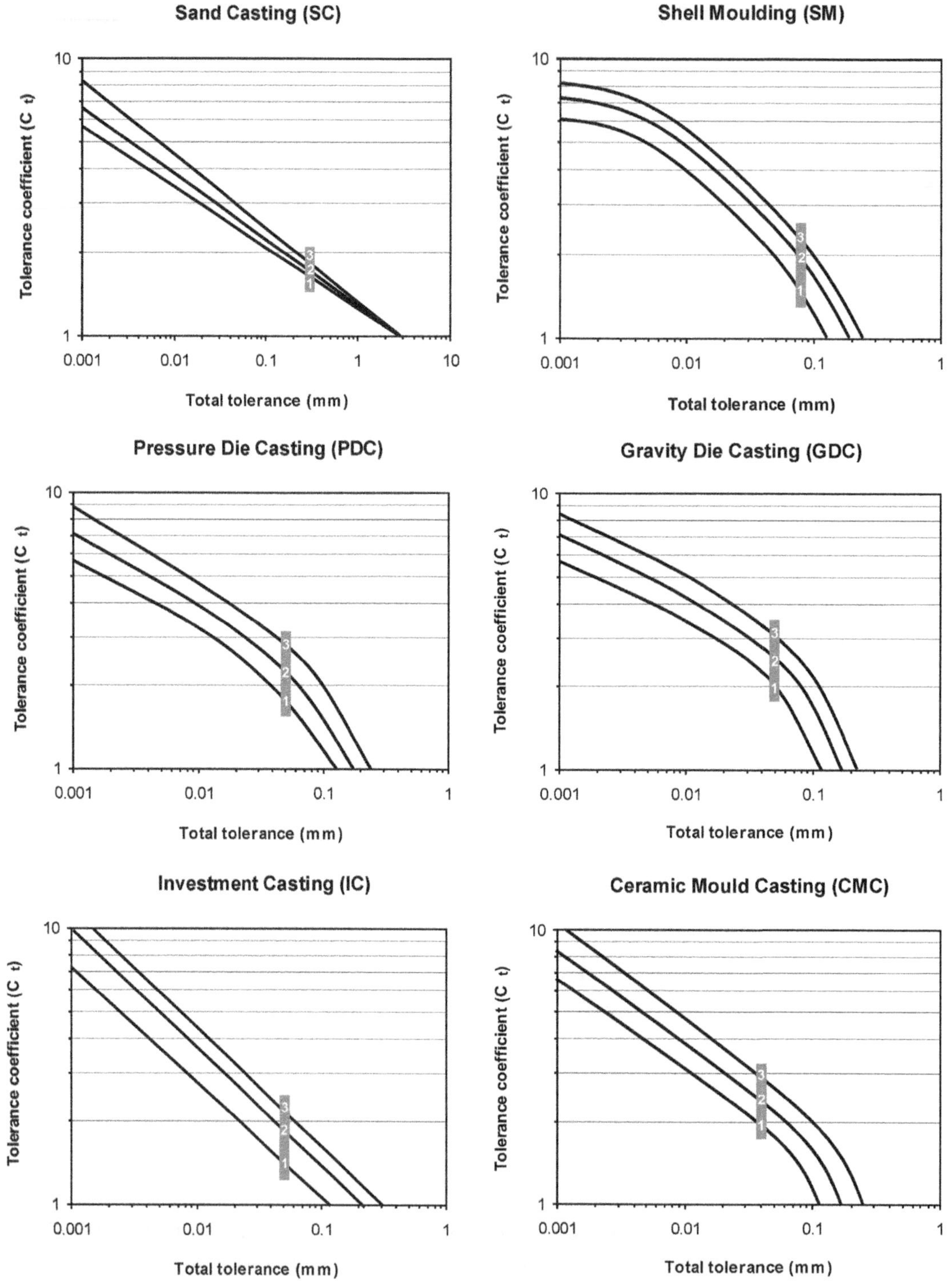

Figure 12.15: Chart Used for the Determination of the Tolerance Coefficient (C_t) for Casting Processes.

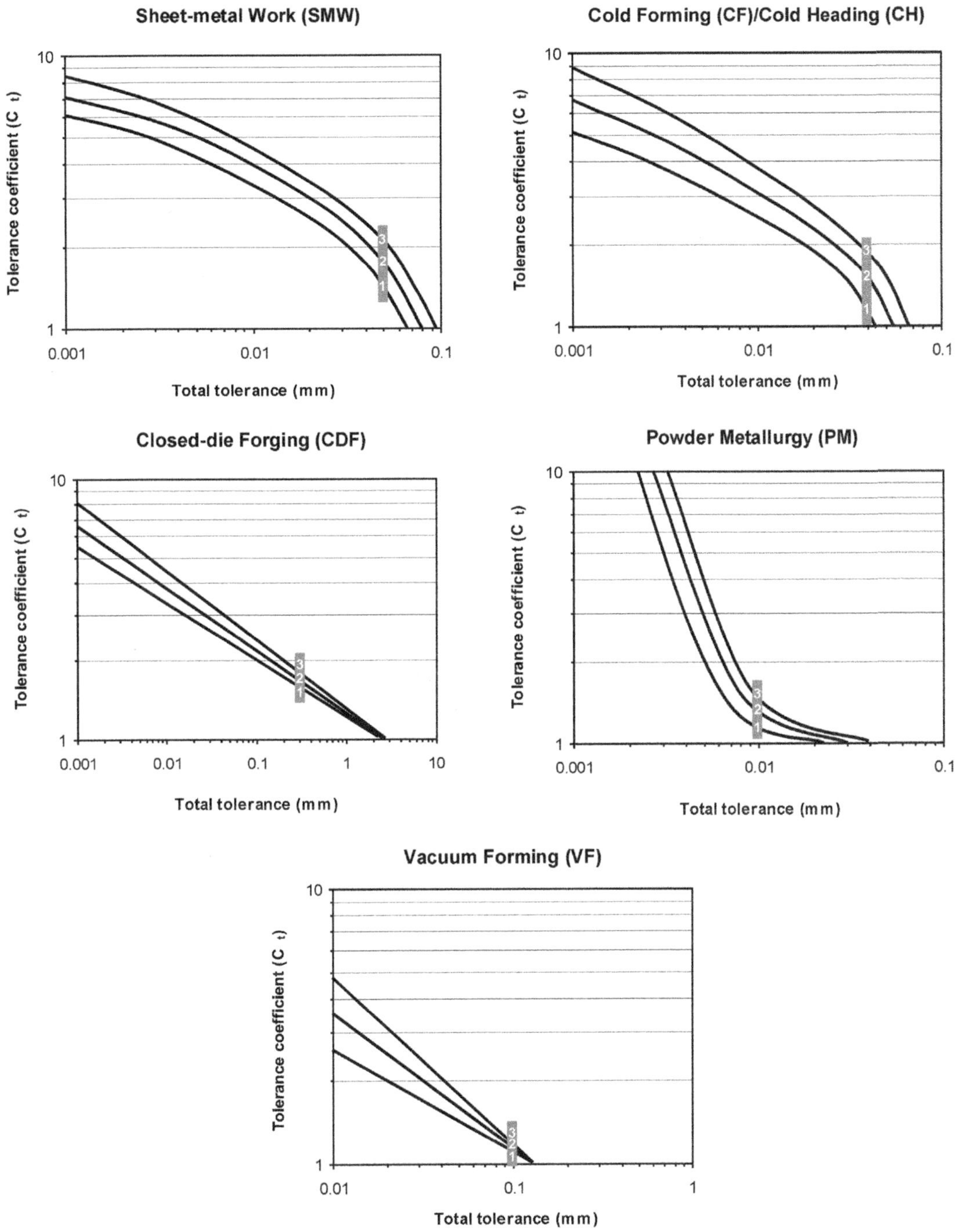

Figure 12.16: Chart Used for the Determination of the Tolerance Coefficient (C_t) for Forming Processes.

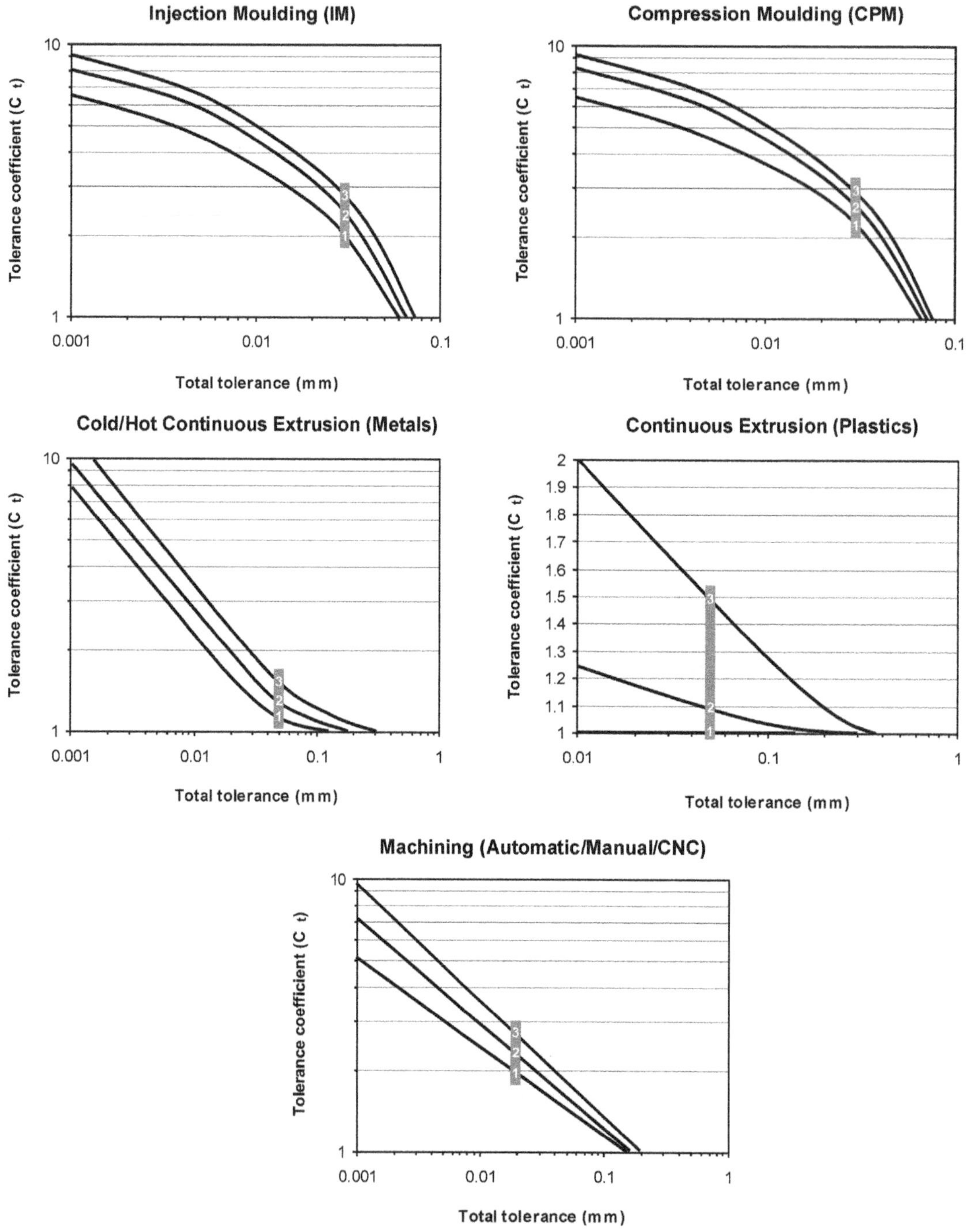

Figure 12.17: Chart Used for the Determination of the Tolerance Coefficient (C_t) for Plastic Moulding, Continuous Extrusion and Machining Processes.

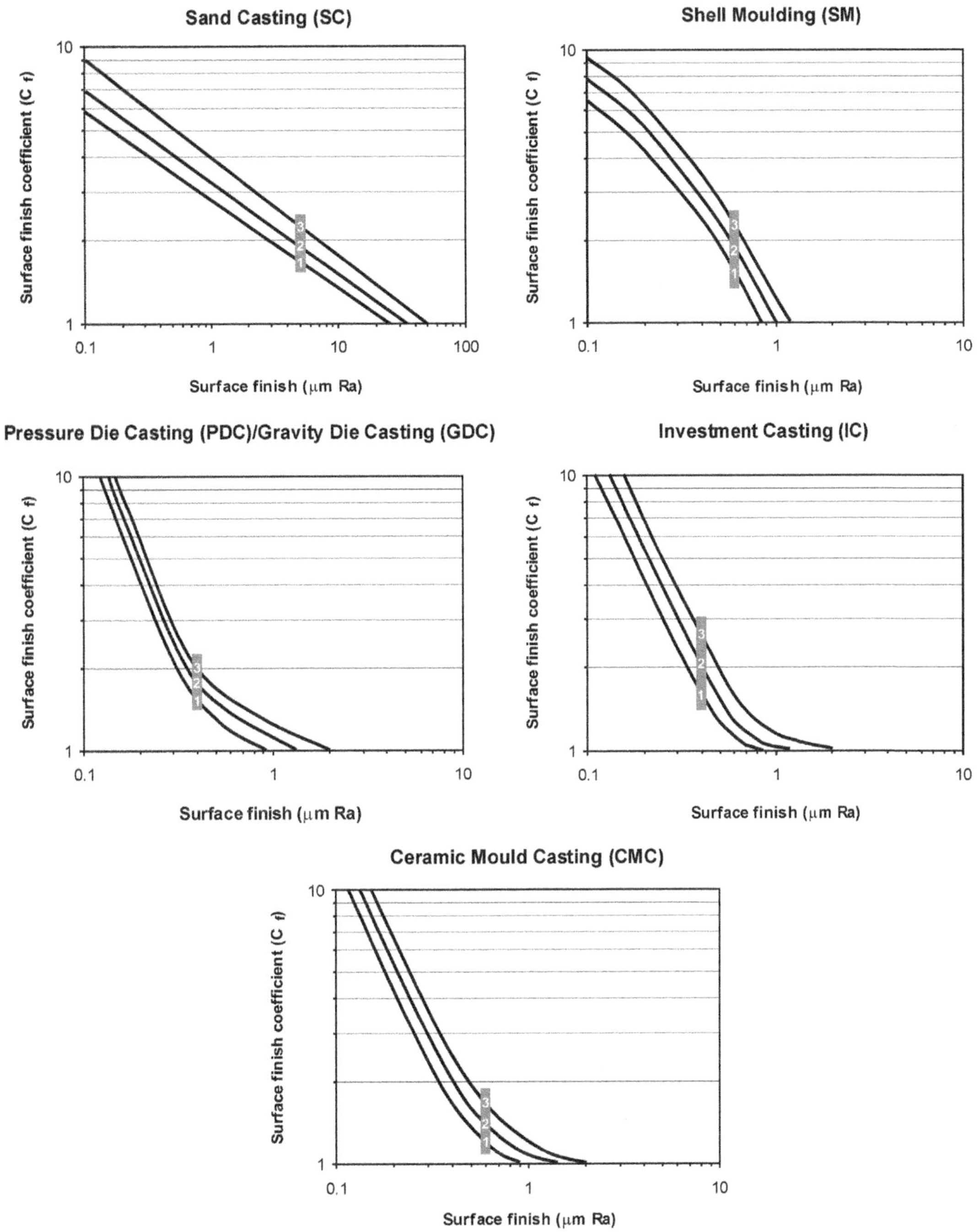

Figure 12.18: Chart Used for the Determination of the Surface Finish Coefficient (C_f) for Casting Processes.

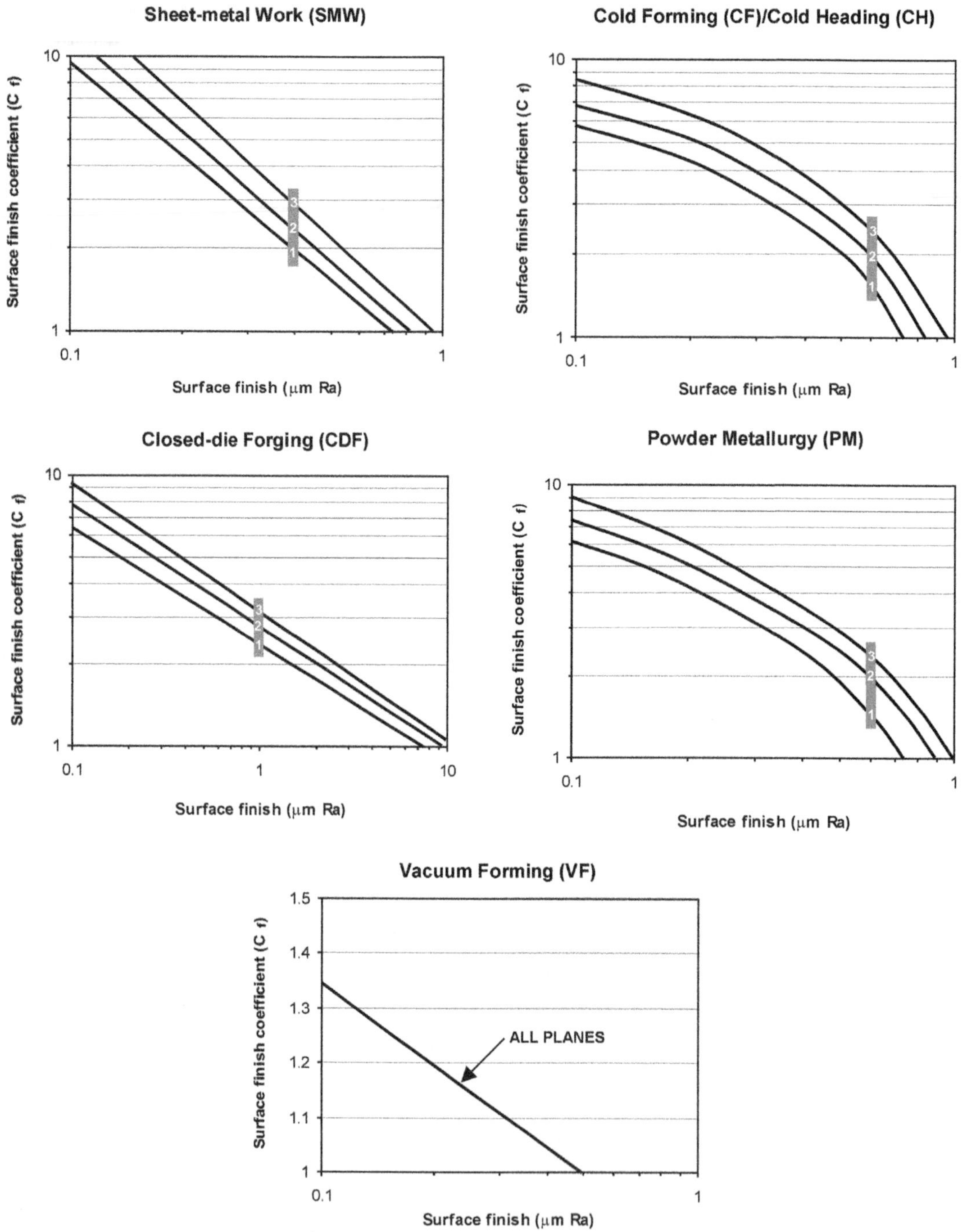

Figure 12.19: Chart Used for the Determination of the Surface Finish Coefficient (C_f) for Forming Processes.

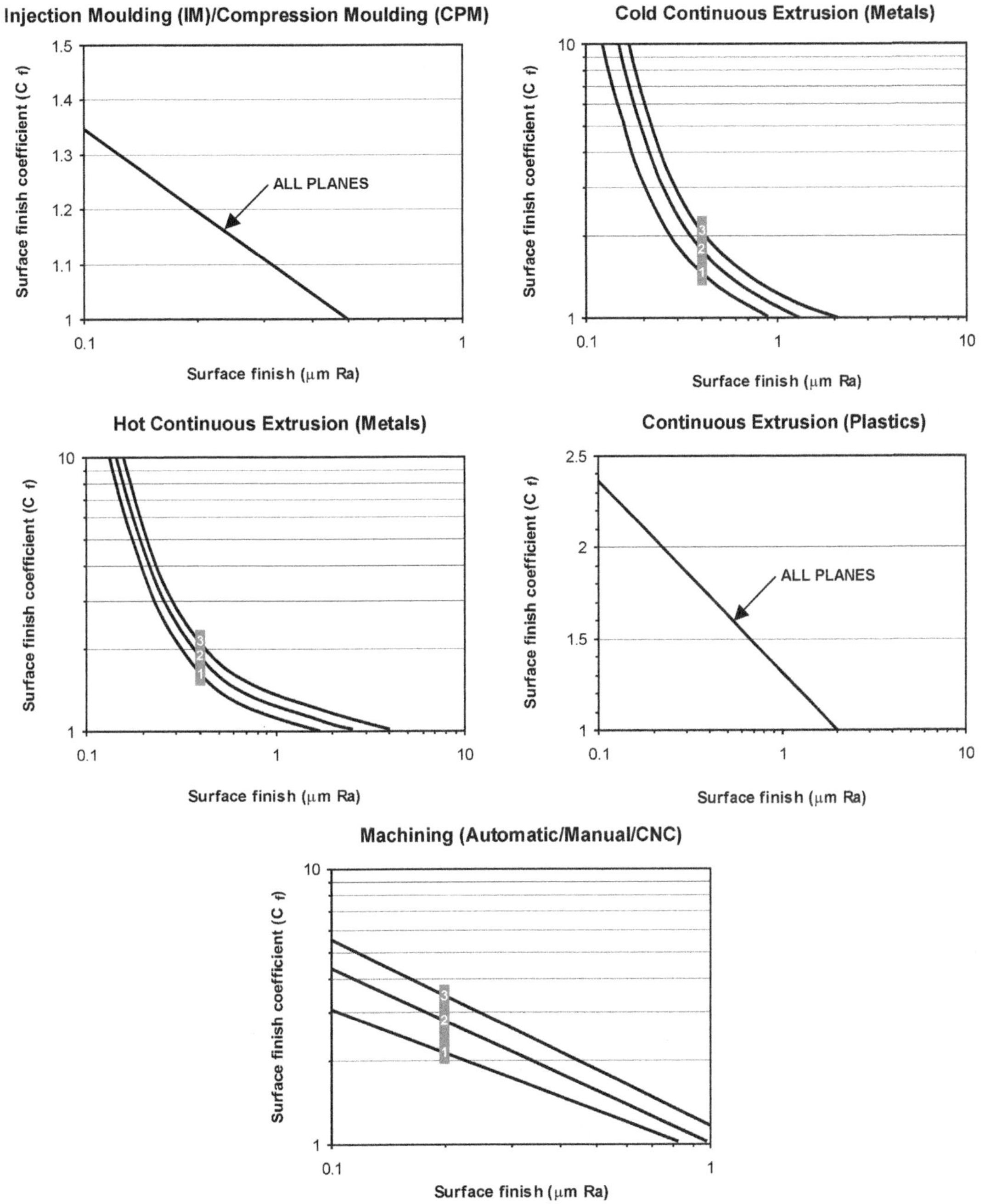

Figure 12.20: Chart Used for the Determination of the Surface Finish Coefficient (C_f) for Plastic Moulding, Continuous Extrusion and Machining Processes.

Table 12.2: Sample Material Cost Values per unit Volume (C_{mt}) for Commonly Used Material Classes.

Material	Material Cost, C_{mt} (pence/mm³)
Cast iron	0.00061
Low carbon steel	0.00052
Alloy steel	0.00196
Stainless steel	0.00250
Copper alloy	0.00400
Aluminium alloy	0.00090
Zinc alloy	0.00147
Thermoplastic - PA, PMMA	0.00052
Thermoplastic - PVC, PE, PS	0.00022
Thermoset	0.00035
Elastomer	0.00021

orthogonal axes or planes (either 1, 2 or 3+) on which the critical tolerances lie, and which cannot be achieved from a single direction using the manufacturing process. Repeat the above process exactly for C_f using the graphs in Figures 12.18–12.20.

$$C_{ft} = C_t \text{ or } C_f, \text{ whichever gives the highest value}.$$

Note that for Chemical Milling (CM2.5 and CM5), $C_{ft} = 1$, as the penalty is taken account of in the formulation of the basic processing cost, P_c.

12.4 Material Cost (M_c)

The material cost, M_c, was defined in Equation (12.1) as the volume of raw material required to process the component, multiplied by the cost of the material per unit volume in the required form, C_{mt}:

$$M_c = V \cdot C_{mt} \tag{12.6}$$

Sample average values for C_{mt} for commonly used material classes can be found in Table 12.2. These values were sourced from Chinese suppliers and for mid-range tonnage values. Company-specific data should be used wherever possible. In many situations the material cost can form a large proportion of the total component cost; therefore, a consistent approach should be taken in the volume calculation if valid comparisons are to be produced. Note that the volume, V, in Equation (12.6) must be worked out in cubic millimetres (mm³). Ref. [9] has relative cost data for a number of material classes that can be used where specific data is not available.

Component manufacture may involve surface coating and/or heat treatments, and have some effect on manufacturing cost. The development of models for this aspect of component manufacturing cost can be found in Refs [10,11].

The volume may be calculated in one of two ways:

1. Using the total volume – if the total volume of material required to produce the component is known (i.e. the volume including any processing waste), then this value is used for 'V' and the waste coefficient, W_c, is ignored.
2. Using the final (finished) volume – if the amount of waste material is not known, then the final component volume may be used. In this case, use the waste coefficient, W_c, which takes into account the waste material consumed by a particular process.

The formulation for 'V' for this method is:

$$V = V_f \cdot W_c \tag{12.7}$$

where V_f = finished volume of the component.

Waste coefficients, W_c, for the sample processes can be found in Table 12.3, relative to the shape classifications provided in Figure 12.8. While in many cases the values quoted can be used with confidence, estimation of the input volume for the process is the approach preferred (method 1 above). In many applications, when calculating the volume of a component, it is not always necessary to go into great detail. Approximate methods are often satisfactory when comparing designs and it can be helpful if a design is broken down into simple shape elements, allowing the quick calculation of a volume. Before looking at the industrial applications of the design costing methodology it should be noted that material and process selection need to be considered together; they should not be viewed in isolation. The analysis presented here does not in any way take into account physical properties such as strength, weight, conductivity, etc.

Note that for Chemical Milling (CM2.5 and CM5), $W_c = 1$, as the penalty is taken account of in the formulation of the basic processing cost, P_c.

12.5 Model Validation

In order to validate the approach, a number of companies were consulted covering a wide range of manufacturing technology and products. Understandably, companies were often reluctant to discuss cost information, even admitting that they had no systematic process or structure to the way new jobs were priced, relying almost exclusively on the knowledge and expertise of one or two senior estimators. However, a number of companies were able to provide both estimated and actual cost data for a sufficient range of components to perform some meaningful validation.

Figure 12.21(a) illustrates the results of a validation exercise in a company producing plastic moulded components. The analysis was performed on a number of products at random and the estimated costs predicted by the evaluation, M_i, have been plotted against the actual

Table 12.3: Waste Coefficient (W_c) for the Sample Processes Relative to Shape Classification Category

Shape Classification	Process																			
	AM	CCEM	CDF	CEP	CF	CH	CMC	CNC	CPM	GDC	HCEM	IC	IM	MM	PDC	PM	SM	SC	SMW	VF
A1	1.6	1	1.1	1	1	1	1.1	1.6	1	1	1	1	1.1	1.6	1	1	1	1.1		1
A2	2	1	1.1	1.1	1	1	1.1	2	1.1	1.1	1	1	1.1	2	1.1	1	1.1	1.1		1.1
A3	2.5	1.5	1.2		1	1	1.2	2.5	1.1	1.1	1.5	1.1	1.1	2.5	1.1	1	1.1	1.2		1.1
A4	3	2	1.2				1.3	3	1.2	1.2	2	1.1	1.1	3	1.2	1	1.2	1.3		1.2
A5	4	3	1.3				1.4	4	1.3	1.3	3	1.2	1.2	4	1.3	1.2	1.3	1.4		1.3
B1	1.7	1	1.1	1	1	1	1.1	1.7	1	1	1	1	1.1	1.7	1	1	1	1.1		1
B2	2.2	1	1.1	1.1	1	1	1.1	2.2	1.1	1.1	1	1	1.1	2.2	1.1	1	1.1	1.1		1.1
B3	2.8	1.5	1.2		1	1	1.2	2.8	1.1	1.1	1.5	1.1	1.1	2.8	1.1	1	1.1	1.2		1.1
B4	4	2	1.2				1.3	4	1.1	1.2	2	1.1	1.1	4	1.2	1	1.2	1.3		1.1
B5	6	3	1.3				1.4	6	1.2	1.3	3	1.2	1.2	6	1.3	1.2	1.3	1.4		1.2
C1	1.8	1	1.1	1	1	1	1.1	1.8	1	1	1	1	1.1	1.8	1.1	1	1.1	1.1	1.2	1
C2	2.4	1	1.1	1.1	1	1	1.2	2.4	1.1	1.1	1	1	1.1	2.4	1.1	1	1.1	1.2	1.2	1.1
C3	4	2	1.1		1	1	1.3	4	1.1	1.1	2	1.1	1.1	4	1.1	1	1.1	1.3	1.4	1.1
C4	6	3	1.2				1.4	6	1.1	1.2	3	1.1	1.2	6	1.2	1	1.2	1.4	1.5	1.1
C5	8	4	1.3				1.6	8	1.2	1.3	4	1.2	1.3	8	1.3	1.2	1.3	1.6	1.6	1.2

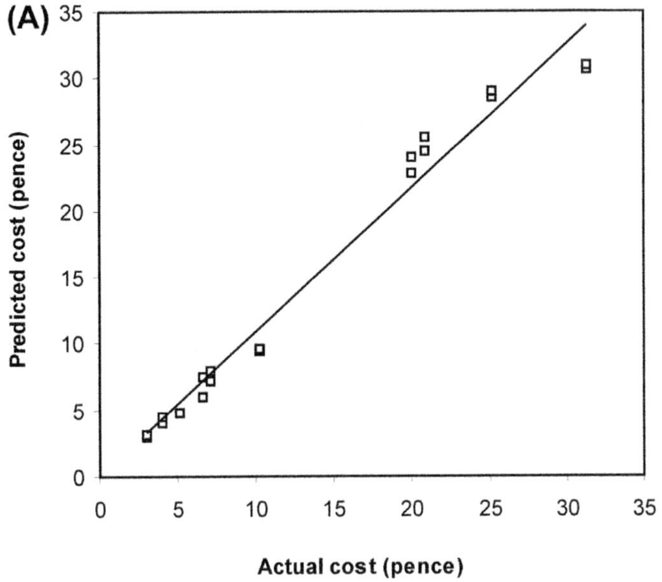

Cost validation results for injection-moulded components

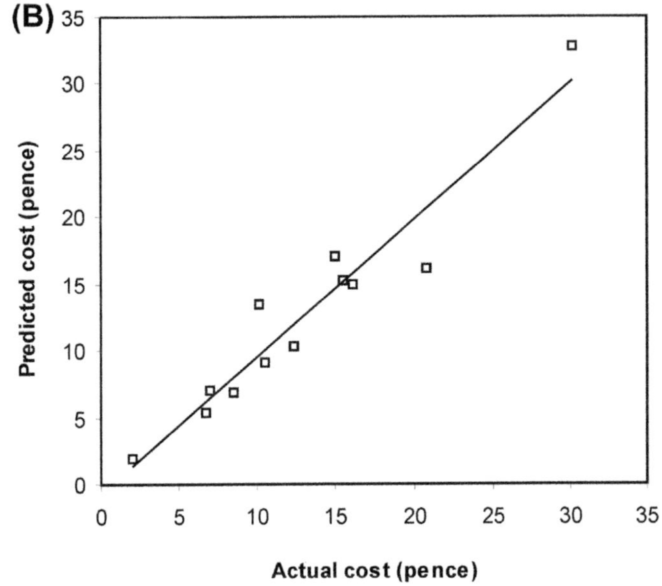

Cost validation results for pressed steel components

Figure 12.21: Costing Methodology Validation Results.

manufacturing costs provided by the company. Figure 12.21(b) illustrates another plot, this time from a company producing pressed sheet-metal parts. Figure 12.22 illustrates some of the components included in the validation studies.

Validation exercises on a range of component types that were carried out by 22 individuals in industry (mechanical, electrical and manufacturing engineers) showed that the main variability encountered was in the calculation of component volume and in the assignment of the shape complexity index [12]. While the determination of component volume is mechanistic, it is recognised that the determination of the most appropriate shape complexity classification requires judgmental skills and experience in the application of the methodology. These problems were largely eliminated when the analysis was carried out in a team environment, where highly consistent and reliable results were produced. In addition, training in the application of the methodology yields considerable improvements in the quality and consistency of the results produced, proving capable of predicting the cost of manufacture of a component to better than 16%. Customising the data to a particular business would significantly enhance the accuracy of the predicted costs obtained from the analysis.

The methodology can be helpful in producing cost estimates where design solutions involve a significant amount of subcontract work. The estimates produced provide support to the make versus buy analysis and can be useful in calibrating supplier quotations. Variations of more than 30% in quotations from subcontractors against identical specifications are common across the range of manufacturing processes. This has been noted by a number of researchers [13]. In this way benefits can be gained whether the methodology is applied as a stand-alone tool during product design/redesign or, more globally, as part of a company's integrated application of simultaneous engineering tools and techniques. The applications of the methodology may be summarised as:

- Determination of component cost in support of DFA.
- Competitor analysis.
- Assistance with make versus buy decisions.
- Cost estimation in concept design with low levels of component detail.
- Support for simultaneous engineering and teamwork.
- Training in design for manufacture.

12.6 Case Studies

12.6.1 Car Headlight Trimscrew

One of the main objectives of DFA is the reduction of component numbers in a product to minimise assembly cost. This tends to generate product design solutions that contain fewer but sometimes more complex components embodying a number of functions. Such an approach is often criticised as being suboptimal; therefore, it is important to know the

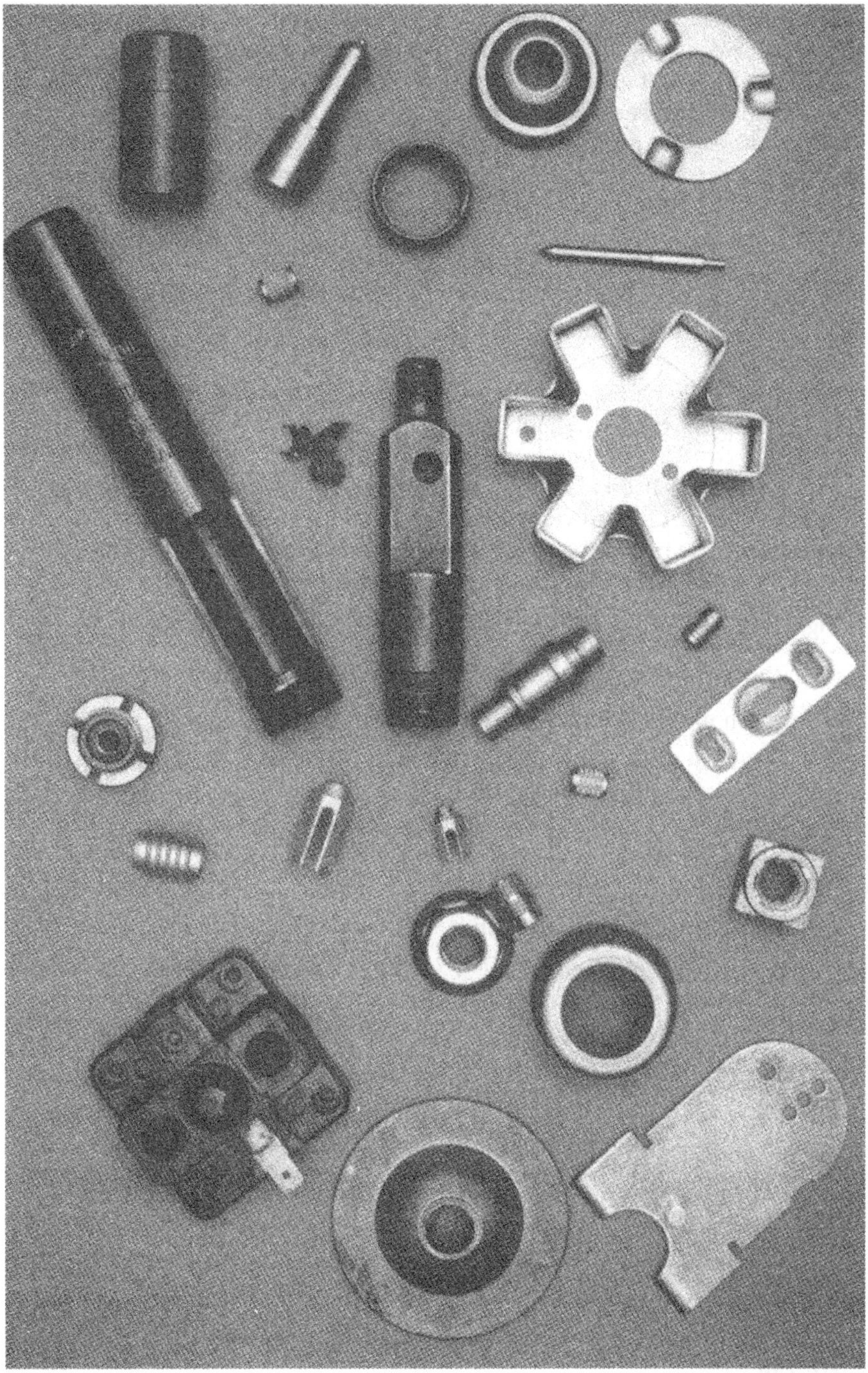

Figure 12.22: Example Components used in the Validation Studies.

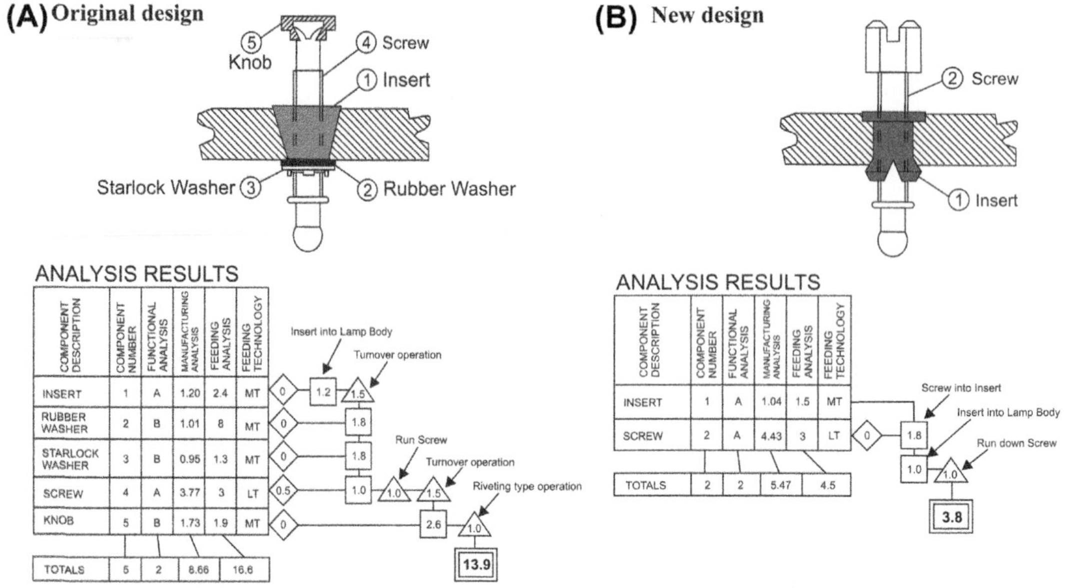

Figure 12.23: Before and After Analysis of a Headlight Trimscrew Design.

consequences of such moves on component manufacturing costs. Product teams should therefore predict the cost of manufacture associated with alternative component design solutions, resulting from the activities of DFA.

An illustration of how the design costing analysis can be used in DFA is given in Figures 12.23 and 12.24 and Table 12.4. Figure 12.23(a) shows the original design of a car headlight trimscrew assembly and Figure 12.23(b) shows the new design. The DFA analyses can also be seen in Figure 12.23(a) and (b), respectively. Notice that these figures include data on manufacturing cost and provide the assembly sequence diagram for each design using the preferred DFA notation. A breakdown of the cost analysis for the two components in the new design of the trim screw is given in Figure 12.24, which also provides links to the charts used to determine the coefficients, etc. Table 12.4 provides a summary of the resulting measures of performance for each design. It can be seen from this that it is possible to fully assess the production cost consequences of each design in terms of both component manufacturer and assembly. Note that the total component manufacturing costs associated with the new design resulting from DFA are less than in the original: this turns out to be the case in many of the DFA studies examined to date by the authors.

12.6.2 Pivot Pin

A simple illustration of a case where the situation is not quite so clear-cut is given in Figure 12.25. The DFA approach drives consideration of the assembly design proposal shown in

PRODUCT NAME: TRIMSCREW

PRODUCT CODE / ID: NEW DESIGN

PRODUCT QUANTITY: 1,000,000 pa

COMPONENT DETAILS				Mc = V x Cmt x [Wc]				Ⓐ		Rc = Cc x Cmp x Cs x [Cft]										Ⓑ	Ⓐ+Ⓑ
PART No. ID	PART DESCRIPTION	MATERIAL	PRIMARY PROCESS	SHAPE COMP.	Volume (mm^3)	Cmt	Wc	Mc	Pc	Cc	Cmp	Section (mm)	Cs	Toler-ance (mm)	Ct	Surface Finish (μm Ra)	Cf	Cft	Rc	(Pc x Rc)	Mi (cost in pence)
1	INSERT	THERMOPLASTIC	INJ. MOULDING	A1	626	0.00022	1.1	0.15	0.9	1	1	1	1	0.1	1	0.8	1	1	1	0.9	1.04
2	SCREW	LOW CARB. STEEL	AUTO. MACHINING	A2	4262	0.00052	N/A	2.22	1.2	1.2	1.4	2	1	0.1	1.1	0.8	1	1.1	1.84	2.21	4.43
																				TOTAL	5.47
				SHAPE CLASSIFICATION (FIGURE 12.8)		MATERIAL COST (TABLE 12.2)	WASTE COEFFICIENT (TABLE 12.3)		BASIC PROCESSING COST (FIGURES 12.1 & 12.3 RESPECTIVELY)	SHAPE COMPLEXITY COEFFICIENT (FIGURE 12.9)	MATERIAL TO PROCESS SUITABILITY COEFFICIENT (TABLE 12.1)		MINIMUM SECTION COEFFICIENT (FIGURE12.14)		TOLERANCE COEFFICIENT (FIGURE 12.17)		SURFACE FINISH COEFFICIENT (FIGURE 12.20)				

Figure 12.24: Cost Analysis for the Manufacture of the Components in the New Headlight Trimscrew Design.

Table 12.4: Resulting Design Measures for Each Trimscrew Assembly.

	Original	New
Total parts count	5	2
Design efficiency	40%	100%
Total manufacturing analysis (pence)	8.66	5.47
Total feeding index	16.6	4.5
Feeding ratio	8.3	2.3
Total fitting index	13.9	3.8
Fitting index	7	1.9

design 'B'. An investigation of the two pivot pin designs using the cost analysis suggests that, from a component manufacturing point of view, design 'A' represents a cost saving. In this example, the same manufacturing process (automatic machining) is used for both designs and the difference in cost results from the different initial material volume requirements. (The values of $P_c = 3$ and $R_c = 2.75$ are the same in each case.) Supplier cost data is used in the case of the standard clip fasteners. Hence, selection on the basis of cost demands a trade-off between assembly and manufacturing cost. Both design solutions are commonly seen in products from various business sectors and product groups.

12.6.3 Spark Plug Body

A case study comparing alternative processing routes is illustrated next. Cold forming and automatic machining processing routes for a spark plug body design and production quantity requirements show significant manufacturing cost variations, as summarised in Figure 1.1. Figure 12.26 presents the detailed cost analysis giving the values obtained from P_c and the individual elements involved in the calculation of R_c, together in the table with details of the design. The benefits of the high material utilisation associated with cold forming mean a large cost saving at the annual production quantity of one million components. (The input volume for the machined component is almost five times that required for cold forming.) However, as the annual production requirement reduces, the processing cost moves more in favour of machining, and at 30,000 per annum the sample data predicts little difference in cost between the two methods of production (as shown in the lower part of Figure 12.26).

12.6.4 Control Valve Sleeve

A case where a material and process change eliminates the need for secondary processing is shown in Figure 12.27. An aluminium pressure die casting is initially considered for the sleeve shown, but secondary processing may be needed to ensure conformance to surface finish requirements as the achievement of 0.4 μm Ra is on the boundary of technical feasibility. An optional design uses injection-moulded polysulphone (PSU). The sample data does not

DESIGN A

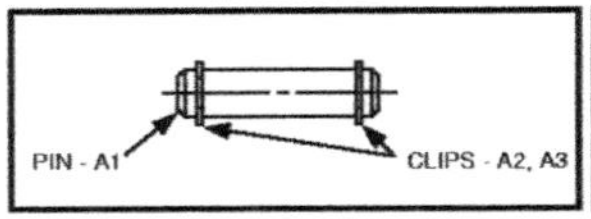

DESIGN B

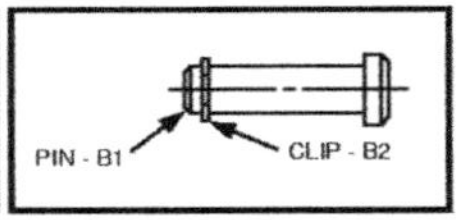

PRODUCT NAME: PIVOT PIN
PRODUCT CODE / ID: DESIGNS A AND B
PRODUCT QUANTITY: 30,000 pa

COMPONENT DETAILS				Mc = V x Cmt x [Wc]				(A)		Rc = Cc x Cmp x Cs x [Cft]										(B)	(A) + (B)
PART No. ID	PART DESCRIPTION	MATERIAL	PRIMARY PROCESS	SHAPE COMP.	Volume (mm^3)	Cmt	Wc*	Mc	Pc	Cc	Cmp	Section (mm)	Cs	Tolerance (mm)	Ct	Surface Finish (µm Ra)	Cf	Cft	Rc	(Pc x Rc)	Mi (cost in pence)
DESIGN A																					
A1	PIN	ALLOY STEEL	AUTO. MACHINING	A1	4850	0.00196	N/A	9.51	3	1	2.5	5	1	0.1	1.1	0.8	1	1	2.75	5.75	15.26
A2	CLIP	STAINLESS STEEL	SHEET METAL-WORK																		1.4*
A3	CLIP	STAINLESS STEEL	SHEET METAL-WORK																		1.4*
																				TOTAL	18.06
DESIGN B																					
B1	PIN	ALLOY STEEL	AUTO. MACHINING	A1	9900	0.00196	N/A	19.40	3	1	2.5	5	1	0.1	1.1	0.8	1	1	2.75	5.75	25.15
B2	CLIP	STAINLESS STEEL	SHEET METAL-WORK																		1.8*
																				TOTAL	26.95

* Supplier quotation

Figure 12.25: Estimated Costs for Alternative Designs of Pivot Pin Components.

PLUG BODY

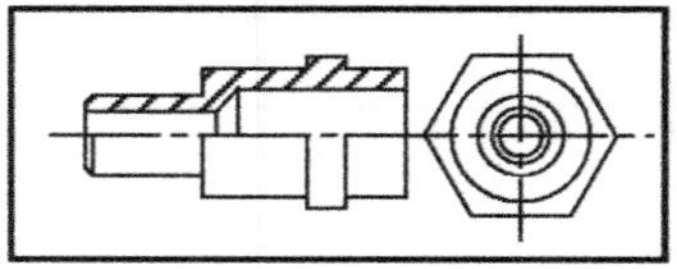

PRODUCT NAME: PLUG BODY
PRODUCT CODE / ID: -
PRODUCT QUANTITY: 1,000,000 pa

COMPONENT DETAILS				Mc = V x Cmt x [Wc]				Ⓐ		Rc = Cc x Cmp x Cs x [Cft]										Ⓑ	Ⓐ+Ⓑ
PART No. ID	PART DESCRIPTION	MATERIAL	PRIMARY PROCESS	SHAPE COMP.	Volume (mm^3)	Cmt	Wc	Mc	Pc	Cc	Cmp	Section (mm)	Cs	Toler-ance (mm)	Ct	Surface Finish (μm Ra)	Cf	Cft	Rc	(Pc x Rc)	Mi (cost in pence)
AT 1,000,000 pa																					
1	PLUG BODY	LOW CARB. STEEL	AUTO. MACHINING	A2	22100	0.00052	N/A	11.49	1.2	1.2	1.4	1.5	1	0.2	1	1.6	1	1	1.68	2.02	13.51
1	PLUG BODY	LOW CARB. STEEL	COLD FORMING	A2	3860	0.00052	1	2	0.8	1	1.3	1.5	1.2	0.2	1	1.6	1	1	1.56	1.25	3.25
AT 30,000 pa																					
1	PLUG BODY	LOW CARB. STEEL	AUTO. MACHINING	A2	22100	0.00052	N/A	11.49	3	1.2	1.4	1.5	1	0.2	1	1.6	1	1	1.68	5.04	16.53
1	PLUG BODY	LOW CARB. STEEL	COLD FORMING	A2	3860	0.00052	1	2	7	1	1.3	1.5	1.2	0.2	1	1.6	1	1	1.56	10.92	12.92

Figure 12.26: Comparison of Automatic Machining and Cold Forming Processes for the Manufacture of a Plug Body.

CONTROL VALVE SLEEVE

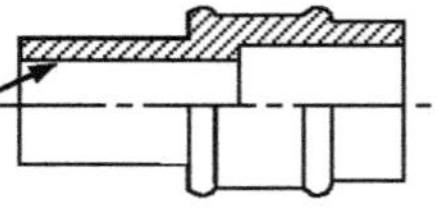

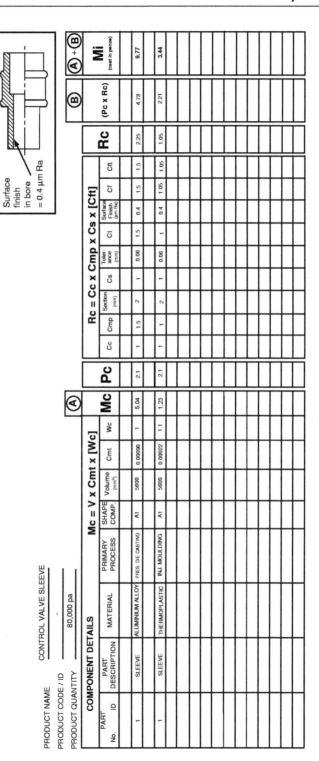

PRODUCT NAME: CONTROL VALVE SLEEVE

PRODUCT CODE / ID: -

PRODUCT QUANTITY: 80,000 pa

COMPONENT DETAILS					Mc = V x Cmt x [Wc]				(A)		Rc = Cc x Cmp x Cs x [Cft]										(B)	(A)+(B)
PART No.	ID	PART DESCRIPTION	MATERIAL	PRIMARY PROCESS	SHAPE COMP.	Volume (mm³)	Cmt	Wc	Mc	Pc	Cc	Cmp	Section (mm)	Cs	Tolerance (mm)	Ct	Surface Finish (µm Ra)	Cf	Cft	Rc	(Pc x Rc)	Mi (cost in pence)
1		SLEEVE	ALUMINIUM ALLOY	PRES. DIE CASTING	A1	5600	0.00090	1	5.04	2.1	1	1.5	2	1	0.06	1.5	0.4	1.5	1.5	2.25	4.73	9.77
1		SLEEVE	THERMOPLASTIC	INJ. MOULDING	A1	5600	0.00022	1.1	1.23	2.1	1	1	2	1	0.06	1	0.4	1.05	1.05	1.05	2.21	3.44

Figure 12.27: Comparison of Pressure Die Casting and Injection Moulding Processes for the Manufacture of a Critical Surface Finish.

differentiate between plastic injection moulding and pressure die casting in terms of basic processing cost. The savings indicated by the cost analysis result from lower material costs and surface finish capability of the injection moulding process, reflected in C_{ft} reduced from 1.5 to 1.05. Adopting injection moulding here removes additional machining and minimises the complexity of the manufacturing layout.

12.7 Bespoke Costing Development

Given the wide range of processes and their variants, and the problems of producing cost estimates from generic data that businesses can believe in, it is necessary to explore how we might go about getting companies to enter their own process knowledge into the component costing methodology presented previously. In this way an organisation can take ownership of the process costing knowledge and its maintenance. The development of this process of 'calibration' will enable a business to tune the data in the system to known component costs and take into account problems of varying material and processing cost in different parts of the world. However, the problem of enabling the user to add new processes to the methodology is rather complex. The main difficulties are associated with the need to collect and represent process knowledge for the calculation of basic processing cost, P_c, and the design-dependent relative cost coefficient, R_c. The adding of new material costs, M_c, and any necessary waste coefficients, W_c, is not considered to be a significant problem. The objective of these notes is to outline a process for the addition of costing information for new processes to the database to facilitate the costing of designs in early stages of the design process. Ref. [11] details an application of the bespoke costing model for a manual tool-polishing application.

12.7.1 Basic Processing Cost (P_c)

In order to determine the basic processing cost, P_c, of a simple or ideal design it is necessary to understand the production factors on which it depends. These are equipment costs including installation, operating costs (labour, supervision and overheads), processing times, tooling costs and component demand. The above variables are taken in account of in the calculation of P_c, using the following equation:

$$P_c = A \cdot T + \frac{B}{N} \tag{12.8}$$

where

A = Total average cost of setting up and operating a specific process, including plant, labour, supervision and overheads, per second; in the chosen country;
T = Average time in seconds for the processing of an ideal design for the process;
B = Average annual cost of tooling for processing an ideal component, including maintenance;
N = Total production quantity per annum.

The above values of A, B and T are based on processing a simple or ideal design well suited to the process in terms of both material and geometry. They are experience-based quantities and should be based where possible on established standards and expertise in companies specialising in the process under consideration.

12.7.2 Addition of P_c Data for a New Manufacturing Process

The steps proposed are as follows:

1. Select a manufacturing process that is currently included in the costing methodology that is nearest to the new process to be added (see Section 12.2). For example, consider the addition of reaction injection moulding to the system. (A similar process would be injection moulding.)
2. Examine the data used for the quantity 'A' for the surrogate process and determine if this can be used as it stands. If not, decide by how much it should be changed. In the first instance, this should be checked with sources including published material (manufacturing books, manuals), manufacturing experts and specialist suppliers. The average operating cost of an injection moulding facility in the UK is taken as 'X'. Obtain a view on a comparative value for reaction injection moulding.
3. Repeat the process in (2) above for the determination of the value for 'T'. The average operating time for a simple design of component in injection moulding is 'Y'. Obtain a view on a comparative figure for reaction injection moulding.
4. Repeat the process in (2) above for the value of 'B'. The average total tooling cost for injection moulding a simple design in the UK is 'Z'. Obtain a view on a comparative figure for reaction injection moulding.
5. The values obtained above are used to calculate 'P_c' for a range of values for 'N'. Produce a plot for reaction injection moulding and compare and discuss.
6. Add the pilot data to the system and represent as such. Add reaction injection moulding data and make as pilot data only.
7. Check the data against known costs for components well suited to the process and calibrate accordingly. Calibrate the new process to known reaction injection moulding case studies.
8. Add data to the main database, coded as a new process. The user should be informed that reaction injection moulding cost estimates are based on new data.
9. Once the data is proven, code as a standard process. The user should be informed as such.

12.7.3 Relative Cost Coefficient (R_c)

The relative cost coefficient is used to determine how much more expensive it will be to produce a component with more demanding characteristics than the 'ideal' design. In order to determine this quantity it is necessary to consider the effects of design-dependent criteria.

These are material to process compatibility; geometry to process suitability; including complexity, size and thickness, tolerance requirements and surface finish.

The equation used for the calculation of the relative cost coefficient, R_c, is as follows:

$$R_c = C_{mp} \cdot C_c \cdot C_s \cdot C_{ft} \tag{12.9}$$

where

C_{mp} = relative cost associated with material–process compatibility when compared with an ideal material process combination;
C_c = relative cost associated with producing different geometries from the ideal for the process under consideration;
C_c = relative cost associated with achieving a section reduction/thickness or size outside the envelope of the ideal design;
C_{ft} = C_t or C_f (whichever is the greater);

with

C_t = relative cost associated with obtaining a specified tolerance;
C_f = relative cost associated with obtaining a specified surface finish.

The combination of C_t and C_f into C_{ft} is based on the assumption that, when a fine surface finish is being produced, fine tolerances can be produced for the same cost and vice versa. This method of comparison and accumulation of costs, based on the product of the above variables, is analogous to the methodologies used by experts in the field of cost engineering and cost estimating. Note that the various elements going to of R_c i.e. C_{mp}, C_c, C_s, C_t, and C_f, need to take account of any secondary processing required to achieve the specified reductions, tolerances and finish, etc. for the component design. For a simple or ideal design of component, each of the relative cost coefficients is unity, but as a component design moves away from that state, the coefficients tend to increase in magnitude, thus increasing the processing cost.

If R_c data is not obtained, any estimate produced will be a lower bound only; the quality of the estimate will improve as more information is represented regarding the effect of the design-dependent factors.

12.7.4 Addition of R_c Data for a New Manufacturing Process

The steps proposed are as follows:

1. Following on from the procedure for P_c, select the process in the database nearest to the new process to be added. Again, let's consider adding reaction injection moulding to the system. (A similar process would be injection moulding.)

2. Examine the data used for the variable 'C_{mp}' for the surrogate process and determine if this can be used directly as it stands. If not, decide by how much it should be changed. In the first instance, this should be checked with sources including published material (manufacturing books and manuals), manufacturing experts and specialist suppliers. Obtain comparative figures for the materials to be considered and tabulate the values.
3. Repeat the process in (2) above for the determination of the value for 'C_c'. Obtain comparative figures against the respective shape categories and plot or tabulate the results. Refer to shape classification charts.
4. Repeat the process in (2) above for the value of 'C_s'. Obtain comparative figures taking account of section reductions/thickness and size. Tabulate the results.
5. Repeat the process in (2) above for the value of 'C_t'. Obtain comparative figures taking account of tolerance requirements. Tabulate the results.
6. Repeat the process in (2) above for the value of 'C_f'. Obtain comparative figures taking account of finish requirements. Tabulate the results.
7. Add the pilot data to the system and represent it as such. Add reaction injection moulding data and make as pilot data only.
8. Check the data against known costs for components well suited to the process and calibrate accordingly. Calibrate the new process to known case studies.
9. Add data to the main database, coded as a new process. The user should be informed that cost estimates are based on new data. Once the data is proven, code as a standard process.

References

[1] M.S. Hundal, Cost models for product design, Proc. ICED'93, The Hague, 17–19 August, 1993 1115–1122.

[2] T. Lenau, J. Haudrum, Cost evaluation of alternative production methods, Materials and Design 15 (4) (1994) 235–247.

[3] G. Pahl, W. Beitz, Engineering Design: A Systematic Approach, second edition, Design Council, London, 1996.

[4] M.M. Farag, Materials Selection for Engineering Design, Prentice Hall, Hemel Hempstead, 1997.

[5] K. Schreve, H.R. Schuster, A.H. Basson, Manufacturing cost estimation during design of fabricated parts, Proc. EDC'98 Brunel University, UK, July, 1998 437–444.

[6] P. Liebl, M. Hundal, G. Hoehne, Cost calculation with a feature-based CAD system using modules for calculation, comparison and forecast, Journal of Engineering Design 10 (1) (1999) 93–102.

[7] A.M.K. Esawi, M.F. Ashby, Cost estimates to guide pre-selection of processes, Materials and Design 24 (2005) 605–616.

[8] M. Ashby, H. Shercliff, D. Cebon, Materials: Engineering, Science, Processing and Design, second edition, Butterworth-Heinemann, Oxford, 2009.

[9] G. Boothroyd, P. Dewhurst, W.A. Knight, Product Design for Manufacture and Assembly, third edition, CRC Press, Boca Raton, 2010.

[10] Electroplating Costs Calculation; http://polynet.dk/ingpro/surface/elecomk.htm, accessed 2 September 2012.

[11] K.J. Fisher, M.J. Lobaugh, R.M. Michael, S.K. Sweeney, P.J. Kuvshinikov, Development of a cost model for manual tool polishing, International Journal of Modern Engineering 6 (2) (2006) 1–17.

[12] G. Hird, Personal communication, Lucas Industries Ltd, Birmingham, 1994.

[13] M. Sealy, P. Berriman, Y.-M. Marti, A practical solution for implementing sustainable improvements in product introduction performance, . Proc. FISITA'92 XXIV Congress, 7–11 June 1992, London, 1992 51–59.

CHAPTER 13

Assembly Costing

13.1 Introduction

This chapter on assembly costing is intended to support the process of assembly-oriented design through the provision of assembly performance metrics. As with conventional DFA approaches, the methodology allows the user to match design features with typical assembly situations (and associated penalties) on charts for each aspect of the assembly analysis. In this way, ambiguity is reduced and the user may identify features that are of high penalty and redesign these where necessary. The assembly cost measure should not strictly be taken as an absolute value. In practice, assembly costs are difficult to quantify and measure, and correlation requires testing a large number of industrial case studies. Nevertheless, the analysis results are useful when used in a relative mode of application. The reader interested in assembly costing models and applications is referred to Refs [1–4].

Many designs are created with complex assembly sequences and fitting and handling operations involving complex and restricted motions, poor stability, difficult orientation and alignment, and simultaneous multiple insertions. The overall effect is reduced ease of assembly, resulting in increased assembly times and cost. To improve the ease of assembly of a design, each operation needs to be carefully considered. Since something like 50% of all labour in the mechanical and electrical industries is involved in assembly, fitting and handling processes must be addressed in proactive DFA. The development of suitable insertion ports and handling features is essential for cost-effective assembly operations. In present DFA methodologies the fitting and handling analyses are used to evaluate insertion processes, which are ranked quantitatively depending on the difficulty of the task. The higher the score, the more inefficient the assembly operation (fitting or handling) is assumed to be, with 1.5 as a threshold value for unacceptable design of an individual operation.

Although the fitting and handling analyses are both well-established means of assessing assembly operations, they are highly judgmental, require training in their application and have no provision for design advice. Within a more proactive DFA methodology such information needs to be provided to the designer in a transparent and intuitive manner. The data should enable the designer to consider the effects of component and assembly port design on

Manufacturing Process Selection Handbook. http://dx.doi.org/10.1016/B978-0-08-099360-7.00013-6

the cost of product assembly. The capability of individual handling and alignment features with respect to their ability to help (or hinder) the assembly operation needs to be presented to the designer.

In order to make progress it is intended to allow the designer to view the data at different levels of detail, ranging from direct comparisons to detailed elements of specific features. The use of different representations will be investigated to make the information user-friendly. One way in which this may be possible is to take a more fundamental approach to the cost/time of component fitting and to use graphical representations of the effects of design geometry to allow for easy comparison at a glance, rather than sorting through tabulated data. In the following, we will consider manual assembly processes only. Manual assembly is by far the most common assembly system used in industry, in spite of the advent of more dedicated, automatic and programmable systems, mainly due to the inherent flexibility of manual or human operations.

13.2 Assembly Costing Model

The total cost of manual assembly comprises the sum of the total handling and fitting times multiplied by the labour rate (includes tooling costs, equipment costs, direct labour, supervision and overheads) in pence per second. The handling analysis below returns a Component Handling Index, H, related to a time factor for handling. Similarly, the time associated with the fitting of components in assemblies is represented by a Component Fitting Index, F, through a straightforward analysis of a component's fitting characteristics. Therefore, the total cost of manual assembly, C_{ma}, is:

$$C_{ma} = C_1 \, (F + H) \tag{13.1}$$

where

H = component handling index (seconds);

F = component fitting index (seconds);

C_1 = labour rate (pence per second).

In order to calculate an assembly cost, two further assumptions must be made:

1. The ideal assembly time for a combined handling and fitting operation is between 2 and 3 seconds. The exact time is dependent on factors such as workplace layout, environment and worker relaxation. In the case where an ideal time of 2 seconds is assumed, then the indices H and F can be taken as values in seconds. If 3 seconds is assumed, it is necessary to multiply the indices by 1.5 to obtain an estimate for the assembly time in seconds.

2. The labour rate, C_l, is calculated based on an annual salary of £22,000 (plus 40% overheads for a semi-skilled worker in the UK, 2012), for a 250-working-day year (5-day week minus statutory holidays) and a 7.5-hour working day. This gives the cost of manual labour as $C_l = 0.46$ pence per second.

13.2.1 Component Handling Analysis

The Component Handling Index, H, can be defined as:

$$H = A_h + \left[\sum_{i=1}^{n} P_{o_i} + \sum_{i=1}^{n} P_{g_i}\right] \tag{13.2}$$

where

A_h = basic handling index for an ideal design using a given handling process;

P_o = orientation penalty for the component design;

P_g = general handling property penalty.

13.2.1.1 Basic Component Handling Indices (A_h)

The basic handling indices, A_h, for a selection of common component handling characteristics are shown in Table 13.1. Select one only.

Table 13.1: Basic Handling Index (A_h) for a Selection of Component Handling Characteristics.

Component Handling Characteristic	Index (A_h)
One hand only	1
Very small aids/tools	1.5
Large and/or heavy (two hands/tools)	1.5
Very large and/or very heavy (two people/hoist)	3

We will now go on to consider the determination of the design-dependent, time-related, penalty indices associated with the geometry and characteristics of the design.

13.2.1.2 Orientation Penalties (P_o)

Orientation penalty indices, P_o, for a selection of common situations are shown in Figure 13.1. Select as appropriate.

13.2.1.3 General Handling Penalties (P_g)

The general handling indices, P_g, for a selection of common situations are shown in Table 13.2. Select as appropriate.

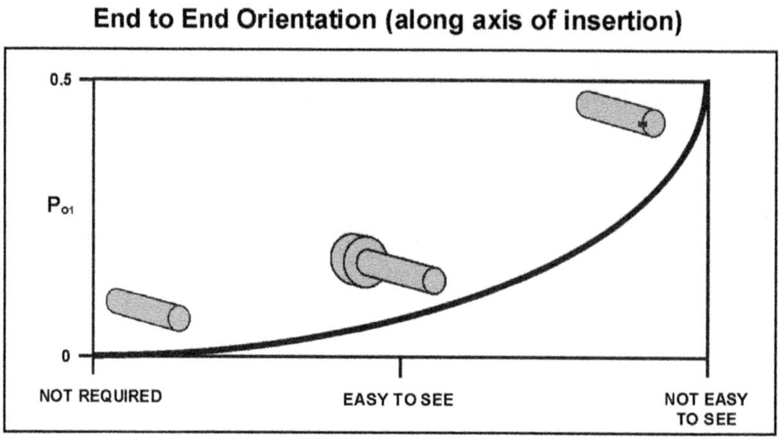

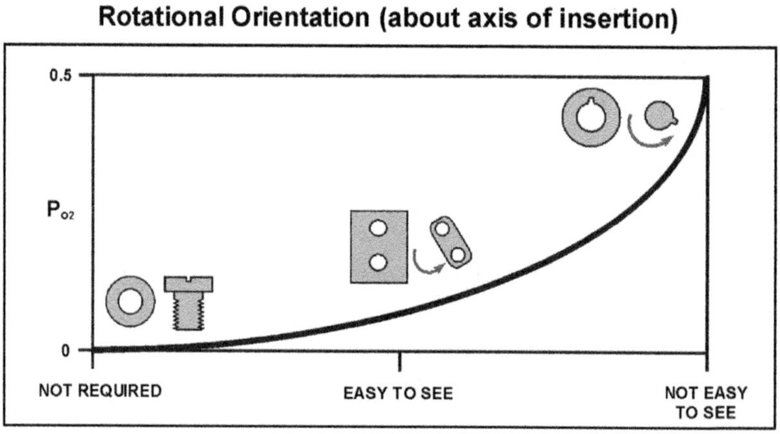

Figure 13.1: Orientation Penalties (P_o).

Table 13.2: Handling Sensitivity Index (P_g) for a Selection of Component Handling Sensitivities.

Component Handling Sensitivity	Index (P_g)
Fragile	0.4
Flexible	0.6
Adherent	0.5
Tangle/severely tangle	0.8/1.5
Severely nest	0.7
Sharp/abrasive	0.3
Hot/contaminated	0.5
Thin (gripping problem)	0.2
None of the above	0

13.2.2 Component Fitting Analysis

The Fitting Index, F, for a particular process in the sequence of assembly is defined as:

$$F = A_{\mathrm{f}} + \left[\sum_{i=1}^{n} P_{\mathrm{f}_i} + \sum_{i=1}^{n} P_{\mathrm{a}_i} \right] \tag{13.3}$$

where

A_{f} = basic fitting index for an ideal design using a given assembly process;

P_{f} = insertion penalty for the component design;

P_{a} = penalty for additional assembly processes on parts in place.

13.2.2.1 Basic Component Fitting Index (A_f)

Fitting indices for a selection of common processes are shown in Table 13.3. Select one only.

We will now go on to consider the determination of the design-dependent, time-related, penalty indices associated with the geometry and characteristics of assembly port designs.

Table 13.3: Fitting Indices (A_f) for a Number of Common Assembly Processes.

Assembly Process	Index (A_f)
Insertion only	1
Snap fit	1.3
Screw fastener	4
Rivet fastener	2.5
Clip fastener (plastic bending)	3
Placement in work holder (P_f and P_a usually not required)	1

13.2.2.2 Insertion Penalties (P_f)

Select all from relevant penalties for insertion operations from Figure 13.2(a) and (b).

13.2.2.3 Additional Assembly Processes (P_a)

Table 13.4 provides the additional assembly process indices, P_a, for a number of assembly processes carried out on components already positioned in the assembly build. Select as appropriate.

13.3 Assembly Structure Diagram

To facilitate an assembly costing analysis, it is essential to understand the structure of the proposed product and an assembly structure diagram is useful in this respect. Through its use,

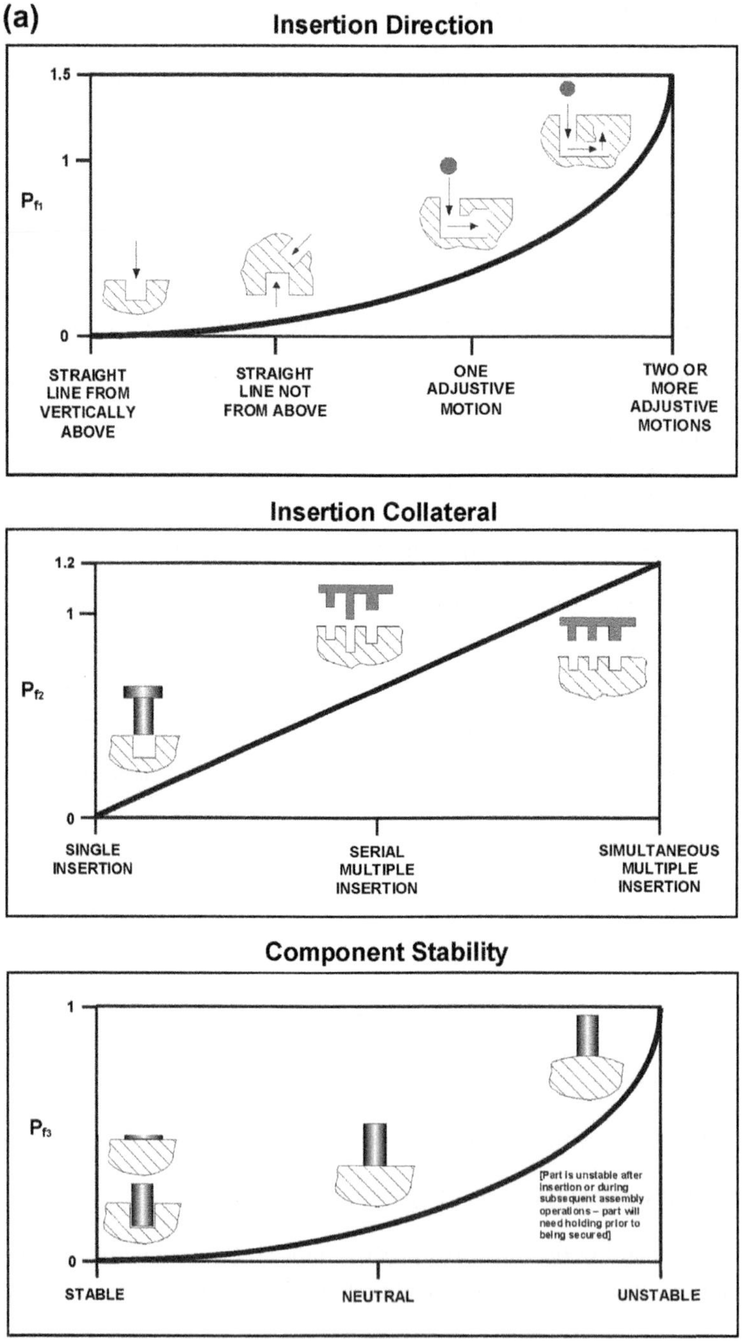

Figure 13.2: (a) Component Insertion Penalties (P_{fi}).

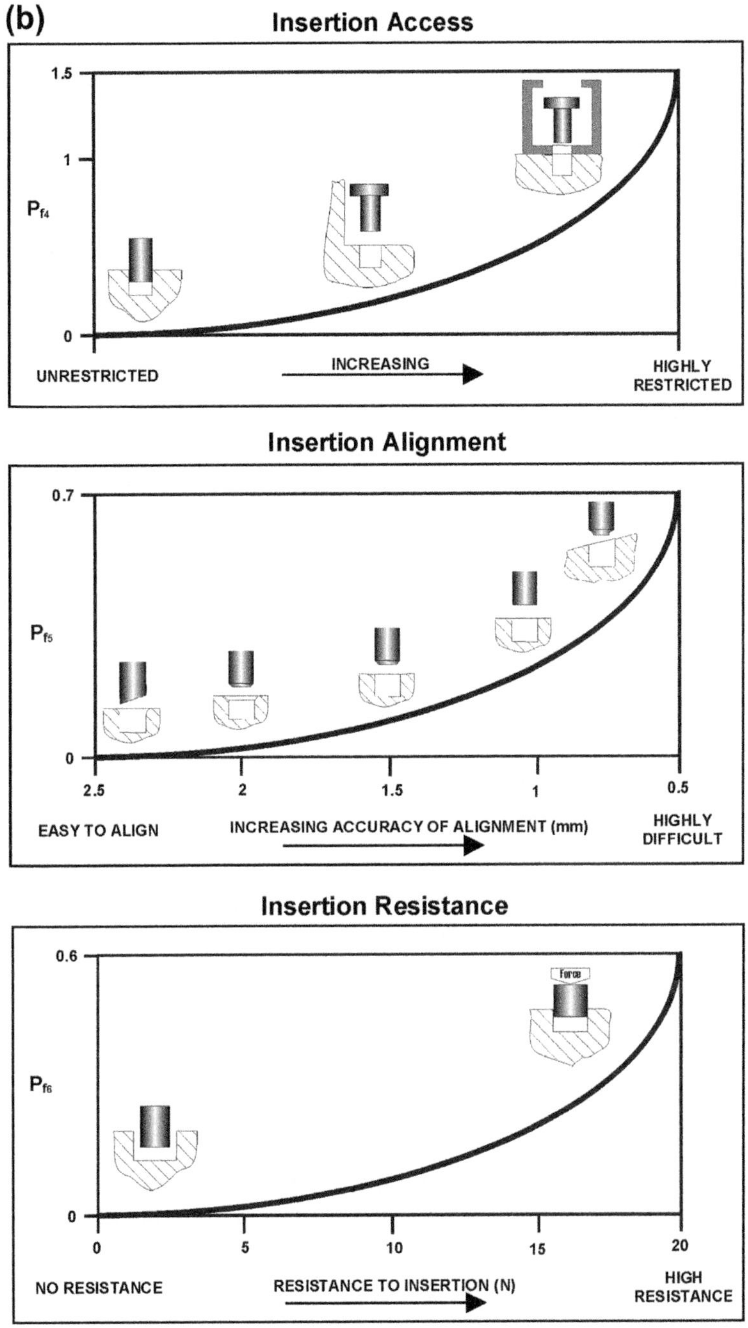

Figure 13.2: (b) Component Insertion Penalties (P_{fi}).

Table 13.4: Additional Assembly Index (P_a) for a Number of Common Assembly Processes.

Additional Assembly Process	Index (P_a)
Additional screw running	4
Later plastic deformation	3
Soldering/brazing/gas welding	6
Adhesive bonding/spot welding	5
Reorientation	1.5
Liquid/gas fill or empty	5
Set/test/measure/other (easy/difficult)	1.5/7.5

components in an assembly are logically mapped and in essence represent the product's disassembly sequence from left to right. Constructing this diagram is seen as a beneficial exercise as it supports an assembly perspective upon the design and compels the designer to focus on each component in the assembly. Individual component costs, M_i, may also be included in the diagram, in addition to the manual assembly cost for each component, C_{ma}, total M_i and C_{ma} for the product and sub-assembly, and component identification labels. An example is shown in Figure 13.3. Note that the inclusion of M_i in the assembly structure diagram is optional. A blank manual assembly costing table is provided in Appendix D to support the costing methodology.

13.4 Case Studies

13.4.1 Staple Remover

The design of a staple remover is shown in Figure 13.4. It is required to find the total production cost of the staple remover, including the cost of manufacturing the components. Figure 13.5 shows the assembly structure diagram for the assembled product, and the assembly costing analysis to support the assembly cost values for each operation is provided in Figure 13.6. The component cost, M_i, has already been determined from the methodology provided earlier. The total cost of the stapler per unit is found to be approximately £0.27 ($0.44), split almost evenly between assembly and manufacturing costs. Of course, a profit margin (typically between 15% and 25%) would be added to this cost, as this is the cost to the company to manufacture and assemble the product. Packaging, shipping and storage could also increase this cost substantially.

Figure 13.7 shows a redesign solution for the staple remover using just a single pressed sheet-metal component made from spring steel. This design eliminates the need for any assembly operations, although the cost of the material and complexity of the press tooling will only be justified if a large volume is produced in order to be competitive.

13.4.2 Floppy Disk

This second case study is concerned with the assembly time and cost of a 1.44 Mb floppy disk for use with a personal computer. Although this product is almost outmoded in this digital

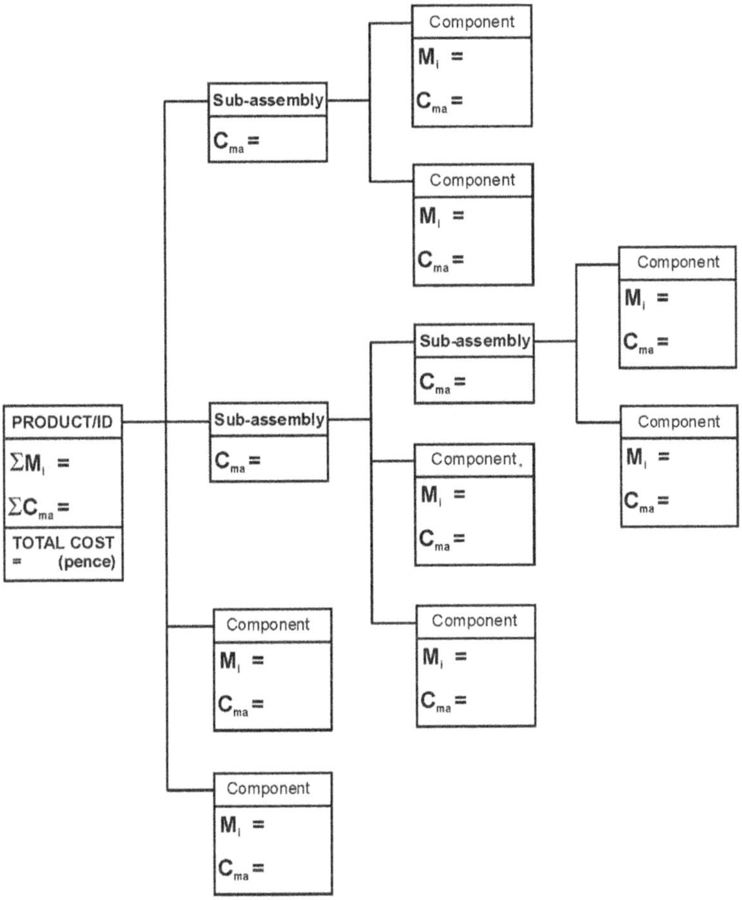

Figure 13.3: Example Format of an Assembly Structure Diagram.

age, it is a useful product line to further demonstrate the methodology for costing assemblies. Figure 13.8 shows the component parts.

The results are shown together with the assembly structure diagram in Figure 13.9 and a full assembly costing analysis is provided in Figure 13.10. The total assembly cost of the floppy disk per unit is found to be approximately £0.24 ($0.39) and the calculated assembly time is approximately 52 seconds. Note that a relaxation is not taken into account and the fact that the operator would be working in a clean environment room wearing protective clothing to stop contamination.

The time contribution of each assembly operation compared to the overall assembly time is shown as a percentage in Figure 13.11. A Pareto chart format is used with the greatest contribution to the total assembly time to the left. As highlighted, locating the front case

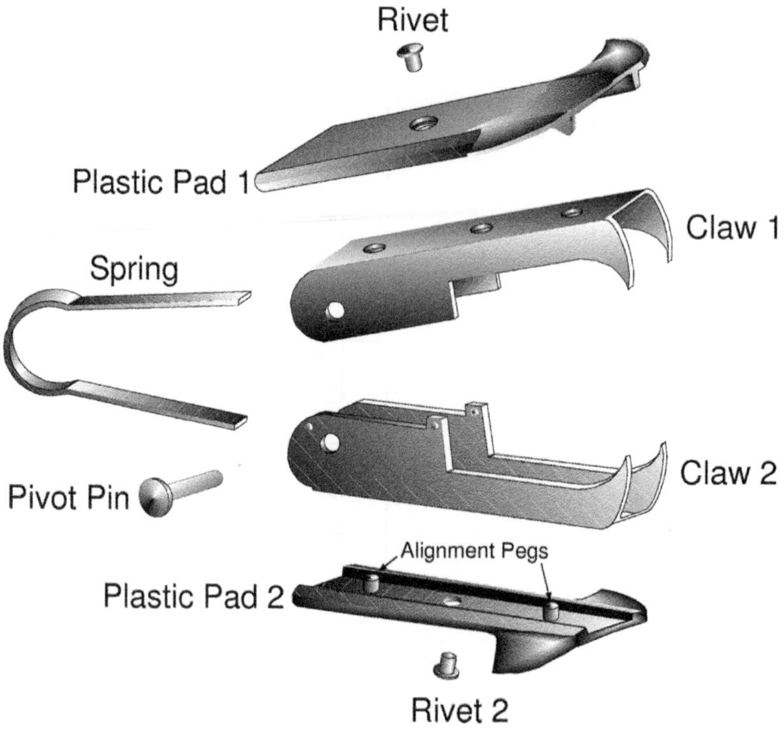

Figure 13.4: Staple Remover Exploded View.

sub-assembly on to the back case sub-assembly whilst the spring is in position is a difficult and time-consuming assembly task. Screen placement and spring fitting are two other operations of a time-consuming nature. In order to improve the ease of assembly of a particular concept design and reduce assembly costs, the use of the metrics in this manner can help identify potentially problematic areas and give guidance on redesign through reference to the charts provided.

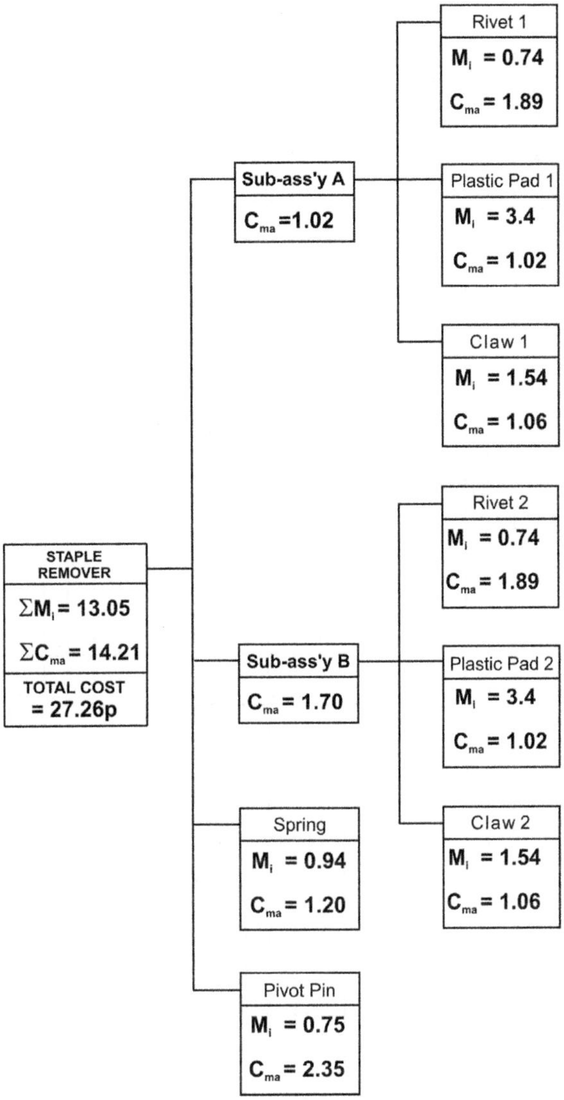

Figure 13.5: Staple Remover Assembly Structure Diagram.

PRODUCT NAME Staple Remover

PRODUCT CODE / ID -

LABOUR RATE (Cl) (pence per second) 0.46

COMPONENT/SUB-ASSEMBLY DETAILS			$H = Ah + [\Sigma Po + \Sigma Pg]$						$F = Af + [\Sigma Pf + \Sigma Pa]$											$Cl[H + F]$
PART/SUB-ASSEMBLY No. ID	PART DESCRIPTION	ASSEMBLY PROCESS	Ah	Po_1	Po_2	ΣPo	ΣPg	**H**	Af	Pf_1	Pf_2	Pf_3	Pf_4	Pf_5	Pf_6	ΣPf	ΣPa	**F**	Line Total	**Cma** (cost in pence)
1	RIVET 1	HAND./FIT.	1.5	0.1	0	0.1	0	1.6	2.5	0	0	0	0	0	0	0	0	2.5	4.1	1.89
2	PLASTIC PAD 1	HAND./FIT.	1	0.1	0.1	0.2	0	1.2	1	0	0	0	0	0	0	0	0	1	2.2	1.02
3	CLAW 1	HAND./FIT.	1	0.1	0.1	0.2	0	1.2	1	0	0	0	0	0.1	0	0.1	0	1.1	2.3	1.06
4	RIVET 2	HAND./FIT.	1.5	0.1	0	0.1	0	1.6	2.5	0	0	0	0	0	0	0	0	2.5	4.1	1.89
5	PLASTIC PAD 2	HAND./FIT.	1	0.1	0.1	0.2	0	1.2	1	0	0	0	0	0	0	0	0	1	2.2	1.02
6	CLAW 2	HAND./FIT.	1	0.1	0.1	0.2	0	1.2	1	0	0	0	0	0.1	0	0.1	0	1.1	2.3	1.06
SUB-ASS'Y A		HAND./FIT.	1	0.1	0.1	0.2	0	1.2	1	0	0	0	0	0	0	0	0	1	2.2	1.02
SUB-ASS'Y B		HAND./FIT.	1	0.1	0.1	0.2	0	1.2	1	0.1	0	1	0	0.3	0.1	1.5	0	2.5	3.7	1.70
7	SPRING	HAND./FIT.	1	0.1	0	0.1	0	1.1	1	0.1	0	0.2	0.2	0	0	0.5	0	1.5	2.6	1.20
8	PIVOT PIN	HAND./FIT.	1	0.1	0	0.1	0	1.1	1	0	0	0	0	0	0	0	3	4	5.1	2.35
																			30.8s	**TOTAL = 14.21 p**

Figure 13.6: Staple Remover Assembly Costing Analysis.

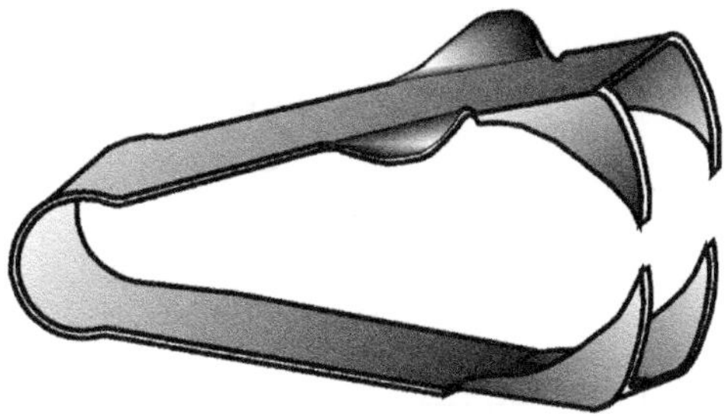

Figure 13.7: Staple Remover Redesign.

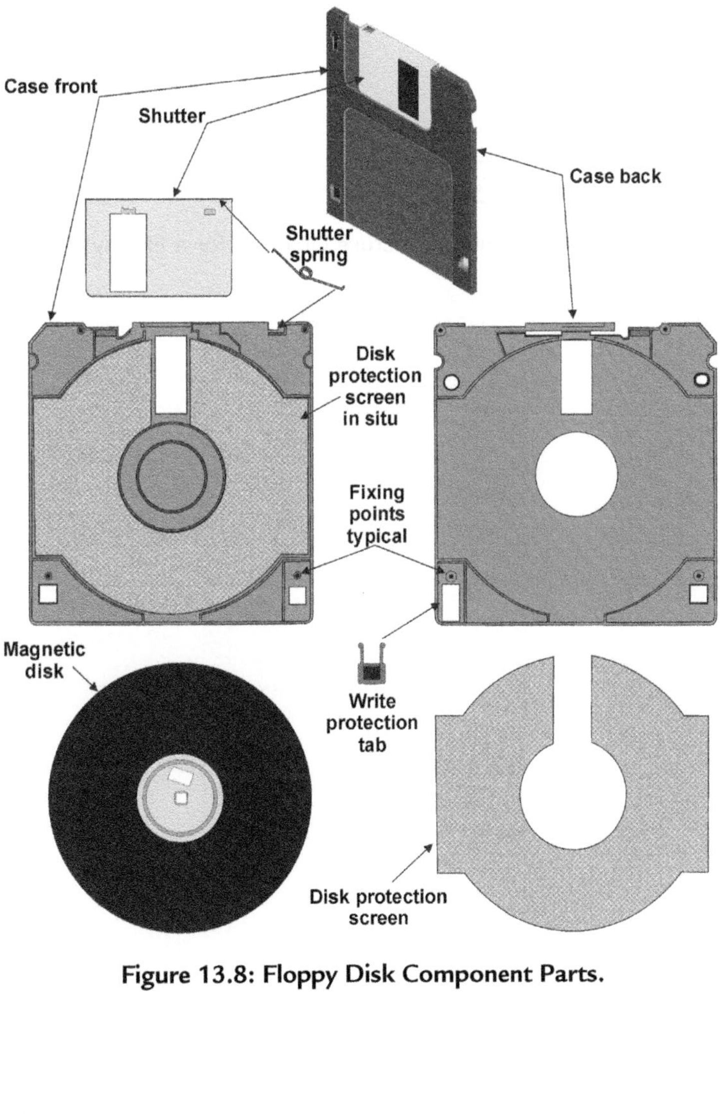

Figure 13.8: Floppy Disk Component Parts.

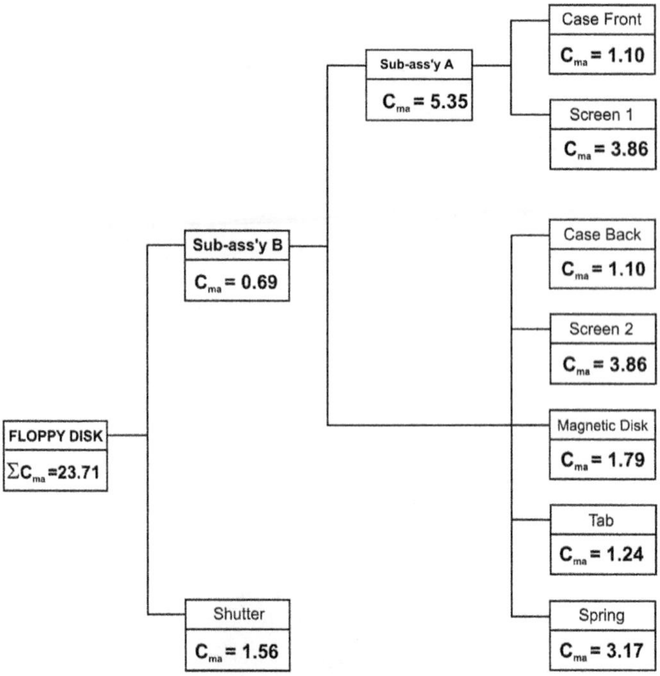

Figure 13.9: Assembly Structure Diagram for a Floppy Disk.

PRODUCT NAME Floppy Disk

PRODUCT CODE / ID -

LABOUR RATE (Cl) (pence per second) 0.46

COMPONENT/SUB-ASSEMBLY DETAILS			H = Ah + [ΣPo + ΣPg]					H	F = Af + [ΣPf + ΣPa]									F	Line Total	Cl[H + F] Cma (cost in pence)
PART/SUB-ASSEMBLY No. ID	PART DESCRIPTION	ASSEMBLY PROCESS	Ah	Po_1	Po_2	ΣPo	ΣPg		Af	Pf_1	Pf_2	Pf_3	Pf_4	Pf_5	Pf_6	ΣPf	ΣPa			
1	CASE FRONT	HAND./FIT.	1	0.1	0.1	0.2	0.2	1.4	1	0	0	0	0	0	0	0	0	1	2.4	1.10
2	SCREEN 1	HAND./FIT.	1	0.1	0	0.1	1.2	2.3	1	0	0	0	0	0.1	0	0.1	5	6.1	8.4	3.86
3	CASE BACK	HAND./FIT.	1	0.1	0.1	0.2	0.2	1.4	1	0	0	0	0	0	0	0	0	1	2.4	1.10
4	SCREEN 2	HAND./FIT.	1	0.1	0	0.1	1.2	2.3	1	0	0	0	0	0.1	0	0.1	5	6.1	8.4	3.86
5	MAGNETIC DISK*	HAND./FIT.	1	0.1	0	0.1	1.7	2.9	1	0	0	0	0	0	0	0	0	1	3.9	1.79
6	TAB	HAND./FIT.	1	0.1	0.1	0.2	0.2	1.4	1	0	0	0	0	0.3	0	0.3	0	1.3	2.7	1.24
7	SPRING	HAND./FIT.	1.5	0.5	0.1	0.6	0.8	2.9	1	1	0.5	1	0	0.5	0	3	0	4	6.9	3.17
SUB-ASS'Y A	CASE FRONT	HAND./FIT.	1	0.1	0.1	0.2	0.2	1.4	1	0	1	1	1.5	0.7	0	4.2	5	10.2	11.6	5.34
SUB-ASS'Y B	CASE FRONT + BACK	RE-ORIENT.															1.5	1.5	1.5	0.69
8	SHUTTER	HAND./FIT.	1	0.1	0.1	0.2	0.4	1.6	1	0.3	0	0	0	0.5	0	0.8	0	1.8	3.4	1.56
																			51.6s	**TOTAL = 23.71p**

* The magnetic disk and central driver are bought-in from a supplier as a single item.

Figure 13.10: Floppy Disk Assembly Costing Analysis.

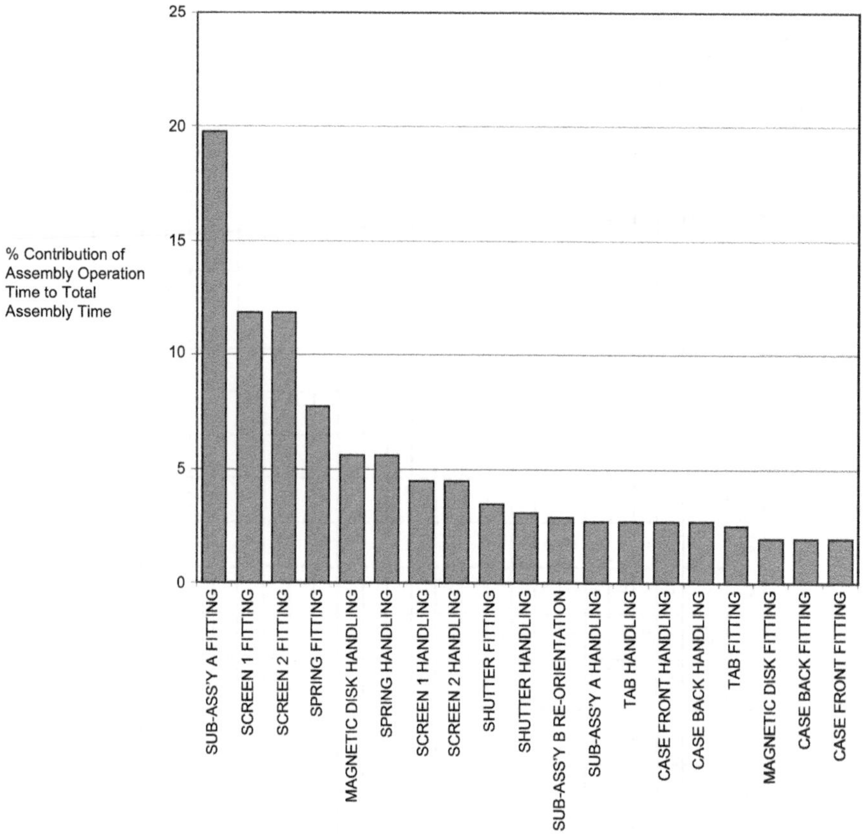

Figure 13.11: Pareto Chart of the Assembly Operation Times for the Floppy Disk.

References

[1] M. Andreasen, S. Kahler, T. Lund, Design for Assembly, second ed., IFS Publications and Springer-Verlag, New York, 1988.

[2] G. Boothroyd, P. Dewhurst, W.A. Knight, Product Design for Manufacture and Assembly, third ed., CRC Press, Boca Raton, 2010.

[3] M. Johnson, R. Kirchain, Quantifying the effects of parts consolidation and development costs on material selection decisions: a process-based costing approach, International Journal of Production Economics 119 (1) (2009) 174–186.

[4] M. Eswaramoorthi, P.S.S. Prasad, P.V. Mohanram, Developing an effective strategy to configure assembly systems using lean concepts, International Journal of Lean Thinking 1 (2) (2010) 15–39.

Appendices

Appendix A – Guidelines for Design for Manufacture (DFM)

A number of rules have been developed to aid designers when thinking about the manufacture of component parts by common processes:

- Ensure maximum simplicity in overall design.
- Be aware of the capabilities of in-house manufacture and outside suppliers.
- Select materials to suit the process, as well as lowest cost and availability.
- It is not desirable to design structures with abrupt changes in section due to stress concentrations. Aim at uniform wall thickness, cross-sections and gradual changes.
- Parts should be designed so that as many operations as possible can be performed without requiring repositioning. This promotes accuracy and minimises handling.
- Dimensions in one plane should all be from a single datum rather than from a variety of points to avoid overlap of tolerances; this facilitates the making of gauges and fixtures.
- Put a price on every tolerance and surface finish. Usually only 20% of the dimensional characteristics in a product are critical.
- The designer should always aim for minimum weight consistent with strength and stiffness requirements. While material costs are minimised by this criterion, there will also usually be a reduction in labour and tooling costs.
- Wherever possible, design to use general-purpose tooling rather than special dies, form cutters, etc. (an exception is high-volume production, where special tooling may be more cost-effective).
- If a component is normally exposed to view, make sure that its appearance is as pleasing as economy in production permits.
- For economic reasons, an attempt should always be made to fulfil several functions with a single function carrier.
- Identify critical characteristics (tolerances, surface finishes).
- Identify factors that influence manufacture of critical characteristics.
- Estimate and minimise component costs.
- Establish maximum tolerances for each characteristic, and avoid tight tolerances.
- Determine process capability of dimensional characteristics early.
- Design parts to be easily inspectable.
- Avoid secondary processes, and minimise number of machined surfaces.

- Minimise number re-orientations during manufacture.
- Use standard manufacturing processes where possible.
- Use generous radii/fillets on castings, mouldings and machined parts.
- Design parts for easy tooling/jigging using standard systems.
- Utilise special characteristics of processes (e.g. moulded inserts, colours).
- Use good detail design for manufacture and conform to drawing standards.

Appendix B – Guidelines for Design for Assembly (DFA)

A number of rules have been developed to aid designers when thinking about the assembly of component parts using common assembly technologies:

- Reduce part-count (and types) by consolidation and integration.
- Eliminate unnecessary joining processes or, when they must be used, reduce number of fasteners to a minimum, or use common, efficient fastening systems.
- Modularise the design, e.g. few but standard interfaces, simple coupling mechanisms and simple architecture/element orientation.
- Design for an optimum assembly sequence.
- Provide a base for assembly to act as a fixture or work carrier.
- Design the assembly process in a layered fashion (from above).
- Keep centre of gravity low.
- Use gravity to aid assembly operations.
- Minimise overall product weight.
- Design parts for multifunctional use where possible.
- Strive to eliminate adjustments (especially blind adjustments/shimming).
- Ensure adequate access and unrestricted vision.
- Use standard components where possible.
- Maximise part symmetry.
- Design parts that cannot be installed incorrectly.
- Minimise handling and re-orientation of parts.
- Design parts for ease of handling from bulk (avoid nesting, tangling).
- Design parts to be stiff and rigid, not brittle or fragile.
- Design parts to be self-aligning and self-locating (use tapers, chamfers, radii, etc.).
- Use good detail design for assembly and conform to drawing standards.
- Remove burrs and flash on component parts.

Appendix B1 – Functional Analysis

Functional Analysis provides a quantitative means of identifying potentially redundant components and those that are candidates for integration with other mating parts within a DFA approach. To carry out a Functional Analysis, itemise every component by name and list the components in a logical product assembly sequence.

- **Stage 1.** Examine the functional requirements of the product.
- **Stage 2.** Decide whether the product can be considered as a whole or as a series of functional subsections. It is best, if possible, to consider the product as a whole to avoid the duplication of parts or features, which may be in adjacent subsections.
- **Stage 3.** Components are divided into two categories:
  - 'A' components – these carry out functions vital to the performance of the product such as drive shafts, insulators, etc.
  - 'B' components – their purpose is not critical to the product function, such as fasteners, spacers etc.

Categorise mating components in a logical progression using the questioning routine in the Functional Criteria Chart until every component has been considered. The objective of this analysis is to determine those components necessary for the function of the product (Category A) and to highlight those that are candidates for elimination or combination with functional parts.

Design Efficiency (E)

Using this methodology, it is possible to assess the design in terms of a Design Efficiency (E):

$$E = \left(\frac{A}{(A + B)} \right) \times 100\,\%$$

When designing a new product, a Design Efficiency as high as possible should be obtained, with 60% being a suggested threshold based on a study of 'good' designs.

To start, pick a major functional component (e.g. rotor shaft) in the product and call it an 'A' part. Use the chart (Figure B1) to rate all other components in order, according to their assembly sequence.

A 'part' is any non-fluid component. It includes every washer, rivet, sticker, etc.

When using the Functional Criteria Chart, note that the questions are arranged into three columns. When answering questions in the 'current design' column, consider the design only as it is and not how it should or might be. For questions in the 'consider specification' column, consider how it is and ask if it has to be. The last column, 'other options', should provoke ideas of how the design could be.

Care should be exercised when rating identical parts such as screws, seals and springs as often they may not perform the same function throughout the product or the function can be achieved with less of them.

If there is any doubt regarding a component's category, default to B.

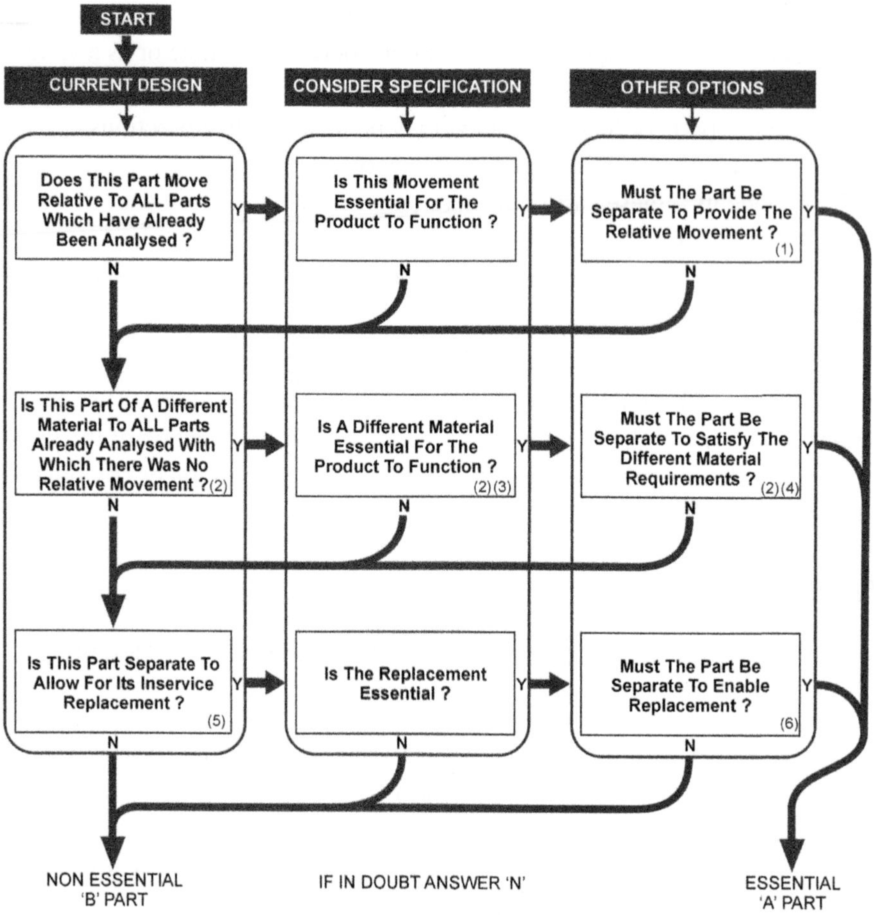

Figure B1: DFA Functional Analysis Chart (adapted from Ref. [1.4]).

Notes on the Functional Analysis Chart (Figure B1)

1. It may be possible to obtain the required movement by flexing the component, for example moulded hinges, diaphragms, bellows, thermal expansion, etc.
2. In some cases, the same materials may have different functional properties, e.g. hot/cold, positive/negative, north/south. In these cases, consider parts as different materials and continue the analysis.
3. A requirement for a different material could be for insulation, wear resistance, sealing, vibration damping, etc.
4. The part and its mating part could be made of the same material, or it may be possible to locally process the component to give the required property, etc.
5. Examples of these parts would be fuses, brush gear, light bulbs, filters, etc.
6. Can adjustments be made by plastically deforming a part?

Appendix C – Weld Joint Design Configurations

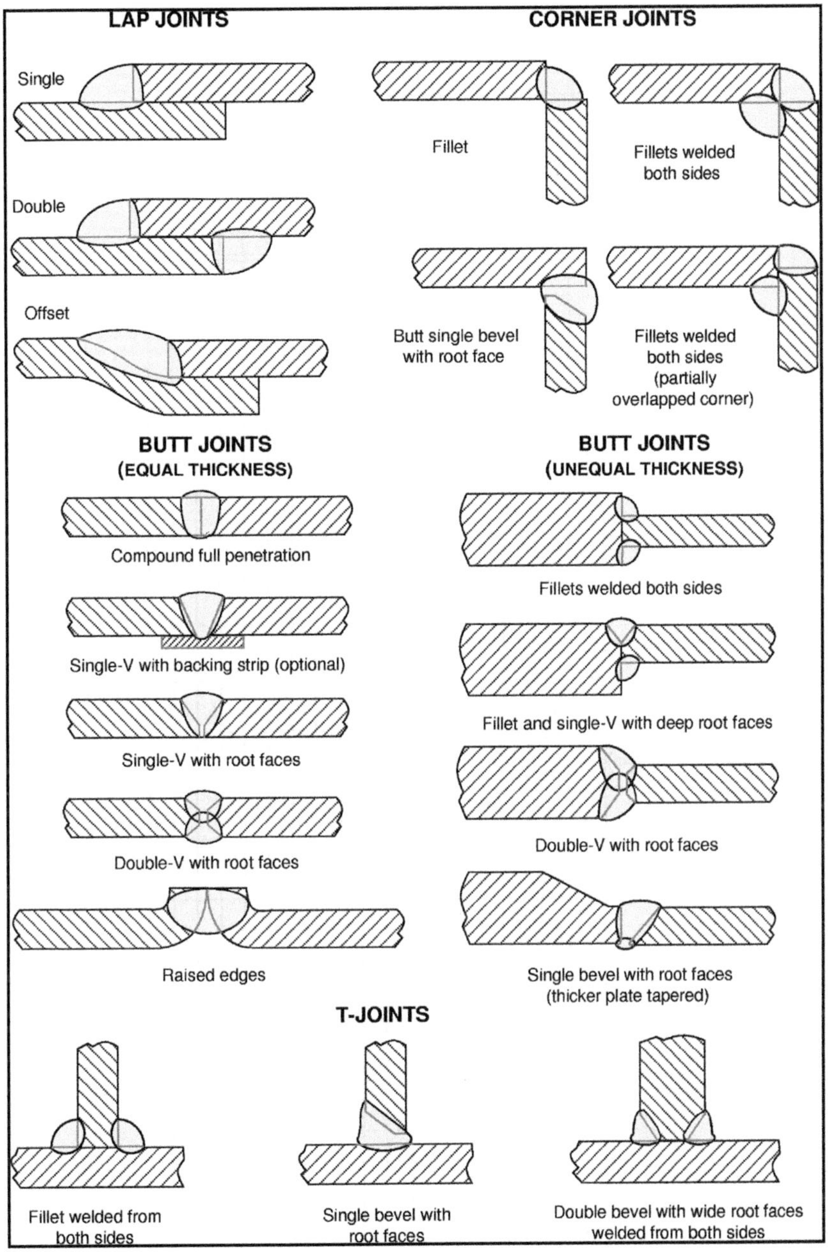

Appendix D – Blank Component Costing Table

PRODUCT NAME ______________

PRODUCT CODE / ID ______________

PRODUCT QUANTITY ______________

COMPONENT DETAILS					Mc = V x Cmt x [Wc]				Ⓐ		Rc = Cc x Cmp x Cs x [Cft]										Ⓑ	Ⓐ+Ⓑ
PART																						
No.	ID	PART DESCRIPTION	MATERIAL	PRIMARY PROCESS	SHAPE COMP.	Volume (mm³)	Cmt	Wc	Mc	Pc	Cc	Cmp	Section (mm)	Cs	Toler-ance (mm)	Ct	Surface Finish (µm Ra)	Cf	Cft	Rc	(Pc x Rc)	Mi (cost in pence)

Appendix E – Blank Assembly Costing Table

PRODUCT NAME ______________

PRODUCT CODE / ID ______________

LABOUR RATE (Cl) (pence per second) ______________

COMPONENT/SUB-ASSEMBLY DETAILS			$H = Ah + [\Sigma Po + \Sigma Pg]$						$F = Af + [\Sigma Pf + \Sigma Pa]$											Cl[H + F]
PART/SUB-ASSEMBLY No. ID	PART DESCRIPTION	ASSEMBLY PROCESS	Ah	Po_1	Po_2	ΣPo	ΣPg	**H**	Af	Pf_1	Pf_2	Pf_3	Pf_4	Pf_5	Pf_6	ΣPf	ΣPa	**F**	Line Total	**Cma** (cost in pence)

Bibliography

M. Ashby, H. Shercliff, D. Cebon, Materials: Engineering, Science, Processing and Design, second ed., Oxford:Butterworth-Heinemann.

R.Bakerjian (Ed.), Design for Manufacturability – Tool and Manufacturing Engineers Handbook, Vol. 6, fourth ed., Society of Manufacturing Engineers, Dearborn, MI, 1992.

R. Black, Design and Manufacture: An Integrated Approach, Macmillan, Basingstoke, 1995.

R.W. Bolz (Ed.), Production Processes: The Productivity Handbook, fifth ed., Industrial Press, New York, 1981.

J.G. Bralla (Ed.), Design for Manufacturability Handbook, second ed., McGraw-Hill, New York, 1998.

J.S. Burnell-Gray, P.K. Datta (Eds.), Surface Engineering Casebook: Solutions to Corrosion and Wear Related Failures, Woodhead, Abington, 1996.

C.K. Chua, K.F. Leong, C.S. Lim, Rapid Prototyping, third ed., World Scientific, Singapore, 2010.

E.P. Degarmo, R.A. Kohser, Materials and Processes in Manufacturing, tenth ed., Wiley, New York, 2007.

G.E. Dieter, tenth edition, ASM Handbook – Materials Selection and Design, Vol. 20, ASM International, Ohio, 1997.

M.P. Groover, Fundamentals of Modern Manufacturing, second ed., Wiley, New York, 2002.

S. Kalpakyian, S.R. Schmid, Manufacturing Engineering and Technology, fourth ed., Prentice Hall, New York, 2009.

D. Koshal, Manufacturing Engineer's Reference Book, Butterworth-Heinemann, Oxford, 1993.

R.O. Parmley (Ed.), The Standard Handbook of Fastening and Joining, McGraw-Hill, New York, 1996.

J. Schey, Introduction to Manufacturing Processes, third ed., McGraw-Hill, New York, 1999.

R.H. Todd, D.K. Allen, L. Alting, Manufacturing Processes Reference Guide, Industrial Press, New York, 1994.

J.M. Walker (Ed.), Handbook of Manufacturing Engineering, Marcel Dekker, New York, 1996.

C. Wick, R.F. Veilleux, Quality Control and Assembly – Tool and Manufacturing Engineers Handbook, Vol. 4, Society of Manufacturing Engineers, Dearborn, MI, 1987.

British Standards (http://shop.bsigroup.com)

BS 1134 (2010) Assessment of Surface Texture.

PD 6470 (1984) The Management of Design for Economic Production.

BS 6615 (1996) Specification for Dimensional Tolerances for Metal and Metal Alloy Castings.

BS 7000-2 (2008) Design Management Systems. Guide to Managing the Design of Manufactured Products.

BS 7010 (1988) Code of Practice for a System of Tolerances for the Dimensions of Plastics Mouldings.

BS 8887-1 (2006) Design for Manufacture, Assembly, Disassembly and End-of-life Processing (MADE).

Design- and Manufacture-related Websites

http://www.assemblymag.com

http://www.custompartnet.com/rapid-process-selector

http://www.dfma.com

http://www.efunda.com

http://www.engineersedge.com

http://www.engineershandbook.com

http://www.eng-tips.com

http://www.lme.com

http://machinedesign.com

http://www.matweb.com
http://www.mistakeproofing.com
http://www.npd-solutions.com
http://www.tec-ease.com/
http://source.theengineer.co.uk
http://www.tolcap.com

Index

Note: Page numbers followed by "f" and "t" indicate figures and tables.

H

I

J

V

W

X

Y

Z

Printed and bound by CPI Group (UK) Ltd, Croydon, CR0 4YY
03/10/2024
01040301-0005